MARXISM AND EDUCATION BEYOND IDENTITY

MARXISM AND EDUCATION

This series assumes the ongoing relevance of Marx's contributions to critical social analysis and aims to encourage continuation of the development of the legacy of Marxist traditions in and for education. The remit for the substantive focus of scholarship and analysis appearing in the series extends from the global to the local in relation to dynamics of capitalism and encompasses historical and contemporary developments in political economy of education as well as forms of critique and resistances to capitalist social relations. The series announces a new beginning and proceeds in a spirit of openness and dialogue within and between Marxism and education, and between Marxism and its various critics. The essential feature of the work of the series is that Marxism and Marxist frameworks are to be taken seriously, not as formulaic knowledge and unassailable methodology but critically as inspirational resources for renewal of research and understanding, and as support for action in and upon structures and processes of education and their relations to society. The series is dedicated to the realization of positive human potentialities as education and thus, with Marx, to our education as educators.

Renewing Dialogues in Marxism and Education: Openings
Edited by Anthony Green, Glenn Rikowski, and Helen Raduntz

Critical Race Theory and Education: A Marxist Response
Mike Cole

Revolutionizing Pedagogy: Education for Social Justice Within and Beyond Global Neo-Liberalism
Edited by Sheila Macrine, Peter McLaren, and Dave Hill

Marxism and Education beyond Identity: Sexuality and Schooling
Faith Agostinone-Wilson

education—*Continued*
   145, 146, 151, 152, 153, 154,
   155, 164, 165, 167, 177, 189,
   190, 191, 192, 194, 197, 198,
   207, 213, 215, 216, 217, 222,
   223, 225, 227, 228, 232, 233,
   234, 235, 236, 237, 240, 241
egalitarian societies, 117, 158, 159,
   163, 165, 166, 181, 182,
   185, 188
Eisenhower, Dwight, 66
Eisenstadt v. Baird, 1972, 171, 175
Elders, Joceyln, 25
empiricism, 4, 142, 147, 148
employees, 54, 66, 135, 192
Employment Non-Discrimination Act,
   2009, 61
   and transgendered, 19
Enlightenment, the, 64, 142, 193
environment, 31, 147, 188, 220,
   222, 224
environmental, 113
equality, 31, 32, 83, 112, 131, 136,
   147, 184, 213, 214
   and reproductive freedom, 28,
      112, 151
   and sexual freedom, 104, 116, 139,
      204, 205, 213
equity feminism, 130, 162,
   163, 164
essentialism, 124, 151, 156, 158
evangelicals, 173
   and Christian right, 101, 164, 182
   and voting patterns, 50
   attitudes toward comprehensive
      sexuality education, 22
   attitudes toward LGBTQ people,
      21, 72
ex-gay movement, 69
ex-gay therapies, 69, 76
exploitation, 4, 87, 104, 111, 121,
   123, 128, 130, 131, 132, 134,
   135, 137, 139, 145, 148, 149,
   150, 157, 158, 159, 160, 161,
   164, 165, 167, 179, 183, 184,
   191, 207, 212, 217

Faith-Based and Community
   Initiatives Program, 58
faith-based initiatives, 23, 24
   and politicians, 58, 183
false consciousness, 3, 5, 117, 137,
   139, 146
Falwell, Jerry, 52, 70, 208
family, 1, 4, 63, 67, 81, 85, 88, 94,
   95, 102, 103, 104, 109, 116, 118,
   123, 132, 139, 141, 152, 155,
   157, 163, 166, 167, 179, 186,
   189, 190, 192, 198, 207, 221,
   222, 223
   and alternative forms, 13, 71, 75,
      92, 100, 102, 117, 141, 166, 167,
      182, 185, 188, 197, 212, 223
   and anarchist theory, 5, 82
   composition of, 13, 92, 94, 95, 97
   and Consanguine form, 114, 115
   and conservative theory, 5, 82
   and materialist theory, 5, 79, 82,
      114, 150
   and materialistic theory, 181
   and Monogamous, 50, 67, 115
   and nuclear, 1, 5, 62, 78, 153, 182,
      197, 212
   and Pairing form, 115
   and poverty, 51, 93, 94, 167, 182,
      186, 192, 193
   and procreation, 60, 74, 86, 107,
      168, 174, 212
   and psychoanalytical theory, 5, 79,
      82, 103
   and Punaluan form, 115
   and socialist feminism, 5
   and two-parent, 33, 70, 73, 81,
      93, 99
Family and Medical Leave Act,
   1993, 185
Family Research Council, 23, 58
family values, 4, 5, 35, 57, 59, 66, 67,
   75, 95, 99, 122, 141, 146, 168,
   183, 201
fascism, 103, 104, 105, 107, 112, 113,
   187, 200
fascist ideology, 105

Craig, Larry, 55
critical race theory, 121, 122, 127
culture, 1, 30, 83, 121, 123, 135, 137, 145, 156, 159, 165, 201, 204, 205
curriculum, 31, 100, 126, 192
  and LGBTQ, 8, 16, 40
  lack of LGBTQ representation at university level, 8, 31, 40
  lack of LGBTQ representation in K-12, 8, 31, 41, 49

Dale, James, 34, 35
Darwin, Charles, 103
Daughters of Bilitis, 197, 202, 204, 213
deconstruction, 121, 122, 125, 131, 143, 146, 148, 149, 199
Defense of Marriage Act, 1966, 14, 16, 23, 59
  percent opposing, 14
Democratic party, 11, 52, 57, 59, 194, 197, 216, 217
Democrats, 3, 11, 20, 21, 50, 127, 155, 194, 197
depression, 46, 47
desire vs. need, 4, 46, 120, 135, 137, 139, 140, 146
developmental rights, 32
dialectical materialism, 4, 49, 90, 122, 148, 165
  critique of postmodernism, 71, 121, 135, 139, 149, 158
disability, 14, 39, 180, 185, 190, 193, 205, 212
discrimination, 8, 34, 56, 133, 140, 152, 162, 165, 167, 171, 175, 181, 197, 209
  and housing, 18, 19, 95, 137
  schools prohibiting, 8, 11, 35, 37, 39, 72
  and sexual orientation, 8, 9, 10, 18, 19, 36, 37, 39, 41, 52, 57, 61, 66, 67, 68, 73, 77, 118, 185, 205, 208
  states prohibiting, 11, 66
  and workplace, 11, 18, 19, 52, 57, 153, 185, 187

divorce, 16, 74, 82, 85, 89, 91, 100, 115, 118, 151, 152, 167, 168, 181, 182, 183
Dobbs, Lou, 106
Doe v. Bellefonte Area School District, 2004, 42
Doe v. Brockton School Community, 2000, 42
domestic labor, 81, 85, 86, 87, 89, 108, 117, 118, 145, 148, 158, 165, 166, 168, 181, 183, 184, 188, 217, 223
domestic partnerships, 13, 39, 141, 190, 206, 214
  and civil unions, 15, 59, 71, 185
  legal benefits of, 15, 140, 169, 185
  process of applying for, 15
domestic sphere, 79, 81, 127, 130, 149, 167
domestic violence, 131, 149, 152, 154, 157, 163, 168
Don't Ask, Don't Tell, 20, 21, 22, 197
  and Boy Scouts, 34, 35
  and military, 20, 35, 42, 65, 66, 213
double standard, 75, 83, 84, 115, 169, 170, 184
DuBois, W.E.B., 65
Dworkin, Andrea, 162
dystopian, 142, 143, 179

economic, 27, 32, 49, 54, 78, 81, 84, 85, 98, 105, 109, 110, 112, 115, 116, 117, 121, 122, 123, 125, 128, 129, 130, 131, 135, 138, 143, 146, 151, 152, 153, 163, 164, 167, 168, 170, 174, 178, 181, 188, 191, 193, 199, 203, 207, 212, 221, 222
economy, 3, 38, 78, 81, 83, 85, 152, 156, 167, 174, 182, 185, 194, 198, 207, 215
education, 7, 8, 16, 22, 23, 24, 25, 26, 27, 28, 31, 32, 33, 38, 40, 41, 44, 46, 49, 69, 81, 83, 85, 86, 87, 89, 96, 104, 118, 121, 125, 127, 129, 130, 137, 142,

church, 3, 63, 72, 73, 74, 79, 80, 81, 82, 83, 99, 102, 109, 170, 174, 195, 207, 217, 218
  and LGBTQ, 1, 21, 45, 46, 51, 72
  support for LGBTQ, 1, 21, 46
Church of Jesus Christ of Latter Day Saints, 58
civil disobedience, 198, 209, 211
civil rights, 10, 21, 51, 53, 55, 57, 66, 68, 70, 73, 122, 129, 131, 147, 152, 200, 202, 207, 208, 209, 219
  and Bible, 208, 210
  and same-sex marriage, 3, 8, 51, 59, 68, 214, 218
  as special rights, 18, 41, 43, 56, 68, 90, 136, 162
class society, 19
clergy, 51, 72, 168, 173, 217, 218
  prohibitions against LGBTQ, 22
Clinton, Bill, 23, 59, 66, 91, 162, 175, 177, 213, 223
Clinton, Hillary, 164
closet, the, 8, 11, 21, 35, 36, 37, 38, 40, 52, 58, 61, 65, 73, 122, 123, 124, 136, 144, 196, 197, 199, 202, 203, 204, 205, 209, 211, 214
  and barriers to organizing, 12, 123, 202, 207, 211
  effects of, 44, 61, 123
  Marxist reading of, 123
Coburn, Tom, 29, 172, 173
Cohn, Roy, 66
collective, 103, 107, 117, 121, 128, 131, 145, 167, 168, 185, 188, 193, 204, 206, 211
collectivist, 61, 103, 211
colonialism, 102, 123, 163
commodification, 86, 132, 135
  and sex, 86, 189
communal, 81, 83, 108, 111, 117, 118, 142, 159, 181, 188, 204
commune, 81, 84
communism, 60, 81, 94, 112, 113, 118
  and homosexuality, 52, 60, 65, 202
communist, 60, 94, 137

compulsory pregnancy, 156, 170, 171, 176
Comstock Law, 1873, 170
Comstock laws
  and contraception, 69, 170
  and reproductive freedom, 69, 170, 171
Concerned Women for America, 29, 58
condoms, 28, 138, 171, 215
  and abstinence-only education, 22, 25, 72, 172
  and anti-contraceptive movement, 23, 72, 168, 172
  failure rates of, 28, 29, 170, 171, 172, 173
  reliability of, 29, 170, 172, 173
conservative familialism, 2, 5, 71, 92, 94, 96, 97, 108, 117, 122, 146
conservatives, 1, 4, 43, 49, 57, 58, 94, 107, 108, 126, 133, 135, 136, 141, 142, 143, 146, 147, 154, 156, 158, 162, 168, 169, 172, 173, 180, 181, 183, 188, 191, 194, 195, 197, 198, 200, 205, 206, 207, 211, 212, 213, 222, 223
  and homophobia, 10, 12, 22, 23, 42, 51, 67, 70, 78, 85, 133, 134, 199
consumerism, 4, 84, 124, 136, 137, 149, 205
  and choice, 89, 139
  and performance, 136
  and postmodernism, 121, 123, 135, 224
contraception, 22, 26, 84, 152, 153, 155, 168, 169, 170, 171, 172, 173, 178, 179, 183, 214
  and Comstock law, 69
  and Comstock laws, 170
  and laws, 168, 169, 170, 171, 172, 173, 174, 175, 177, 180, 181
  and sterilization, 175
cooperatives, 167, 204
corporal punishment, 167
Cory, Donald Webster, 76, 202
Council of Christian Colleges and Universities (CCCU), 208

INDEX 245

Boy Scouts of America v. Dale, 2000, 34
Branch Davidian, 113
Briggs Initiative, 1978 (CA), 19, 42, 205, 216
Brigham Young University, 211
Broekhuizen, Elsa, 58
Brown, Judy, 172
Brown v. Board of Education, 1954, 162
Bruderhof, 107
Bryant, Anita, 52, 66, 68
Buchanan, Patrick, 75, 113, 114
Buffett, Warren, 3
bullying, 12, 36, 40, 73
 and school policies, 32, 37, 38, 39, 40, 41, 42, 48
 types of, 37, 38
Burke, Edmond, 70
Bush, George H. W., 23, 154, 169, 171, 172, 223
Bush, George W., 23, 24, 29, 58, 70, 75, 129, 154, 172, 179, 182, 222, 223
Bush, Jeb, 180

Canady, Charles, 178
capitalism, 1, 4, 43, 57, 71, 81, 83, 86, 87, 94, 95, 97, 110, 111, 112, 113, 114, 116, 122, 123, 124, 126, 127, 134, 135, 136, 139, 142, 144, 146, 147, 148, 156, 157, 160, 162, 165, 166, 167, 184, 186, 191, 192, 193, 197, 200, 201, 204, 206, 207, 209, 213, 216, 220, 223, 224
 and alienation, 42, 82, 88, 91, 118, 167
 and commodified sex, 86, 91, 128, 136, 138, 167, 189
 contradictions of, 59, 94, 193, 212
 and homophobia, 9, 55, 56, 59, 62, 78, 86, 93, 130, 131, 144, 190, 205, 212, 214
 and nuclear family, 2, 5, 9, 42, 55, 62, 68, 78, 79, 81, 84, 91, 93, 114, 118, 123, 152, 153, 157, 159, 160, 161, 167, 181, 183, 184, 190, 205
 and overproduction, 138
 and postmodernism, 5, 113, 121, 135
 and production, 56, 82, 114, 132, 142
 and property relations, 78, 138
 and racism, 92, 94, 130, 131, 132, 144, 193, 212
 and religion, 154, 193, 208
 and reproduction, 56, 82, 86, 114, 123, 152, 167, 174
 and sexism, 55, 130, 131, 132, 143, 144, 152, 153, 157, 159, 163, 167, 187, 190, 193, 200, 212
 and sexual harassment, 42, 187, 188
Carey v. Population Services International, 1978, 28
Carpenter, Edward, 64
Carter, Jimmy, 21, 223
censorship, 23, 116
 and LGBTQ, 8, 30, 32, 33, 34, 119, 122, 189, 207
Cercle Hermaphroditos, 201
character education, 31
charity, 18, 58, 94, 99, 184, 220
Cheney, Richard, 96, 107
Chicago Society for Human Rights, 201
child care, 69, 71, 81, 84, 86, 87, 93, 98, 110, 111, 117, 118, 142, 146, 152, 153, 156, 158, 160, 163, 167, 179, 181, 183, 188, 217
childhood, 77
children, 2, 5, 13, 73, 74, 80, 81, 82, 83, 85, 90, 91, 95, 96, 98, 101, 104, 105, 106, 115, 117, 119, 120, 140, 155, 164, 166, 167, 179, 181, 182, 188, 189, 197, 200
Children's Health Insurance Program (CHIP), 179
Christian Coalition, 23
Christian Reconstructionism, 58
Christian right, 208
 and evangelicals, 101, 164, 182
 voting patterns, 50
Christianity, 44, 45, 46, 48, 55, 60, 68, 73, 74, 158, 164, 167, 207, 208

activist groups, 8, 10, 125, 147, 223
  conflict within, 68, 132, 133, 134, 191, 193, 194, 199, 211
  and organizing, 12, 130, 131, 191, 198, 202, 218, 220, 224
adaptation, 70
Adkisson, Jim, 1
adolescence, 18, 26, 45, 84, 91, 119, 152, 173, 177, 219
Adolescent and Family Life Act, 1981, 23
adoption, 53, 155, 180
  states prohibiting LGBTQ parents, 13, 67, 71
affirmative action, 113, 162
African-Americans, 19, 26, 44, 51, 70, 86, 99, 105, 124, 133, 134, 140, 143, 162, 175, 180, 189, 200, 203, 220, 221
  and marriage, 50
  and nuclear family, 83
  and sexuality, 21, 65, 83, 134
  stereotypes of, 3, 21, 27, 93, 136
Ahmanson Jr., Howard, 58
alienation, 88, 90, 91, 188, 221
  and capitalism, 5, 42, 82, 88, 167
  and homophobia, 11, 61, 210
  and nuclear family, 5, 44, 82, 89, 117, 167
  and relationships, 44
Alinski, Saul, 177
American Baptist Convention, 217
American Life League, 172
American Medical Association, 171, 174
American Psychiatric Association, 49, 55, 66, 75, 77
anarchism, 79, 107, 108, 111, 112, 114, 127
  Marx's critique of, 112
anthropology, 142, 158, 159
  and family, 157
  and feminism, 157
anti-contraceptive movement, 24, 28, 168, 218
  and abortion, 25, 154, 177
  and pharmacists, 171, 177
  and Plan B, 29
assimilation, 52, 61, 164, 166, 197, 199, 202, 203
authoritarian, 105, 106, 107
authoritarianism, 103, 104, 106
authority, 103, 106, 107, 119, 185

backlash, 22, 26, 42, 58, 98, 101, 149, 154, 168, 170, 171, 187, 197, 204
  and anti-gay legislation, 15, 41, 52, 66, 94, 194, 196, 201
  and definition, 56
  and homophobia, 3, 18, 49, 52, 56, 57, 65, 66, 77, 94, 147, 196, 205
Baird, William, 171
Banks, Tyra, 163
Bawer, Bruce, 19, 147
Beatie, Thomas, 13
Beck, Glenn, 1, 48, 107
Bennett, William, 70
bestiality, 71, 72
Bible, 21, 29, 42, 45, 46, 52, 58, 103, 171
  and civil rights, 208, 210
  and "clobber verses," 22, 209
  and gender roles, 53, 154, 164
  and nuclear family, 75, 82
  and support of homophobia, 74, 208
bigamy, 67
birth control, 23, 146, 168, 170, 171, 177
  as abortion, 154, 169, 170, 172, 173
  and laws, 69, 170, 171
birth control pills, 171, 172
birth rates, 26, 168, 170, 181, 182
bisexual, 7, 17, 23
  percentage of people identifying as, 78
bisexuality, 159
  definition of, 10
bourgeois, 97, 112, 116, 122, 141, 165, 194
Bowers v. Hardwick, 1986, 50, 60, 67, 206
Boy Scouts of America, 34, 35, 41
Boy Scouts of America Equal Access Act, 2002, 36

Index

abortion, 57, 112, 116, 156, 164, 168, 169, 170, 176, 180, 199, 214, 215, 217, 220
   access to, 28, 152, 153, 154, 170, 173, 175, 176, 180, 181, 183, 213, 215
   and anti-contraceptive movement, 2, 28, 169
   and back alley, 154, 170, 174, 175, 216
   and breast cancer link, 24, 178
   and counseling, 28, 168, 176, 177, 178, 220
   and gag rules, 28, 176, 178
   history of, 170, 173, 174, 175, 177, 216
   international rates of, 28, 174
   and late-term, 28, 175, 176, 177, 178
   and laws, 220
   and parental notification laws, 28, 176, 177
   and "partial birth," 176, 178
   and "partialbirth," 168
   percent anti-choice, 154, 155, 169, 173
   percent pro-choice, 154, 155, 169, 173
   pre-Roe, 155, 174, 175, 179
   and privacy, 171, 172, 173, 175, 176, 177, 179, 196, 220
   and reproductive rights, 28, 152, 154, 213, 219
   and sexuality, 169
   U.S. rates of, 25, 29, 169, 173
   and youth access to, 28, 154, 176, 177

About Face Youth Theatre (Chicago), 25
abstinence, 23, 68, 69, 70, 95, 170, 173
abstinence-only education, 2, 4, 23, 154
   and condoms, 22, 25, 72
   and curriculum, 24, 25, 95
   and funding, 22, 23, 25, 69, 175
   and homophobia, 8, 25
   and laws, 22, 23, 24
   and monogamy, 25, 69
   percent of public schools requiring, 24
   philosophies of, 24, 25
   policies of, 207
   and purity movement, 25, 69
   and religion, 26
   success of, 24
abuse, 37, 39, 44, 101, 105, 113, 152, 163
ACT UP, 206, 215, 221
activism, 4, 8, 31, 34, 41, 51, 54, 67, 76, 124, 125, 133, 135, 136, 137, 142, 143, 144, 162, 165, 173, 175, 177, 179, 181, 188, 189, 191, 195, 197, 198, 199, 202, 204, 223
   and sexuality, 6, 8, 45, 65, 68, 147, 149, 190, 192, 198, 200, 201, 203, 204, 205, 206, 207, 213, 214, 216, 217, 221, 223
   and strategy, 18, 43, 68, 85, 87, 132, 139, 150, 190, 200, 204, 206, 209, 210, 213, 218, 221
   and tactics, 6, 56, 68, 150, 190, 200, 204, 206, 209, 210, 214, 216, 218

This page intentionally left blank

Udry, C.A. (2008, September-October). Balance sheet of U.S. imperialism. *International Socialist Review, 61*, 36–41.

U.S. Department of Health and Human Services. (2006). Curriculum requirements for abstinence-only education. Retrieved February 23, 2009, from: http://aspe.hhs.gov/hsp/08/AbstinenceEducation/report.pdf

Waller, M. (2008, September 24). LaBruzzo: Sterilization plan fights poverty. *The Times-Picayune.* Retrieved April 7, 2009, from: http://www.nola.com/news/t-p/capital/index.ssf?/base/news-6/122223371288730.xml&coll=1

Wallerstein, J. (1999). Clues. In W. Horn, D. Blankenhorn, and M. Pearlstein, (Eds.), *The fatherhood movement: A call to action (77–82).* New York: Lexington Books.

Wang, H. (2001). Women's 'oppression' and property relations: From Sati and bride-burning to late capitalist 'domestic labor' theories. In M. Zavarzadeh, T. Ebert, and D. Morton, (Eds.), *Marxism, queer theory, gender (pp.217–250).* Syracuse, NY: The Red Factory.

Warner, J. (2006). *Perfect madness: Motherhood in the age of anxiety.* New York: Riverhead Books.

Warnick, B. (2009, April). Student speech rights and the special characteristics of the school environment. *Educational Researcher, 38(3),* 200–215.

Weiner, L. (2008). Building the international movement we need: Why a consistent defense of Democracy and equality is essential. In M. Compton and L. Weiner, (Eds.), *The global assault on teaching, teachers, and their unions: Stories for resistance (pp.251–268).* New York: Palgrave-MacMillan.

Whitehead, B. (1999). Clueless generation. In W. Horn, D. Blankenhorn, and M. Pearlstein, (Eds.), *The fatherhood movement: A call to action (65–75).* New York: Lexington Books.

Wilhelm, S. (2009). A crisis in abortion funding. *Socialist Worker.* Retrieved July 7, 2009, from: http://socialistworker.org/print/2009/07/07/crisis-in-abortion-funding

Willett, C. (2002). Parenting and other human casualties in the pursuit of academic excellence. In A. Superson and A. Cudd, (Eds.), *Theorizing backlash: Philosophical reflections on the resistance to feminism (pp.119–131).* New York: Rowan & Littlefield Publishers, Inc.

Wilson, J.Q. (1996, March). Against homosexual marriage. *Commentary.* In A. Sullivan, (Ed.), *Same-sex marriage pro & con: A reader (pp.160–169).* New York: Vintage Books.

Wolf, S. (2009a, January-February). Stonewall: The birth of gay power. *ISR, 63,* 40–45.

Wolf, S. (2009b, January-February). Teamsters and trannies, unite! *ISR, 63,* 6–7.

Wolf, S. (2009c). *Sexuality and socialism: History, politics, and theory of lgbt liberation.* Chicago, IL: Haymarket Books.

Xiridou, M., Geskus, R., de Wit, J., Coutinho, R., and Kretzschmar, M. (2003, May 2). The contribution of steady and casual partnerships in the incidence of HIV infection among homosexual men in Amsterdam. *AIDS, 17(7),* 1029–1038.

Yalom, M. (2001). *A history of the wife.* New York: Harper Collins Publishers.

Stewart, C. (2008). Work on Aboriginal education in a social justice union: Reflections from the inside. In M. Compton and L. Weiner, (Eds.), *The global assault on teaching, teachers, and their unions: Stories for resistance (pp.173–176)*. New York: Palgrave-MacMillan.

Suber, T. (2007). Where is the movement? Retrieved January 18, 2009 from: http://richgibson.com/wheremovement.htm.

Suber, T. (2009). Whither the anti-war movement? Retrieved April 7, 2009, from: http://www.richgibson.com/rouge_forum/newspaper/spring2009/index.html#wither

Sullivan, A., (Ed.). (2004). *Same-sex marriage pro & con: A reader*. New York: Vintage Books.

Sullivan, A. (1996). *Virtually normal: An argument about homosexuality*. New York: Vintage Books.

Sunshine, S. (2008, Winter). Rebranding fascism: National-anarchists. *The Public Eye, 23(4)*, 1, 12–19.

Superson, A. (2002). Welcome to the boys' club: Male socialization and the backlash against feminism in tenure decisions. In A. Superson and A. Cudd, (Eds.), *Theorizing backlash: Philosophical reflections on the resistance to feminism (pp.89–117)*. New York: Rowan & Littlefield Publishers, Inc.

Sutherland, K. (2005, January). Marx and MacKinnon: The promise and perils of Marxism for feminist legal theory. *Science & Society, 69(1)*, 113–132.

Suskind, R. (2003, January 1). Why are these men laughing? *Esquire*. Retrieved March 27, 2009, from: http://74.125.47.132/search?q=cache:LrNZ0qrL2wkJ:www.ronsuskind.com/newsite/articles/archives/000032.html+john+diiulio+mayberry&cd=4&hl=en&ct=clnk&gl=us.

Sycamore, M.B. (2008). There's more to life than platinum. In M.B. Sycamore, (Ed.), *That's revolting! Queer strategies for resisting assimilation (pp.1–7)*. New York: Soft Skull Press.

Sykes, H. (2004, January/February). Pedagogies of censorship, injury, and masochism: Teacher responses to homophobic speech in physical education. *Journal of Curriculum Studies, 36(1)*, 75–99.

Thoreau, H. (2005). *Walden*. England: Digireads.com

Tierney, W. (1997). *Academic outlaws: Queer theory and cultural studies in the academy*. New York: Sage Publications.

Tilly, L. (1986). Paths of proletarianization: Organization of production, sexual division of labor, and women's collective action. In E. Leacock and H. Safa, (Eds.), *Women's work (25–40)*. MA: Bergin & Garvey Publishers.

Tinker v. Des Moines. (1969). Retrieved November 1, 2009, from: http://www.oyez.org/cases/1960-1969/1968/1968_21

Thweatt, D. and Rose, S.E. (2009, May 18). Showdown in D.C. over marriage equality. *Socialist Worker*. Retrieved May 20, 2009, from: http://socialistworker.org/2009/05/18/showdown-over-marriage-equality

Toppo, G. (2009, March 17). Gay themed film cost Oklahoma teacher her job. *USA Today*. Retrieved November 8, 2009, from: http://www.usatoday.com/news/education/2009-03-16-teacher-laramie_N.htm

Torosyan, R. (2007). Public discourse and the Stewart model of critical thinking. In J. Holt, (Ed.), *The Daily Show and philosophy (pp.107–120)*. Malden, MA: Blackwell Publishing.

Trahair, R.C.S. (1999). *Utopia and utopians: An historical dictionary*. Westport, CT: Greenwood Press.

Trotsky, L. (1970). *Women and the family*. New York: Pathfinder Press.

Sears, A. (2005, January). Queer anti-capitalism: What's left of lesbian and gay liberation? *Science & Society, 69(1)*, 95–112.

Seidman, L.M. (2008, Fall). The Dale problem: Property and speech under the regulatory state. *University of Chicago Law Review, 75(4)*, 1541–1600.

Sexual Minority Assessment Research Team. (2009). *Best practices for asking questions about sexual orientation on surveys*. The Williams Institute. Retrieved November 5, 2009, from: http://www.law.ucla.edu/williamsinstitute/pdf/SMART_FINAL_Nov09.pdf

Shashahani, S. (1986). Mamasani women: Changes in the division of labor among a sedentarized pastoral people of Iran. In E. Leacock and H. Safa, (Eds.), *Women's work (111–121)*. MA: Bergin & Garvey Publishers.

Shepard, B. (2008). Sylvia and Sylvia's children. In M.B. Sycamore, (Ed.), *That's revolting! Queer strategies for resisting assimilation (pp.123–140)*. New York: Soft Skull Press.

Shepard, J. (2009). *The meaning of Matthew*. New York: Hudson Street Press.

Siegel, E. (2009, October 8). Oklahoma abortion law: Details to be posted online. *Huffington Post*. Retrieved December 26, 2009, from: http://www.huffingtonpost.com/2009/10/08/oklahoma-abortion-law-det_n_313779.html

Silver, N. (2009, October 22). Arguments against gay marriage literally stop making sense. *FiveThiryEight*. Retrieved October 26, 2009, from:
http://www.fivethirtyeight.com/2009/10/arguments-against-gay-marriage.html

Slack, M. (2009, May 9). Joe the plumber slurs gay people. *Huffington Post*. Retrieved May 9, 2009. From: http://www.huffingtonpost.com/2009/05/04/joe-the-plumber-queer-mea_n_196116.html

Smith, E. (2008). Stripping for the movement. In M.B. Sycamore, (Ed.), *That's revolting! Queer strategies for resisting assimilation (pp.296–304)*. New York: Soft Skull Press.

Smith, S. (2005). *Women and socialism: Essays on women's liberation*. Chicago, IL: Haymarket Books.

Smith, S. (2009). Needed: A new movement for abortion rights. *Socialist Worker*. Retrieved June 4, 2009, from: http://socialistworker.org/2009/05/28/new-movement-for-abortion-rights

Socarides, Ch. (1995). *Homosexuality: A freedom too far*. Roberkai Publishers.

Sommers, C.H. (2005). *Who stole feminism? How women have betrayed women*. New York: Simon & Schuster.

Somoza, L. (2009, June 1). This Mensa moment is brought to you by G. Gordon Liddy. *The Huffington Post*. Retrieved July 17, 2009, from: http://www.huffingtonpost.com/lora-somoza/this-mensa-moment-is-brou_b_209899.html

Spade, D. (2008). Fighting to win. In M.B. Sycamore, (Ed.), *That's revolting! Queer strategies for resisting assimilation (pp.47–53)*. New York: Soft Skull Press.

Spong, J.S. (1990). Blessing gay and lesbian commitments. In A. Sullivan, (Ed.), *Same-sex marriage pro & con: A reader (pp.67–71)*. New York: Vintage Books.

Spring, J. (2008). *The American school*. New York: MacMillan.

Stan, A. (2009, Spring). New tactics and coalitions take aim at Planned Parenthood. *The Public Eye, 24(1)*, 1; 8–12.

Stanton, E. (1898). *The woman's bible*. Retrieved July 21, 2009, from: http://www.sacred-texts.com/wmn/wb/

Steen Fenrich. (n.d.). *The LGBT Hate Crimes Project*. Retrieved October 27, 2009, from: http://www.lgbthatecrimes.org/doku.php/steen-fenrich#the-motive

Stein, B. (2006, November 26). In class warfare, guess which class is winning? *New York Times*. Retrieved June 30, 2010, from: http://www.nytimes.com/2006/11/26/business/yourmoney/26every.html

Rotello, G. (2007, October 4). If ENDA doesn't protect the transgendered, it doesn't protect me. *The Huffington Post*. Retrieved June 6, 2009, from: http://www.huffingtonpost.com/gabriel-rotello/if-enda-doesnt-protect-th_b_67202.html

Rousseau, H. (1762). *The social contract: Or principles of political right.* Retrieved September 2, 2009, from: http://www.constitution.org/jjr/socon.htm

Rowland, D. (2004). *The boundaries of her body: The troubling history of women's rights in America*. Naperville, IL: Sphinx Publishing.

Ruiz, N. (2006, Fall). The illusory nation: Why Pat Buchanan's America never has or will exist. *Public Resistance, 2 (1)*. Retrieved February 4, 2009, from: http://web.mac.com/publicresistance/iWeb/1.2/1.2%20Contents.html

Rust v. Sullivan. (1991). Retrieved July 23, 2009, from: http://www.oyez.org/cases/1990-1999/1990/1990_89_1391

Rustgi, S., Doty, M., and Collins, S. (2009). Women at risk: Why many women are forgoing needed health care. *The Commonwealth Fund*. Retrieved May 31, 2009, from: http://www.commonwealthfund.org/~/media/Files/Publications/Issue%20Brief/2009/May/Women%20at%20Risk/PDF_1262_Rustgi_women_at_risk_issue_brief_Final.pdf

Safa, H. (1986). Runaway shops and female employment: The search for cheap labor. In E. Leacock and H. Safa, (Eds.), *Women's work (58–71)*. MA: Bergin & Garvey Publishers.

Sagarin, E. (1969). *Odd man in: Societies of deviance in America*. New York: Quadrangle Books.

Santinover, J. (1996). *Homosexuality and the politics of truth*. Grand Rapids, MI: Baker Book House Co.

Sargent, L. (2008, December). Waiting in the wings. *Z Magazine, 21(12)*, 7–8.

Scalia's Dissent. (1995). *Romer v. Evans*. Retrieved May 23, 2009, from: http://74.125.95.132/search?q=cache:fD5RaYTXRMcJ:www.law.cornell.edu/supct/html/94-1039.ZD.html+romer+v.+evans+scalia+dissent&cd=1&hl=en&ct=clnk&gl=us&client=firefox-a.

Schaller, T.F. (2008). *Whistling past Dixie: How Democrats can win without the south*. New York: Simon and Schuster.

Schell, J. (2009, May 28). Torture and truth. *The Nation*. Retrieved July 7, 2009, from: http://www.cbsnews.com/stories/2009/05/28/opinion/main

Scher, A. (2008, Winter). Post-Palin feminism. *The Public Eye, 23(4)*, 1; 20–24.

Schroeder, S. (2008). Queer parents: An oxymoron? In M.B. Sycamore, (Ed.), *That's revolting! Queer strategies for resisting assimilation (pp.100–104)*. New York: Soft Skull Press.

Schulman, S. (2008). That incredible exhilaration. In M.B. Sycamore, (Ed.), *That's revolting! Queer strategies for resisting assimilation (pp.237–248)*. New York: Soft Skull Press.

Schulte, E. (2009a, January 9). Why is Rick Warren on Obama's guest list? *Socialist Worker*. Retrieved January 12, 2009, from: http://socialistworker.org/2009/01/09/rick-warren-on-obama-guest-list

Schulte, E. (2009b, February 11). Another challenge to DOMA. *Socialist Worker*. Retrieved February 11, 2009, from: http://socialistworker.org/2009/02/11/another-challenge-to-doma

Schulte, E. (2009c, March 18). The stark facts about violence against women. *Socialist Worker*. Retrieved March 23, 2009, from: http://socialistworker.org/print/2009/03/18/facts-about-violence-against-women

Schulte, E. (2009d, May 27). Protests erupt over California Prop 8 ruling. *Socialist Worker*. Retrieved May 29, 2009, from: http://socialistworker.org/2009/05/27/protests-erupt-over-prop-8

Planned Parenthood of Central Missouri v. Danforth. (1976). Retrieved October 20, 2009, from: http://www.law.cornell.edu/supct/html/historics/USSC_CR_0428_0052_ZS.html

Plato (360 B.C.E.). *Laws.* Retrieved May 23, 2009, from: http://classics.mit.edu/Plato/laws.8.viiii.html.

Popenoe, D. (1999). Challenging the culture of fatherlessness. In W. Horn, D. Blankenhorn, and M. Pearlstein, (Eds.), *The fatherhood movement: A call to action (17–24).* New York: Lexington Books.

Prager, D. (1990, April/June). Homosexuality, the bible, and us: A Jewish perspective. *Ultimate Issues.* In A. Sullivan, (Ed.), *Same-sex marriage pro & con: A reader (pp.61–66).* New York: Vintage Books.

Queen, C. (2008). Never a bridesmaid, never a bride. In M.B. Sycamore, (Ed.), *That's revolting! Queer strategies for resisting assimilation (pp.105–112).* New York: Soft Skull Press.

Rampton, S. (2008, December 3). Up close and personal: Mormon homophobia. *CounterPunch.* Retrieved December 10, 2008, from: http://www.counterpunch.org

Rauch, J. (1996, May 6). For better or worse? *The New Republic.* In A. Sullivan, (Ed.), *Same-sex marriage pro & con: A reader (pp.170–181).* New York: Vintage Books.

Ray, I. (2005). The Huntington High project. In K. Jennings, (Ed.), *One teacher in 10: LGBT educators share their stories* (pp.231–239). Los Angeles: Alyson Books.

Record, M. (2005). One more person to love them. In K. Jennings, (Ed.), *One teacher in 10: LGBT educators share their stories* (pp.255–260). Los Angeles: Alyson Books.

Redmond, H. and Colson, N. (2008, November 24). Desperate and nowhere to turn. *Socialist Worker.* Retrieved November 24, 2008, from: http://socialistworker.org/print/2008/11/24/desperate-and-nowhere-turn

Reich, W. (1971). *The invasion of compulsory sex-morality.* New York: Farrar, Straus, and Giroux.

Reich, W. (1970). *The mass psychology of fascism.* New York: Farrar, Straus, and Giroux.

Rikowski, G. and McClaren, P. (2002). Postmodernism in educational theory. In D. Hill, P. McClaren, M. Cole, and G. Rikowski, (Eds.), *Marxism against postmodernism in educational theory (pp.3–13).* New York: Oxford Books.

Roberts, G. (2005). Transitioning in the school system. In K. Jennings, (Ed.), *One teacher in 10: LGBT educators share their stories* (pp.170–177). Los Angeles: Alyson Books.

Robertson, S.L. (2008). "Remaking the world": Neoliberalism and the transformation of education and teachers' labor. In M. Compton and L. Weiner, (Eds.), *The global assault on teaching, teachers, and their unions: Stories for resistance (pp.11–27).* New York: Palgrave-MacMillan.

Roe v. Wade (1973). Retrieved May 23, 2009, from: http://www.oyez.org/cases/1970–1979/1971/1971_70_18

Roiphe, K. (1994). *The morning after: Sex, fear, and feminism.* New York: Back Bay Books.

Rosado, J.A.. (2008). Corroding our quality of life. In M.B. Sycamore, (Ed.), *That's revolting! Queer strategies for resisting assimilation (pp.317–328).* New York: Soft Skull Press.

Rosenbury, L.A. (2007, April). Between home and school. *University of Pennsylvania Law Review, 155(4),* 833–898.

Ross, L. (2009, June 2). Repeal Hyde: Even republicans know it's wrong to politick with women's lives. *On the Issues.* Retrieved July 21, 2009, from: www.ontheissuesmagazine.com/cafe2/article/46

Ross, T. (2008, September 26). Nebraska lawmakers consider revising 'safe-haven' law. *Yahoo!News.* Retrieved September 26, 2008, from: www.http:/news.yahoo.com/s/ap/20080926/ap_on_re_us/children_safe_haven

Nemec, B. and SalMonella. (2008). Riding radio to choke the IMF. In M.B. Sycamore, (Ed.), *That's revolting! Queer strategies for resisting assimilation (pp.162–173)*. New York: Soft Skull Press.

Nicholson, L. (1992, Summer). Feminism and the politics of postmodernism. *Boundary2, 19(2)*, 53–70.

Nicolari, P.A. (2005). If Ellen can do it, so can I. In K. Jennings, (Ed.), *One teacher in 10: LGBT educators share their stories* (pp.16–24). Los Angeles: Alyson Books.

Nightlight Christian adoptions website. (n.d.). Retrieved July 24, 2009, from: http://www.nightlight.org/adoption-services/snowflakes-embryo/

Nowlan, B. (2001). Post-Marxist queer theory and 'the politics of AIDS'. In M. Zavarzadeh, T. Ebert, and D. Morton, (Eds.), *Marxism, queer theory, gender (pp.115–154)*. Syracuse, NY: The Red Factory.

O' Brien, D. (Director). (2008). *Equality U*. [Documentary]. U.S.A.: Eye Think Pictures. (Available from: http://eyethinkpictures.com/EYETHINK_PICTURES.html).

O'Brien, M. (2008). Stayin' alive: Trans survival and struggle on the streets of Philadelphia. In M.B. Sycamore, (Ed.), *That's revolting! Queer strategies for resisting assimilation (pp.305–311)*. New York: Soft Skull Press.

Ognibene, R. (2005). The importance of discomfort. In K. Jennings, (Ed.), *One teacher in 10: LGBT educators share their stories* (pp.223–230). Los Angeles: Alyson Books.

Omnibus appropriations bill. (2009). Retrieved October 19, 2009, from: http://appropriations.house.gov/pdf/2009_Con_Bill_DivF.pdf

Page, C. (2006). *How the pro-choice movement saved America: Freedom, politics, and the war on sex*. New York: Basic Books.

Pearlstein, M. (1999). Fatherhood and language. In W. Horn, D. Blankenhorn, and M. Pearlstein, (Eds.), *The fatherhood movement: A call to action (127–131)*. New York: Lexington Books.

Pedersen, N. (2009, Summer). Lammas Day. *Renaissance, 14(4)*, 26.

Pelham, B. and Crabtree, S. (2009, March 10). *Religiosity and perceived intolerance of gays and lesbians*. Retrieved November 2, 2009, from: http://www.gallup.com/poll/116491/religiosity-perceived-intolerance-gays-lesbians.aspx

Perroti, J. and Westheimer, K. (2001). *When the drama club is not enough: Lessons from the Safe Schools Program for gay and lesbian students*. Boston, MA: Beacon Press.

Personal responsibility and work opportunity reconciliation act. (1996). Retrieved October 19, 2009, from: http://thomas.loc.gov/cgi-bin/query/z?c104:H.R.3734.ENR:

Peterson, L. (2009, Winter). Turning the tables. *Bitch, 42*, 35–40.

Petr, B. (2005). Going back to where I grew up gay. In K. Jennings, (Ed.), *One teacher in 10: LGBT educators share their stories* (pp.199–202). Los Angeles: Alyson Books.

Petrovic, J.E. (2002, July). Promoting democracy and overcoming heterosexism: And never the twain shall meet? *Sex Education, 2(2)*, 145–154.

PFLAG website. (n.d.). Retrieved June 30, 2010, from: http://community.pflag.org/Page.aspx?pid=194&srcid=-2

Pilger, J. (2009, November/December). Power, illusion, and America's last taboo. *ISR, 68*, 24–27.

Pinar, W. (1998). *Queer theory in education*. New York: Lawrence Erlbaum.

Pinar, W., Reynolds, W.M., Slattery, P., and Taubman, P.M. (1995). *Understanding curriculum*. New York: Peter Lang.

Piven, F. and Cloward, R. (1993). *Regulating the poor: The functions of public welfare*. New York: Random House.

Planned Parenthood v. Casey. (1992). Retrieved July 24, 2009, from: http://caselaw.lp.findlaw.com/scripts/getcase.pl?court=US&vol=505&invol=833

Meacham, J. and Quinn, S. (2007, July 9). The religion-industrial complex. *The Washington Post*. Retrieved June 30, 2010, from: http://newsweek.washingtonpost.com/onfaith/georgetown/2007/07/the_religionindustrial_complex.html

Mecca, T.A. (2008). It's all about class. In M.B. Sycamore, (Ed.), *That's revolting! Queer strategies for resisting assimilation (pp.29–38)*. New York: Soft Skull Press.

Meyer, E.J. (2009). *Gender, bullying, and harassment: Strategies to end sexism and homophobia in schools*. New York: Teachers College Press.

Mincy, R. and Pouncy, H. (1999). There must be 50 ways to start a family. In W. Horn, D. Blankenhorn, and M. Pearlstein, (Eds.), *The fatherhood movement: A call to action (83–104)*. New York: Lexington Books.

Minge, W. (1986). The industrial revolution and the European family: "Childhood" as a market for family labor. In E. Leacock and H. Safa, (Eds.), *Women's work (13–24)*. MA: Bergin & Garvey Publishers.

Moeller, C. (2002). Marginalized voices: Challenging dominant privilege in higher education. In A. Superson and A. Cudd, (Eds.), Theorizing backlash: Philosophical reflections on the Resistance to feminism (pp.155–180). New York: Rowan & Littlefield Publishers, Inc.

Montgomery v. Independent School District No. 709. (2000). Retrieved November 1, 2009, from: http://www.lexisnexis.com.library.aurora.edu/us/lnacademic/search/casessubmitForm.do

Mooney, C. (2005). *The republican war on science*. New York: MJF Books.

Morgan, L.H. (1877). *Ancient society*. New York: Cornell University.

Morris, A. (1984). *The origins of the civil rights movement: Black communities organizing for change*. New York: The Free Press.

Morton, D. (2001). Pataphysics of the closet: Queer theory as the art of imaginary solutions for unimaginary problems. In M. Zavarzadeh, T. Ebert, and D. Morton, (Eds.), *Marxism, queer theory, gender (pp.1–70)*. Syracuse, NY: The Red Factory.

Moynihan, D.P. (1965). *The negro family: A case for national action*. Retrieved September 4, 2009, from: http://www.dol.gov/oasam/programs/history/webid-meynihan.htm

Mullings, L. (1986). Uneven development: Class, race, and gender in the United States before 1900. In E. Leacock and H. Safa, (Eds.), *Women's work (41–57)*. MA: Bergin & Garvey Publishers.

Murphy, K.A. (2008). Teachers and their unions: Why social class "counts." In M. Compton and L. Weiner, (Eds.), *The global assault on teaching, teachers, and their unions: Stories for resistance (pp.75–79)*. New York: Palgrave-MacMillan.

Murray, C. (1994). *Losing ground: American social policy 1950–1980*. New York: Basic Books.

Murray, C. (2006). *In our hands: A plan to replace the welfare state*. Washington, D.C.: AEI Press.

Murray, C. and Herstein, R. (1996). *The bell curve: Intelligence and class structure in American life*. New York: Free Press Paperbacks.

Murray, S.A. (2008, October). Abstinence-only sex education and the democratic deficit. *Z Magazine, 21(10)*, 32–35.

Murrow, D. (2005). *Why men hate going to church*. Nashville, TN: Thomas Nelson.

Nabozny v. Podlesny. (1996). Retrieved November 1, 2009, from: http://www.lexisnexis.com.library.aurora.edu/us/lnacademic/search/casessubmitForm.do

National school climate survey. (2005). *Executive Summary- GLSEN*. Retrieved October 11, 2009, from: http://www.glsen.org/binary-data/GLSEN_ATTACHMENTS/file/582-2.pdf

Neill, A.S. (1960). *Summerhill: A radical approach to child rearing*. New York: Hart Publishing Co.

Lakoff, G. (2002). *Moral politics: How liberals and conservatives think.* Chicago, IL: University of Chicago Press.

Lapon, G. (2009, May 18). Why women need single payer. *Socialist Worker.* Retrieved May 31, 2009, from: http://socialistworker.org/print/2009/05/18/why-women-need-single-payer

Lawrence v. Texas 478 U.S. 186 (2003). Retrieved May 23, 2009, from: http://www.law.cornell.edu/supct/html/02-102.ZS.html

Lazarus, M.E. (1852). *Love vs. marriage.* New York: Fowler and Wells.

Leacock, E. (1981). *Myths of male dominance: Collected articles on women cross-culturally.* Chicago: Haymarket Books.

Leacock, E. (1986). Postscript: Implications for organization. In E. Leacock and H. Safa, (Eds.), *Women's work (253–265).* MA: Bergin & Garvey Publishers.

Levine, B.E. (2008, October). Psycho-pharmaceutical industrial complex. *Z Magazine, 21(10),* 28–31.

Linkins, J. (2009, May 1). Judy Shepard responds to "hoax" comments. Huffington Post. Retrieved May 19, 2009, from: http://www.huffingtonpost.com/2009/05/01/judy-shepard-responds-to_n_194483.html

Lipow, G. (2008, July/August). Cooling a fevered planet. *Z Magazine, 21(7/8),* 43–47.

Loewen, J. (2005). *Sundown towns: A hidden dimension of American racism.* New York: The New Press.

Lowry, L. (1993). *The giver.* New York: Bantam Books.

Luttrell, W. (2003). *Pregnant bodies, fertile minds: Gender, race, and the schooling of pregnant teens.* New York: Routledge.

Lynd, S. and Grubacic, A. (2008). *Wobblies and Zapatistas: Conversations on anarchism, Marxism, and radical history.* Oakland, CA: PM Press.

Lyons, P. (2005). Feeling all right on the religious left. In K. Jennings, (Ed.), *One teacher in 10: LGBT educators share their stories* (pp.68–78). Los Angeles: Alyson Books.

MacDonald, D. (2009, March 10*). Gay sex is downright dangerous and abstinence won't kill you: I should know.* Free Republic Website. Retrieved March 31, 2009, from: http://74.125.95.132/search?q=cache:mxM2Ker_epsJ:www.freerepublic.com/focus/f-news/2204723/posts+gays+should+be+abstinent&cd=15&hl=en&ct=clnk&gl=us

Mann, S.A. and Huffman, D.J. (2005, January). The decentering of second wave feminism and the rise of the third wave. *Science & Society, 69(1),* 59–61.

Marriage Protection Week, full text. (2003). Retrieved May 29, 2009, from: http://marriage.about.com/cs/newsandviews/f/marprotfaq2.htm

Martin, J. (2006). Gender and education. In M. Cole, (Ed.), *Education, equality and human rights: Issues of gender, race, sexuality, disability, and social class (pp.22–42).* London: Routledge.

Marx, K. (1998). *The German ideology.* New York: Bantam Books.

Marx, K. (1977). *Capital: Volume I.* New York: Random House.

Marx, K. (1973). *Grundrisse.* London: Penguin Books.

Maybee, J. (2002). Politicizing the personal and other tales from the front lines. In A. Superson and A.Cudd, (Eds.), *Theorizing backlash: Philosophical reflections on the resistance to feminism (pp.133–152).* New York: Rowan & Littlefield Publishers, Inc.

McClaren, P. and Farahmandpur, R. (2002). Breaking signifying chains: A Marxist position on postmoderninsm. In D. Hill, P. McClaren, M. Cole, and G. Rikowski, (Eds.), *Marxism against postmodernism in educational theory (pp.35–66).* New York: Oxford Books.

McGowan, D. (2001, Spring). Making sense of Dale. *Constitutional Commentary, 18(1),* 121–176.

Johnson, D. (1990, April 9). Ryan White dies of AIDS at age 18. *The New York Times*. Retrieved June 13, 2009, from: http://www.nytimes.com/1990/04/09/obituaries/ryan-white-dies-of-aids-at-18-his-struggle-helped-pierce-myths.html

Johnson, R. (2009, September-October). Why we shouldn't wait. *International Socialist Review, 67*, 2-3.

Kahn, Y.H. (1989). The kedushah of homosexual relationships. In A. Sullivan, (Ed.), *Same-sex marriage pro & con: A reader (pp.71–77)*. New York: Vintage Books.

Kaplan, L. (1995). *The story of Jane: The legendary underground feminist abortion service*. Chicago, IL: The University of Chicago Press.

Kelly, J. (2006). Women thirty-five years on: Still unequal after all this time. In M. Cole, (Ed.), *Education, equality and human rights: Issues of gender, race, sexuality, disability, and social class (pp.7–21)*. London: Routledge.

Kelly, J. (2002). Women, work, and the family: Or why postmodernism cannot explain the links. In D. Hill, P. McClaren, M. Cole, and G. Rikowski, (Eds.), *Marxism against postmodernism in educational theory (pp.211–235)*. New York: Oxford Books.

Kelly, M. (1986). Introduction. In E. Leacock and H. Safa, (Eds.), *Women's work (1–10)*. MA: Bergin & Garvey Publishers.

Khan, H. and Tapper, J. (2009, May 27). Newt Gingrich on Twitter: Sotomayor 'racist,' should withdraw. *ABC News*. Retrieved July 17, 2009, from: http://abcnews.go.com/Politics/SoniaSotomayor/story?id=7685284&page=1

Khan, S. (2009, Spring). Tying the not: How the right succeeded in passing Proposition 8. *The Public Eye, 24(1)*, 3–7.

Khanlarzadheh, M. (2009, February). Iranian women and economic sanctions. *Z Magazine, 22(2)*, 38–40.

Kinsey, A. (1998). *Sexual behavior in the human male*. Bloomington, IN: Indiana University Press.

Kivel, P. (1990). *The men's movement, the Promise Keepers, and other forms of male backlash*. Retrieved July 25, 2009, from: http://www.paulkivel.com

Kornegay, J. (2009, April 24). Bullied to death. *Socialist Worker*. Retrieved June 4, 2009, from: http://socialistworker.org/2009/04/24/bullied-to-death

Kosciw, J. and Diaz, E. (2008). *Involved, invisible, ignored: The experiences of lesbian, gay, bisexual, and transgender parents and their students in our nation's k-12 schools*. New York: GLSEN.

Kosciw, J., Diaz, E., and Greytak, E. (2008). *2007 national school climate survey: The experiences of lesbian, gay, bisexual, and transgender youth in our nation's schools*. New York: GLSEN.

Kotulski, D. (2004). *Why you should give a damn about gay marriage*. Los Angeles, CA: Advocate Books.

Krauthammer, C. (1996, July 22). When John and Jim say 'I do.' *Time*. In A. Sullivan, (Ed.), *Same-sex marriage pro & con: A reader (pp.282–285)*. New York: Vintage Books.

Kuban, K. and Grinnell, C. (2008). More Abercrombie than activist? In M.B. Sycamore, (Ed.), *That's revolting! Queer strategies for resisting assimilation (pp.74–86)*. New York: Soft Skull Press.

Kuehn, L. (2008). "The education world is not flat: Neoliberalism's global project and teacher uinons' transnational resistance. In M. Compton and L. Weiner, (Eds.), *The global assault on teaching, teachers, and their unions: Stories for resistance (pp.53–72)*. New York: Palgrave-MacMillan.

Kupfer, R. (2005). Living with the possibilities. In K. Jennings, (Ed.), *One teacher in 10: LGBT educators share their stories (pp.288–294)*. Los Angeles: Alyson Books.

Hammam, M. (1986). Capitalist development, family division of labor, and migration in the Middle East. In E. Leacock and H. Safa, (Eds.), *Women's work (158–173)*. MA: Bergin & Garvey Publishers.

Haraway, D. (1990). *Simians, cyborgs, and women: The reinvention of nature*. New York: Routledge.

Hardisty, J. (2008, Spring). Pushed to the altar: The right-wing roots of marriage promotion. *The Public Eye, 23(1)*, 1, 21–27.

Harris Interactive & GLSEN. (2005). *From teasing to torment: School climate in America, a survey of students and teachers*. New York: GLSEN.

Havenworks. (n.d.). Retrieved April 7, 2009, from: http://www.havenworks.com/people/a-z/a/adkisson-jim-david/

Hayden, D. (2009, July/August). Redesigning the American dream. *Z Magazine, 7/8*, 47–52.

The healthcare costs of having a baby. (2007, June). *March of Dimes*. Retrieved May 31, 2009, from: http://www.marchofdimes.com/files/Thomson.pdf

Henkle v. Gregory. (2001). Retrieved November 1, 2009, from: http://www.lambdalegal.org/in-court/cases/henkle-v-gregory.html

Henry, G. (1948). *Sex variants: A study of homosexual patterns*. New York: Paul Hoeber, Inc.

Henry, R. (1999). After the divorce. In W. Horn, D. Blankenhorn, and M. Pearlstein, (Eds.), *The fatherhood movement: A call to action (105–116)*. New York: Lexington Books.

Heywood, E.H. (1876). *Yours or mine: An essay to show the true basis of property*. Princeton, MA: Co-Operative Publishing Co.

Hickey, T. (2006). Multitude or class: Constituencies of resistance, sources of hope. In M. Cole, (Ed.), *Education, equality and human rights: Issues of gender, race, sexuality, disability, and social class (pp.180–201)*. London: Routledge.

Hill, D., Sanders, M. & Hankin, T. (2002). Marxism, class analysis, and postmodernism. In D. Hill, P. McClaren, M. Cole, & G. Rikowski, (Eds.), *Marxism against postmodernism in educational theory (pp.159–194)*. New York: Oxford Books.

Hillebrand, R. (n.d.). The Oneida community. *New York History Net*. Retrieved August 15, 2009, from: http://www.nyhistory.com/central/oneida.htm

Holt, A. (1977). *Selected writings of Alexandra Kollontai*. New York: W.W. Norton & Company.

Horn, W. (1999). Did you say "movement?" In W. Horn, D. Blankenhorn, and M. Pearlstein, (Eds.), *The fatherhood movement: A call to action (1–16)*. New York: Lexington Books.

H.R. 1592. (2007). *The Matthew Shepard and James Byrd Jr. Act*. Retrieved October 27, 2009, from: http://www.govtrack.us/congress/billtext.xpd?bill=h110-1592

Hypolitio, A.M. (2008). Educational restructuring, democratic education, and teachers. In M. Compton, and L. Weiner, (Eds.), *The global assault on teaching, teachers, and their unions: Stories for resistance (pp.149–160)*. New York: Palgrave-MacMillan.

Jennings, K. (Ed.). (2005). *One teacher in 10: LGBT educators and their stories*. Los Angeles: Alyson Books.

Jill. (2007). The company you keep. *Feministe*. Retrieved July 21, 2009, from: http://www.feministe.us/blog/archives/2007/07/18/the-company-you-keep/

Jindal, P. (2008). Sites of resistance or sites of racism? In M.B. Sycamore, (Ed.), *That's revolting! Queer strategies for resisting assimilation (pp.39–46)*. New York: Soft Skull Press.

John, J. (1993). Creation and natural law. In A. Sullivan, (Ed.), *Same-sex marriage pro & con: A reader (pp.78–81)*. New York: Vintage Books.

Forrest, S. and Ellis, V. (2006). The making of sexualities. In M. Cole, (Ed.), *Education, equality and human rights: Issues of gender, race, sexuality, disability, and social class (pp.89–110).* London: Routledge.

Frank, T. (2004). *What's the matter with Kansas? How conservatives won the heart of America.* New York: Henry Holt and Co.

Fraser, S. (2005). *Ruling America: A history of wealth and power in a democracy.* Harvard UniversityPress.

Freedom of Choice Act. (2009). Retrieved July 24, 2009, from: http://www.prochoiceamerica.org/assets/files/Abortion-Access-to-Abortion-FOCA.pdf

Freeman, R. (2009, March 15). The U.S. is facing a Weimar moment. *Common Dreams.* Retrieved March 27, 2009, from: http://www.commondreams.org/view/2009/03/15.

Freedman, J.M. (2005). A good day. In K. Jennings, (Ed.), *One teacher in 10: LGBT educators share their stories* (pp.7–15). Los Angeles: Alyson Books

Gallagher, M. (1999). The importance of being married. In W. Horn, D. Blankenhorn, and M. Pearlstein, (Eds.), *The fatherhood movement: A call to action (57–64).* New York: Lexington Books.

Gandossy, T. (2009, October 28). Census will report same-sex couples, gay groups see opportunity. *CNN.com.* Retrieved November 5, 2009, from: http://www.cnn.com/2009/US/10/28/same.sex.census/index.html

Gasper, P. (2004, November-December). Is biology destiny? *ISR Magazine, 38.* Retrieved June 13, 2009, from: http://www.isreview.org/issues/38/genes.shtml

Gates, G. (2006). Same-sex couples and the gay, lesbian, bisexual Population: New estimates from the American Community Survey. *The Williams Institute.*Retrieved November 5, 2009, from: http://escholarship.org/uc/item/8h08t0zf

Geier, J. (2009, July/August). The free fall is over, but..., *ISR Magazine, 66,* 29–32.

Gilder, G. (1986). *Men and marriage.* Gretna, LA: Pelican Publishing Company.

Gilder, G. (1981). *Wealth & poverty.* New York: Bantam Books.

Gilman, C.P. (1994). *Women and economics: A study of the economic relation between women and men.* Amherst, NY: Prometheus Books.

Gimenez, M.E. (2005, January). Capitalism and the oppression of women: Marx revisited. *Science & Society, 69(1),* 11–32.

GLSEN and Harris Interactive. (2008). *The principal's perspective: School safety, bullying, and harassment, a survey of public school principals.* New York: GLSEN.

GLSEN website. (n.d.). Retrieved June 30, 2010, from: http://www.glsen.org/cgi-bin/iowa/all/home/index.html

Gold, S. (2005). The best way out was through. In K. Jennings, (Ed.), *One teacher in 10: LGBT educators share their stories* (pp.261–263). Los Angeles: Alyson Books.

Goldman, E. (1911). Marriage and love. *Anarchy Archives.* Retrieved August 11, 2009, from: http://dwardmac.pitzer.edu/Anarchist_Archives/goldman/aando/marriageandlove.html

Goldstein, R. (2003). *Homocons: The rise of the gay right.* New York: Verso.

Goode, E. (2003, April 13). Certain words can trip up AIDS grants, scientists say. *The New York Times,* p.A10.

Griswold v. Connecticut. (1965). Retrieved May 23, 2009, from: http://www.oyez.org/cases/1960-1969/1964/1964_496/

Grubacic, A. (2008, July/August). Roots of resistance: An interview with Roxanne Dunbar-Ortiz. *Z Magazine, 21(7/8),* 51–55.

Gutmann, D. (1999). The species narrative. In W. Horn, D. Blankenhorn, and M. Pearlstein, (Eds.), *The fatherhood movement: A call to action (133–145).* New York: Lexington Books.

Eisenstadt v. Baird (1972). Retrieved July 21, 2009, from: http://www.oyez.org/cases/1970-1979/1971/1971_70_17

Ellis, V. and High, S. (2004, April). Something more to tell you: Gay, lesbian or bisexual young people's experiences of secondary schooling. *British Educational Research Journal, 30(2),* 213–225.

Elshtain, J.B. (1991, October 22). Against gay marriage. *Commonweal.* In A. Sullivan, (Ed.), *Same-sex marriage pro & con: A reader (pp.52–60).* New York: Vintage Books.

Employment Non-Discrimination Act (H.R.3017). (2009). Retrieved November 2, 2009, from: http://edlabor.house.gov/hearings/2009/09/hr-3017-employment-non-discrim.shtml

Engels, F. (1972). *The origin of the family, private property, and the state.* New York: International Publishers.

The Episcopal Church. (n.d.). *Religioustolerance.org.* Retrieved November 2, 2009, from: http://www.religioustolerance.org/hom_epis.htm

Erwin, P. (2008, September 15). Helicopter parents on the job search. *Career Builder.* Retrieved September 24, 2009, from: http://www.careerbuilder.com/Article/CB-1011-Job-Search-Helicopter-Parents-on-the-Job-Search/undefined?ArticleID=1011&cbRecursionCnt=1&cbsid=7f10d38824434786b231ae32cb7ad3f0-307137865-wt-6&ns_siteid=ns_us_g_helicopter_parents_mo

Eskridge, W.N. (2008). *Dishonorable passions: Sodomy laws in America (1861–2003).* New York:Viking.

Ettelbrick, P. (1989, Fall). Since when is marriage a path to liberation? *OUT/LOOK National Gay and Lesbian Quarterly.* In A. Sullivan, (Ed.), *Same-sex marriage pro & con: A reader (pp.122–128).* New York: Vintage Books.

Exodus International website. (n.d.). Retrieved June 6, 2009, from: http://www.exodus-international.org/

Federal Equal Access Act, 20 U.S.C. 4071–74. (1984). Retrieved November 1, 2009, from: http://www.religioustolerance.org/equ_acce.htm

Federal Marriage Amendment, H.J. Res 56. (2003). Retrieved May 29, 2009, from: http://usgovinfo.about.com/cs/usconstitution/a/marriage.htm.

Feldt, G. (2004). *The war on choice: The right wing attack on women's rights and how to fight back.* New York: Bantam Books.

Figueroa, L. (2009, February 6). Gay woman fights over hospital visitation rights in Miami court. *Miami Herald.* Retrieved February 11, 2009, from: http://www.miamiherald.com/living/v-print/story/982447.html

Filipovic, J. (2009, June 1). The Tiller murder wasn't a lone killer's sick plot. *AlterNet.* Retrieved July 4, 2009, from: http://www.alternet.org/reproductivejustice/140387/the_tiller_murder_wasn%27t_a_lone_killer%27s_sick_plot%3B_it_came_out_of_the_radical_anti-abortion_movement/

Flores v. Morgan Hill. (2003). Retrieved November 1, 2009, from: http://www.lexisnexis.com.library.aurora.edu/us/lnacademic/results/docview/docview.do?docLinkInd=true&risb=21_T7751635682&format=GNBFI&sort=BOOLEAN&startDocNo=1&resultsUrlKey=29_T7751635685&cisb=22_T7751635684&treeMax=true&treeWidth=0&csi=6320&docNo=1

Ford, J. and Jasinski, J.L. (2006, March). Sexual orientation and substance use among college students. *Addictive Behaviors, 31(3),* 404–413.

Fone, B. (2000). *Homophobia: A history.* New York: Picador USA.

Forrest, S. (2006). Challenges in teaching and learning about sexuality and homophobia in schools. In M. Cole, (Ed.), *Education, equality and human rights: Issues of gender, race, sexuality, disability, and social class (pp.111–133).* London: Routledge.

Dave Thomas Foundation. (2002, June 19). *Landmark study shows vast majority of Americans support adoption.* Retrieved November 2, 2009, from: http://statistics.adoption.com/information/foster-care-statistics.html
Davis, A. (1981). *Women, race, and class.* New York: Vintage Books.
Davis, B. (2005). The marriage question. In K. Jennings, (Ed.), *One teacher in 10: LGBT educators share their stories* (pp.25–33). Los Angeles: Alyson Books.
Defense of Marriage Act. (1996). Retrieved November 5, 2009, from: http://frwebgate.access.gpo.gov/cgi-bin/getdoc.cgi?dbname=104_cong_public_laws&docid=f:publ199.104
D'Emilio, J. (1983). *Sexual politics, sexual communities: The making of a homosexual minority in the United States, 1940–1970.* Chicago, IL: University of Chicago Press.
D'Emilio, J. (1992). *Making trouble: Essays on gay history, politics, and the university.* New York: Routledge.
D'Emilio, J. (2003). *The world turned: Essays on gay history, politics, and culture.* Durham, N.C.: Duke University Press.
D'Emilio, J. (2009, May/June). LGBT liberation: Build a broad movement. *International Socialist Review,* 65, 21–23.
D'Emilio, J. and Freedman, E.B. (1997). *Intimate matters: A history of sexuality in America.* Chicago, IL: The University of Chicago Press.
Dickinson, T. (2008, December 11). Same-sex setback. *Rolling Stone,* 45–47.
Dickinson, T. (2009, May 28). The GOP jihad. *Rolling Stone,* 40–45.
Digby. (2008, July 28). *Last year's fashion.* Retrieved May 12, 2009, from: http://74.125.95.132/search?q=cache:1EQt0Vc8-msJ:digbysblog.blogspot.com/2008/07/last-years-fashion-by-digby-one-of-most.html+digby+religion+industrial+complex&cd=1&hl=en&ct=clnk&gl=us&client=firefox-a
Diggs, J.R. (2002). The health risks of gay sex. *Corporate Resource Council.* Retrieved June 13, 2009, From: http://www.corporateresourcecouncil.org/white_papers/Health_Risks.pdf
Doe v. Bellefonte Area School District. (2004). Retrieved November 1, 2009, from: http://www.lexisnexis.com.library.aurora.edu/us/lnacademic/search/casessubmitForm.do
Doe v. Brockton School Community. (2000). Retrieved November 1, 2009, from: http://www.lexisnexis.com.library.aurora.edu/us/lnacademic/results/docview/docview.do?docLinkInd=true&risb=21_T7751655927&format=GNBFI&sort=BOOLEAN&startDocNo=1&resultsUrlKey=29_T7751655930&cisb=22_T7751655929&treeMax=true&treeWidth=0&csi=7682&docNo=2
Dolor, U. (2008). Homophobia in St. Lucian schools: A perspective from a select group of teachers. In M. Compton and L. Weiner, (Eds.), *The global assault on teaching, teachers, and their unions: Stories for resistance (pp.169–172).* New York: Palgrave-MacMillan.
Eberly, D. (1999). No democracy without dads. In W. Horn, D. Blankenhorn, and M. Pearlstein, (Eds.), *The fatherhood movement: A call to action (25–33).* New York: Lexington Books.
Ebert, T. (2005, January). Rematerializing feminism. *Science & Society, 69(1),* 33–35.
Ebert, T. (2001). The spectral concrete: Bodies, sex work and (some notes on) citizenship. In M. Zavarzadeh, T. Ebert, and D. Morton, (Eds.), *Marxism, queer theory, gender (pp.275–303).* Syracuse, NY: The Red Factory.
The Editors. (1996, January 6). Let them wed. *The Economist.* In A. Sullivan, (Ed.), *Same-sex marriage pro & con: A reader* (pp.182–185). New York: Vintage Books.
Eggan, F. (2008). Dykes and fags want everything. In M.B. Sycamore, (Ed.), *That's revolting! Queer strategies for resisting assimilation (pp.11–18).* New York: Soft Skull Press.
Eisenbach, D. (2006). *Gay power: An American revolution.* New York: Carroll & Graf Publishers.

Chambers, A. (2006). *God's grace and the homosexual next door: Reaching the heart of the gay men and women in your world*. Eugene, OR: Harvest House Publishers.

Charron, J. and Skylstad, W. (1996, July 19). Statement on same-sex marriage. In A. Sullivan (Ed.), *Same-sex marriage pro & con: A reader (pp.52–53)*. New York: Vintage Books.

Cloud, D. (2001). Queer theory and 'family values': Capitalism's utopias of self-invention. In M. Zavarzadeh, T. Ebert, and D. Morton, (Eds.), *Marxism, queer theory, gender* (pp.71–114). Syracuse, NY: The Red Factory.

Coates, D. (1999). Beyond government. In W. Horn, D. Blankenhorn, and M. Pearlstein, (Eds.), *The fatherhood movement: A call to action (116–126)*. New York: Lexington Books.

Cole, M. (2009). *Critical race theory and education: A Marxist response*. New York: Palgrave Macmillan.

Cole, M. and Hill, D. (2002). 'Resistance postmodernism': Progressive politics or rhetorical left posturing? In D. Hill, P. McClaren, M. Cole, and G. Rikowski, (Eds.), *Marxism against postmodernism in educational theory (pp.89–107)*. New York: Oxford Books.

Colson, N. (2009, April 7). Gay marriage ban overturned in Iowa. *Socialist Worker*. Retrieved April 8, 2009, from: http://socialistworker.org/2009/04/07/gay-marriage-ban-overturned

Colson, N. (2009, February 27). Turning an egg into a "person." *Socialist Worker*. Retrieved March 4, 2009, from: http://socialistworker.org/2009/02/27/an-egg-into-a-person

Compton, M. and Weiner, L. (2008). The global assault on teachers, teaching, and teacher unions. In M. Compton and L. Weiner, (Eds.), *The global assault on teaching, teachers, and their unions: Stories for resistance (pp.3–9)*. New York: Palgrave-MacMillan.

Coontz, S. (2005). *Marriage, a history: How love conquered marriage*. New York: Penguin Books.

Coontz, S. (1997). *The way we really are: Coming to terms with America's changing families*. New York: Basic Books.

Cory, D.W. (1951). The *homosexual in America: A subjective approach*. New York: Greenberg.

Cotter, J. (2001). Sexual harassment as/and (self) invention: Class, sexuality, pedagogy, and (creative) writing. In M. Zavarzadeh, T. Ebert, and D. Morton, (Eds.), *Marxism, queer theory, gender (pp.155–216)*. Syracuse, NY: The Red Factory.

Craig, T. and Boorstein, M. (2009, November 12). Catholic church gives D.C. ultimatum. *The Washington Post*. Retrieved November 12, 2009, from: http://www.washingtonpost.com/wp-dyn/content/article/2009/11/11/AR2009111116943_pf.html

Croll, E. (1986). Rural production and reproduction: Socialist development experiences. In E. Leacock and H. Safa, (Eds.), *Women's work (224–252)*. MA: Bergin and Garvey Publishers.

Cudd, A. (2002a). Analyzing backlash to progressive social movements. In A. Superson and A. Cudd, (Eds.), *Theorizing backlash: Philosophical reflections on the resistance to feminism (pp.3–16)*. New York: Rowan & Littlefield Publishers, Inc.

Cudd, A. (2002b). When sexual harassment is protected speech: Facing the forces of backlash in academe. In A. Superson and A. Cudd, (Eds.), *Theorizing backlash: Philosophical reflections on the resistance to feminism (pp.217–243)*. New York: Rowan & Littlefield Publishers, Inc.

Dailing, P. (2009, October 13). Geneva high school teacher accused of anti-gay remark. *Sun-Times News Group*. Retrieved November 6, 2009, from: http://www.suntimes.com/news/metro/1821529,CST-NWS-teacher13.article

Dallas, J. (2003). *Desires in conflict: Hope for men who struggle with sexual identity*. Eugene, OR: Harvest House Publishers.

Brooks, D. (2003, November 22). The power of marriage. *The New York Times.* In A. Sullivan, (Ed.), *Same-sex marriage pro & con: A reader,* (pp.196–198). New York: Vintage Books.

Brown, B. (2008, November 8). Antipathy toward Obama seen as helping Arkansas limit adoption. *New York Times.* Retrieved May 29, 2009, from: http://www.nytimes.com/2008/11/09/us/politics/09arkansas.html?fta=y

Browning, F. (1996, April 17). Why marry? *The New York Times.* In A. Sullivan, (Ed.), *Same-sex marriage pro & con: A reader (pp.132–134).* New York: Vintage Books.

Bruce, T. (2004, February 25). Respecting marriage and equal rights. *FrontPage Magazine.* In A. Sullivan, (Ed.), *Same-sex marriage pro & con: A reader (pp.199–203).* New York: Vintage Books.

Buckley, W. (2005, February 19). Killers at large. *National Review Online.* Retrieved June 13, 2009, from: http://www.nationalreview.com/buckley/wfb200502191155.asp

Burke, P. (1990). *The French historical revolution: The Annales School, 1929–1989.* California: Stanford University Press.

Burroway, J. (2006, October 1). The heterosexual agenda: Exposing the myths. *Box Turtle Bulletin.* Retrieved November 5, 2008, from: http://www.BoxTurtleBulletin.com/Articles/000,015.htm

Burroway, J. (2008, September 30). Straight from the source: What the Dutch study really says about gay couples. *Box Turtle Bulletin.* Retrieved June 13, 2009, from: http://www.boxturtlebulletin.com/Articles/000,003.htm

Burton-Rose, D. (2008). Queering the underground. In M.B. Sycamore, (Ed.), *That's revolting! Queer strategies for resisting assimilation (pp.19–28).* New York: Soft Skull Press.

Bynum, R. (2009, May 16). Michael Steele: Gay marriage is bad for small businesses. *Huffington Post.* Retrieved May 20, 2009, from: http://www.huffingtonpost.com/2009/05/16/michael-steele-gay-marriage

California Federal Savings and Loan v. Guerra. (1987). Retrieved July 16, 2009, from: http://caselaw.lp.findlaw.com/scripts/getcase.pl?court=US&vol=479&invol=272

Cameron, P. (1989). *Medical consequences of what homosexuals do.* Brochure

Cameron, P., Cameron, K. and Proctor, K. (1989, June). Effect of homosexuality upon public health and social order. *Psychological Reports, 64(3),* 1167–1179.

Canedy, D. (2003, May 15). Governor Jeb Bush to seek guardian for fetus of rape victim. *The New York Times.* Retrieved July 24, 2009, from: http://www.nytimes.com/2003/05/15/us/gov-jeb-bush-to-seek-guardian-for-fetus-of-rape-victim.html

Canfield, K. (1999). Promises worth keeping. In W. Horn, D. Blankenhorn, and M. Pearlstein, (Eds.), *The fatherhood movement: A call to action (43–55).* New York: Lexington Books.

Carey v. Population Services International. (1977). Retrieved October 20, 2009, from: http://www.law.cornell.edu/supct/html/historics/USSC_CR_0431_0678_ZS.html

Carpenter, E. (1894). *Marriage in free society.* Manchester: The Labor Press Society Limited.

Cavanagh, S.L. (2008, September). Sex in the lesbian teacher's closet: The hybrid proliferation of queers in school. *Discourse Studies in the Cultural Politics of Education, 29(3),* 387–399.

Chamallas, M. (2002). The backlash against feminist legal theory. In A. Superson and A. Cudd, (Eds.), *Theorizing backlash: Philosophical reflections on the resistance to feminism (pp.67–86).* New York: Rowan & Littlefield Publishers, Inc.

Chamberlain, P. (2008, Fall). Abstaining from the truth: Sex education as ideology. *The Public Eye 23(3),* 1; 21–27.

Bawer, B. (1994). *A place at the table*. New York: Touchstone Books.

Beach, B. (2005). Safe schools and voices. In K. Jennings, (Ed.), *One teacher in 10: LGBT educators share their stories* (pp.212–219). Los Angeles: Alyson Books.

Bebel, A. (1910). *Women and socialism*. Retrieved July 28, 2009, from: http://www.marx.org/archive/bebel/index.htm

Bell, D. (1986). Central Australian Aboriginal women's love rituals. In E. Leacock and H. Safa, (Eds.), *Women's work (75–95)*. MA: Bergin & Garvey Publishers.

Beneria, L. and Sen, G. (1986). Accumulation, reproduction, and women's role in economic development: Boserup revisted. In E. Leacock and H. Safa, (Eds.), *Women's work (141–157)*. MA: Bergin & Garvey Publishers.

Bennett, W. (1996, June 3). *Newsweek*. In A. Sullivan, (Ed.), *Same-sex marriage pro & con: A reader (pp.274–275)*. New York: Vintage Books.

Berkowitz, B. (2009, January). Heritage Foundation to "fight back." *Z Magazine, 22(1)*, 11.

Berkowitz, B. (2008, November). Bush frantically seeks a legacy. *Z Magazine, 21(11)*, 11–13.

Berlet, C. (2009, January). Brownshirt anarchism, bogus journalism. *Z Magazine, 22(1)*, 17–18.

Best, R. (2003, August 5). The utopian visions of Frances Wright. Ethical Inquiry. Retrieved August 15, 2009, from: http://rbest.ethicalmanifold.net/archives/000125.html

B.J.L. (2008, April 28). CNN's Bash, Roberts, and Phillips ignored Hagee's comments linking Hurricane Katrina to gay pride parade. *Media Matters for America*. Retrieved May 23, 2009, from: http://74.125.95.132/search?q=cache:53Kr-lzbS0IJ:mediamatters.org/research/200804280010+jerry+falwell+hurricane+katrina+gays&cd=5&hl=en&ct=clnk&gl=us&client=firefox-a

Blumenthal, M. (2009, April 7). What a killer was watching. *Blogs & Stories*. Retrieved April 8, 2009, from: http://www.thedailybeast.com/blogs-and-stories/2009-04-07/what-a-killer-was-watching/full/

Blumenthal, M. (2006, November 1). Rick Santorum's beastly politics. *The Nation*. Retrieved May 29, 2009, from: http://74.125.95.132/search?q=cache:-vWqO8NurbAJ:www.thenation.com/doc/20061113/santorum+man+on+dog+santorum&cd=1&hl=en&ct=clnk&gl=us&client=firefox-a

Bourne, J. (2002). Racism, postmodernism and the flight from class. In D. Hill, P. McClaren, M. Cole, and G. Rikowski, (Eds.), *Marxism against postmodernism in educational theory (pp.195–210)*. New York: Oxford Books.

Bowers v. Hardwick 478 U.S. 186. (1986). Retrieved May 23, 2009, from: http://www.law.umkc.edu/faculty/projects/ftrials/conlaw/bowers.html

Bowlby, J. (1983). *Attachment*. New York: Basic Books.

Boy Scouts of America v. Dale. (2000). Retrieved November 1, 2009, from: http://www4.law.cornell.edu/supct/html/99-699.ZS.html

Bowman, B. (1999, September). Art teacher censorship of student produced art in Georgia's public high schools. *CultureWork, 3(3)*. Retrieved November 23, 2009, from: http://aad.uoregon.edu/culturework/culturework11.html

Boy Scouts of America Equal Access Act. (2002). *Sec. 9525 No Child Left Behind Act*. Retrieved October 10, 2009, from: http://www.ed.gov/policy/elsec/leg/esea02/pg112.html

Breen, T. (2009, March 26). States consider drug tests for welfare recipients. *Yahoo!News*. Retrieved April 7, 2009, from: http://news.yahoo.com/s/ap/20090326/ap_on_bi_ge/states_welfare_with_strings

Brooks, C. (2005). Dancing with the issues. In K. Jennings, (Ed.), *One teacher in 10: LGBT educators share their stories* (pp.240–248). Los Angeles: Alyson Books.

References

A boy and his flag. (2009, November 5). *Arkansas Times*. Retrieved December 3, 2009, from: http://www.arktimes.com/articles/articleviewer.aspx?ArticleID=2f5d7a3b-c72a-446b-8d20-3823aa79c021

Aarons, L. (1995). *Prayers for Bobby*. New York: Harper One.

About Face Youth Theatre website. (n.d.). http://aboutfacetheatre.com/AFYT/

Adkins, K.J. (2009, August 5). *George Sodoni felt slighted by interracial romance*. Retrieved September 10, 2009, from: http://www.theroot.com/blogs/president-obama/george-sodoni-felt-slighted-interracial-romance.com

Adkisson, J. (2008). Church manifesto. Retrieved April 6, 2009, from: http://web.knoxnews.com/pdf/021009church-manifesto.pdf

Adolescent family life act. (1981). Retrieved October 19, 2009, from: http://www.hhs.gov/opa/espanol/about/legislation/xxstatut.pdf

Anders, C. (2008). Choice cuts. In M.B. Sycamore, (Ed.), *That's revolting! Queer strategies for resisting assimilation (pp.87–91)*. New York: Soft Skull Press.

Adorno, T., Frenkel-Brunswik, E., Levinson, D., and Sanford, R.N. (1950). *The authoritarian personality*. New York: W.W. Norton and Company, Inc.

Andrews, S.P. (1889). *Love, marriage, and divorce and the sovereignty of the individual*. Retrieved August 15, 2009, from: http://praxeology.net/HJ-HG-SPA-LMD.htm.

Antonio, R. (2000, July). After postmodernism: Reactionary tribalism. *American Journal of Sociology, 106 (1)*, 40–87.

Anyon, J. (1994, Summer). The retreat of Marxism and socialist feminism: Postmodern and poststructural theories in education. *Curriculum Inquiry, 24(2)*, 115–133.

Arkes, H. (1993, July 5). The closet straight. *National Review*. In A. Sullivan, (Ed.), *Same-sex marriage pro & con: A reader (pp.155–159)*. New York: Vintage Books.

Arizpe, L. and Aranda, J. (1986). Women workers in the strawberry agribusiness in Mexico. In E. Leacock and H. Safa, (Eds.), *Women's work (174–193)*. MA: Bergin & Garvey Publishers.

Arterburn, S. and Stoeker, F. (2001). *Every man's marriage*. Colorado Springs, CO: Waterbrook Press.

Atwood, M. (1998). *The handmaid's tale*. New York: Anchor Books.

Bader, E. (2008, Fall). American Life League's pill kills day links birth control and abortion. *The Public Eye, 23(3)*, 3–5.

Bageant, J. (2007). *Deer hunting with Jesus: Dispatches from America's class war*. New York: Three River Press.

Bailey, M., Kandaswamy, P., and Richardson, U. (2008). Is gay marriage racist? In M.B. Sycamore, (Ed.), *That's revolting! Queer strategies for resisting assimilation (pp.113–119)*. New York: Soft Skull Press.

reconcile capitalist-level consumption with environmental sustainability or even survival itself. This has to underlie any potential proposals on organizing an alternative society. In writing about the sense of defeat that many younger activists might be experiencing, Pilger (2009) sounds the alarm: "In the 1950s, we never expected the great wind of the 1960s to blow. Feel the breeze today" (p. 27).

Conclusion

Hopefully, after reading this volume, it will be apparent that sexuality is a pivotal point of organization for educators and activists. Sexuality impacts everyone of every background. Far from being an irrelevant or simply a personal construct, issues impacting the family (health care, education, reproductive rights, workplace inequality, sexual harassment, homophobia) affect all workers in tandem with race and gender. If Marxists do not grapple with sexuality and the family, socialism will not succeed, as was shown in the early 1900s after the Russian Revolution. There, the unevenness and expense of socializing domestic labor was a hard lesson to learn. With our overabundance of material goods, scarcity doesn't have to be the logical outcome of socialism.

Sexuality carries with it a sense of urgency, but not limited to the postmodern dream of desire over human need. As Neill (1960) once declared, "the dream is not good enough.... I want to see heaven on earth, not in the clouds" (p. 211). In a similar manner, we need to move beyond "someday" kinds of dreams to begin to make Marx relevant once again for the working class. The way to do this is through addressing the family and the possibilities of what might happen once capitalism is defeated. Most likely there would be a variety of family forms since marriage would no longer carry property-based privileges. LGBTQ and single-parent families would no longer be socially and economically ostracized. Schools would respond by fully acknowledging the rights of K–12 students to a developmentally appropriate, comprehensive sexuality education free from homophobia or sexism.

This book is especially aimed at the educators who were born after the 1960s and who might feel that the best era in terms of activism has already passed. Those of us who were born into a Nixon administration, followed by Ford, then Carter, then Reagan, then Bush, then Clinton...and then another Bush, have seen the majority of their lives dominated by conservative rhetoric and economic decline. The Obama administration seems equally unlikely to take a stand that is much different from that of his predecessors. We also face something more overwhelming than was realized in the 1960s; the possibility of environmental destruction. There is no longer any way to was an added insult, as 8,500 people die of AIDS every day (p. 299). If countries were allowed to cancel their debts, the money could go toward medication and health care. As Smith commented about this action, "We were telling the naked truth about Bush's global AIDS policies" (p. 304).

Issues of the family offer perhaps the ultimate means of building global constituencies. No matter what part of the world one is talking about, the family touches on gender and sexism, patriarchy, pay differentials, aging, and marriage rights. It involves sexuality, race, and the material realities of housing, health care, education, and work. The United States is in the midst of a growing environmental and economic crisis brought on by a rabid conservative ascendency, as infrastructures collapse and military budgets expand, creating conditions in which people will eventually have to fight for the most basic services. In this context, LGBTQ people have an interest in the family even if some are opposed to the concept of state-sanctioned relationships via marriage (Wolf 2009c). D'Emilio (2003) offers an intriguing possibility that challenges the notion of marriage as oppressive. Could it be that as our concept of sexuality continues to expand, that the very meaning of marriage and the family will be transformed? As D'Emilio proposes, "we are reinventing a very long and old American tradition in which family and community are deeply connected to one another.... If this is true, then the fight for the recognition of our families is a fight that has meaning for everyone" (p. 190).

Windows and Doors occupation in Chicago, similar to Harvey Milk's organizing of the Coors Beer strike in solidarity with the Teamsters in the 1970s (D'Emilio 1992). Spade (2008) envisions transactivism reaching out to different constituencies by tying transgender identity and issues to the overall dismantling of heteronormativity and its attendant hierarchies, which can only benefit working families.

When the philosophy of identity politics takes precedence, activists miss out on the power of cultivating allies. The major weakness of a focus on the narrowness and exclusivity of identity is that it makes it impossible for groups to see common sources of oppression; that is, only blacks can understand and fight racial oppression, only gays can understand and fight homophobia, and so forth. Wolf (2009c) found that some of the LGBTQ activism in the wake of AIDS missed out on opportunities to connect with labor as Harvey Milk did with much success in the 1970s. Often the filter for identity movement activism is constant rage, as D'Emilio (1992) found with militant wings of the Black Panther and feminist movements. While decided and focused anger is an important component that can lead to action, the identity focus of these groups blocked them from building global coalitions and moving forward. Each faction felt that they were alone in the fight (since only X group could understand and fight their particular form of oppression) and dealing with the overwhelming problems of racist and sexist violence began to wear activists down. Even the tried and true tactics of street action are for naught if those involved don't move forward and advance a better strategy (D'Emilio 2003).

The activism surrounding AIDS is an important case example of the need for global coalition building (D'Emilio 1992; Goldstein 2003). When the disease was first discovered, it was solely linked to gay men. AIDS is now a pandemic and the connection between economics, the majority world, and access to resources is all too apparent. Schulman (2008) believed that the original association of HIV/AIDS with gay males allowed the government and the public to justify not taking action. Unfortunately, in response, the gay community tended to narrow its vision to primarily securing expensive drugs for treatment without linking activism to larger social concerns, like intravenous drug use, the War on Drugs, the prison-industrial complex, and poverty, and ignored populations at risk of contracting the virus, such as minority women.

While ACT UP was able to carry out several successful zaps of the pharmaceutical industry, this approach only worked for a few years. It wasn't until ACT UP began to express the message that health care is a universal human right that more long-lasting coalitions began to form around the issue. In 2004 ACT UP connected the AIDS struggle to debt imposed on impoverished countries when 12 activists stepped into the street in front of the Republican National Convention (RNC) and dropped their clothes to send the message: Drop the debt, stop AIDS (Smith 2008). The fact that the city was cleansed of the presence of homeless people in preparation for the RNC overwhelmed with requests as word of mouth spread and the realization hit that this was a health issue affecting not only a small segment of poor women, but all strata of society. Although there were several close calls that forced Jane to consider closing down the service, they felt they couldn't abandon the many women who were in desperate situations:

> What they were doing wasn't charity. It wasn't some do-gooder project meant to lift up the less fortunate. They were working for the liberation of women. They wanted every woman who called them to see herself as an active participant, to take responsibility for her decision...the high cost of the abortions made it possible and logical to talk about abortion in terms of economic oppression, that coming up with the money added greater stress for women whose resources were minimal. (p. 36)

Within the organization, a core group of 4 to 5 women ensured that new members had to be screened and trained in order to maintain the anonymity of the group along with the women who were seeking health services. A strict protocol had to be followed as part of the ethics involved. This caused some internal stress to the group, some members of whom wanted to have a more open environment in which to work. Whether one served as a call screener, performed the abortion, or did post-op counseling, the leadership required to hold things together was immense. After abortion was legalized, Jane transferred its services to the creation of low-cost and free clinics for women. The clinics also offered classes in union organizing and sought to create a curriculum for K–12 schools. Ultimately, Kaplan (1995) concluded that leadership in Jane was represented by ordinary women learning what they were capable of while working together.

Build Global Coalitions Rather Than Identity Politics

The only viable framework for activism worth considering is a global mindset, fighting against homophobia as well as racism, sexism—and capitalism (D'Emilio 1992; Goldstein 2003; Johnson 2009; O'Brien 2008; Rosado 2008). Mecca (2008) describes the heart of this strategy, which hearkens back to the GLF days, as worth reconsidering today: "One week we might be marching for affordable housing with poor black mothers, another week we might be demanding the passage of the Equal Rights Amendment with middle-class white women" (p. 31). Lesbian groups were among the leaders in forming women's health clinics that served minority populations cut off from health care (D'Emilio 1992; Kaplan 1995). In 1970 a meeting was convened where many radical groups attended, called the Revolutionary People's Constitutional Convention (Eskridge 2008). The groups drafted a letter connecting LGBTQ and women's liberation interests. Wolf (2009b) outlines how LGBTQ marriage rights protestors joined in support of the Republic emotions. The marchers were parent surrogates upon whom each observer could project the longing for reconciliation, the yearning to relive the experience of unconditional acceptance. (p. 2)

In discussing the life of civil rights leader Bayard Rustin, D'Emilio (2003) creates a moving portrait of a man who was the embodiment of leadership at a time when being openly gay was a high-risk proposition. Most are not aware that in 1964 Rustin led the largest civil rights protest of the era involving the New York public schools. At that time, nearly half of the city's students cut classes in support of minimum wage increases, public funding for jobs, and job training. Rustin was a master of coalition building and illustrated the importance of the role of leadership and organizations in sustaining a coherent movement ready for mobilization. The following recount by D'Emilio is worth representing in full due to the impact that one leader can make on those hungry for a different world, even if one never meets a leader face-to-face:

I find myself time traveling emotionally to an earlier me. I want Rustin in the life of the fragile adolescent who came to this campus more than a generation ago and found himself confronted by unexpected challenges, without any of the usual moorings. An adolescent who needed a mentor and needed, badly, to borrow someone's courage. I want him with me at the one and only meeting of Students for a Democratic Society that I ever attended, across the street in Fayerweather Hall, and that I fled feeling stupid and inadequate. I want him with me as I sat down in the middle of West 47th St., protesting a speech by the Secretary of State, praying that the charging police horses will stop before they reach me. I want him with me as I leafletted army recruits in front of the Whitehall Induction Center and an angry crowd gathered across the street. I want him as a reassuring presence during the hair-raising, earsplitting fights with my God and country, Cardinal Spellman style Catholic parents over my decision to file as a conscientious objector. I want him near as I sat quivering in front of the members of my draft board on Arthur Avenue in the Bronx, trying to persuade them to grant me a conscientious objector status while they signed induction orders. (pp. 239–240)

Kaplan's (1995) account of the operation of Jane, an underground abortion service, is a case study of the importance of leadership. The women who founded Jane were a combination of activists and housewives who all shared a common vision of reproductive freedom. Because of the dangers they faced if they were caught, these women had to employ several leadership strategies in order to keep Jane both anonymous and afloat. Kaplan describes Jane's journey from initially working with a doctor to whom they had to pay inflated rates in order to perform the abortions. Then the members realized they could learn the procedure themselves. The Jane membership was soon concerns with congregants. Others, including Catholics, worked to repeal anti-contraception laws.

D'Emilio's (1983) sets the church's unlikely involvement with LGBTQ issues in the 1960s in the context of the civil rights movement impacting all institutions, both religious and secular. Poverty and social issues began to take priority, including serving younger populations. The Glide Memorial Methodist Church in San Francisco was targeting runaways who were starting to hit the streets in larger numbers in the 1960s. Many of these young people were gay or transgender with nowhere to go, turning to prostitution to survive. Ted McIlvenna, a Methodist minister, approached the local chapter of the Mattachine Society for advice on how to reach this population with health services and counseling. The post-Marxist Mattachine Society refused to get involved because of fears with being linked to youth under the age of 21. So, on his own, McIlvenna and other ministers and congregants created a "gay line" tour of San Francisco, hitting bars, drag shows, clubs, and activist meetings. This placed the church face to face with LGBTQ youth, marking the first time many church members interacted with someone who was openly gay or lesbian. When the police busted a gay social event with the ministers present, they held a press conference condemning the harassment. By the church becoming involved, the police were put in an immediate dilemma. If they spoke out against the church, they would look irreligious. But if they supported the right of homosexuals to meet, they would be considered their allies.

Leadership Is Necessary

Far from a traditional hierarchical concept of the all-knowing leader, movements today still require people to take on more visible and responsible roles within organizations. The most important reason for leadership is that, for a variety of reasons, not everyone is in a position to step outside the pack and place themselves on the line. Leadership can create conditions where the more vulnerable members of society can still effectively participate until they are at a place of better awareness or less immediate risk (Eisenbach 2006). Leadership also enables people to learn from each other as they apply a combination of insider strategies (elections, courts, lobbying) and outsider strategies (demonstrations, clinic defenses, peer teaching) (Feldt 2004; Kaplan 1995). As Bageant (2007) points out, leaders "bring themselves as models of empowerment. And if they are good at what they do, they bring backbone" (p. 13). Aarons (1995) recounts the power of leadership encompassed in PFLAG contingents of public demonstrations, as people in less gay-friendly regions can feel less alone:

> Projected on these marchers were tens of thousands of child-parent promises: love, pain, rejection, joy, denial, fear, remorse—the entire palette of

Kelly (2002) recommends trade union membership as a key strategy in the short term for women to gain essential social benefits like maternity leave and pay equity. Democratically run unions could eventually extend these basic rights to include universal child care, fully funded reproductive health services, and paid leave for families. Kelly also argues that unions can organize resistance to privatization efforts in the areas of education and welfare "to ensure that women are not forced into the role of unpaid careers" (p. 231). LGBTQ people also benefit from unionization provided the union reaches out to those who aren't formally organized in their workplaces or even the unemployed (Sears 2005). Because women and LGBTQ people are vulnerable in capitalist society, unions are often one of the key ways to protect themselves from exploitation, which is one reason politicians and the media seek to bust education unions (Murphy 2008; Robertson 2008). The ability of unions to mobilize strikes, building occupations, teach-ins, marches, and media events is even more critical for groups who might not have the means to do these things by themselves.

Unions are fraught with problems, the most prominent being the corporatization of the major education unions in the United States. Membership can also reflect racist and homophobic values, making it difficult to craft an unqualified platform in support of sexuality rights (Weiner 2008). Country-firstism produces a narrow view (i.e., returning manufacturing jobs to the U.S.) when the only viable strategy in late-stage capitalism is global solidarity across nations and job types (Compton and Wiener 2008). A majority-rules approach can often reinforce backward policies, in which portions of the membership have to continually bargain away their concerns so that the group as a whole can move forward. Lynd and Grubacic (2008) recommend that "solidarity unionism among any group of the oppressed requires ordinary people to act for themselves, but beyond that, that they interpret their action and on that basis project future actions" (p. 146). If this principle is not part of a democratically run union, leadership can easily steer the rank and file in corporatist directions.

Though they are often a barrier to liberation movements, many churches and religious individuals have acted on the side of sexuality rights. Kotulski (2004) describes how in Connecticut the Unitarian Universalist Association, Presbyterian Church, Reformed Judaism, and the United Church of Christ banded together in 2003 to refuse to sign the marriage licenses of heterosexual couples. Their action sent the message that as long as gays and lesbians were not allowed to marry, to continue to recognize heterosexual marriages in the church would be contributing to the strength of an unjust law. Some of the earliest reproductive rights movement activism came from pro-choice clergy in the late 1960s (Kaplan 1995). Howard Moody, a Baptist minister, helped refer women to safe, underground abortion services. The American Baptist Convention passed a resolution in 1968 asking clergy to address these of a group of 14 men was allowed to testify as an expert witness when New York was holding hearings on the legalization of abortion (Kaplan 1995). Denied official access, women outside of the hearing talked to the press about their own experiences with illegal abortions. A network of alternative clinics and women's centers was created to respond to urgent health care needs. Abortion activism also had an educational component as those using these services learned about the historical and social context of women's oppression and its connection to capitalism.

D'Emilio (2003) reminds us that though the 1960s had many disappointments, it was not a failed decade no matter what media pundits claim. In many ways, the sexual liberation of the 1960s has had the most widespread influence. Even during the most conservative periods, such as during the Reagan administration when some of the boldest waves of activism occurred around the AIDS epidemic. In fewer than 10 years, community-based organizations were created to serve those afflicted with the virus; a mobilization of this size had not been seen since the Great Depression. One could argue that activism around sexuality has kept protest alive. Whether the rhetoric surrounding sexual liberation is reform-minded or revolutionary, its impact cannot be denied.

Consider Unions and Churches

Democratically run unions represent one of the best ways to organize around issues that impact K–12 settings. Weiner (2008) explains:

> We knew our ideas and actions counted because we could influence the union's exercise of institutional power and the union could project our collective voice and power...union principles of collective action and solidarity contradict neoliberalism's core principles, reification of individual effort and competition...unions, when they are doing what they should, encourage notions and practices that run counter to neoliberalism's central tenets. (p. 252)

Teachers have organized globally to fight for students and families, particularly in Argentina, Mexico, and Brazil (Hypolitio 2008). Their core platform has been education as a human right, based on the tenets of critical pedagogy. If unions could extend this same kind of support to sexuality rights, this would be a powerful message with wide outreach. Harvey Milk tapped into the organizing might of the California Teachers Association to fight the Briggs Initiative, resulting in a mass mailing of more than 2 million "No on 6" voter education cards across the state (Wolf 2009c). The presence of strong teaching unions can also encourage student activism in the form of building occupations and walkouts (Moeller 2002).

many of us remain hidden or afraid to declare our intentions. However, acts of public protest threaten our anonymity, and some of us are more vulnerable than others if our identity is known (D'Emilio 2002). This was the case with abortion. Women finally refused to be intimidated or shamed and came forward to talk about their experiences with being denied access to reproductive freedom (Kaplan 1995). By conducting speak-outs at hearings about their own back alley abortions, they created awareness that resonated with the public. Their visible presence broke the silence that reflected the shame that surrounded acknowledgment of women's sexuality. Translating this type of visibility in often oppressive and surveillance-ridden K–12 environments can be a challenge, especially in a precarious economy. As D'Emilio (2003) notes, visibility creates targets. As more activists declare an uncompromising stance about sexuality rights, everyday life in America changes little by little. This would not be the case if we retained our outsider status.

Having organizations in which to strategize and from which to gain support is essential. Kupfer (2005) relates how straight allies were essential in his decision as an educator to take a chance and form an LGBTQ student group. In 1987, ACTUP mobilized its first action against Burroughs Wellcome, which held the patent for the AIDS drug AZT costing upwards of $10,000 annually (Eisenbach 2006, p. 304). ACTUP members, dressed in conservative suits, walked into the drug company headquarters and chained themselves to the pillars inside. After a few months, a similar action was committed on the floor of the New York Stock Exchange where five activists cuffed themselves to a balcony, emphasizing the connection between corporate greed and health care. The company responded by lowering the cost of the medication, though it remains prohibitively expensive for many today. Rosado (2008) describes the activist work of FIERCE, a group of LGBTQ poor and working-class youth who organized in response to Mayor Rudolf Giuliani's militarization of the police force in New York City. FIERCE focused their actions against gentrification in the West Village where more wealthy gays and lesbians were allying with the police against poor and transgendered youth whom they viewed as a menace. FIERCE'S mission is as follows:

> We are concerned with more important things like being able to hang out in the Village without getting harassed or arrested and getting the condoms, food, medical, mental health, housing, and jobs we need. We're fighting for our right to exist." (p. 328)

When Planned Parenthood in upstate New York was under attack by an anti-choice group, a coalition of various community groups quickly mobilized to make sure that public hearings were well represented by the reproductive rights perspective (Feldt 2004). As more allies joined in, the support system for Planned Parenthood was able to both fight off the anti-choice group and expand community health services. In 1969 only one woman out begins in preschool and continues through high school and college, so that students can make informed choices. Women must have absolute access to reproductive health services, from contraception to abortion. LGBTQ people are entitled to the same marriage rights as straights and shouldn't have to make do with the second-class status of "domestic partnership." This means framing sexuality issues as civil rights as opposed to claim rights, which tend to look at benefits to the majority of people and is often used as a rationale for denying rights to a minority (Eskridge 2008). For example, repealing sodomy statutes became a flashpoint for activism when court decisions held up "community norms" and the like as reasons to deny marriage rights.

Taking a civil rights stance also means ceasing to put the rights of transgendered individuals on hold as part of a bargaining strategy for gay and lesbian rights or excluding women's health services as a condition for passing a national health care plan as occurred with the Stupak Amendment. As long as one segment of the population is being oppressed, no one is free. This can be a tricky issue as many activists are troubled by the notion of supporting seemingly conservative causes like marriage or equality of gays in the military. However, Kotulski (2004) has a compelling reply to this concern as she quotes Beth Robinson, a civil rights attorney:

> And if somebody said that gays couldn't have hunting licenses, I'd be out there fighting for the right of gays and lesbians to carry guns and hunt. And then, after all the gays and lesbians had access to hunting licenses, I'd work to eradicate hunting. (p. 101)

Of course relying on the courts and politicians is a dead-end proposition that puts activists in a constant state of dependence and hoping for open-mindedness of judges and legislators (D'Emilio 1992). Compared to full liberation, equality is far short of a socialist vision. The early gay liberation movement launched attacks at the very categories of gay and straight and linked those categories to capitalism. Instead, these early activists began to examine sexuality as a more open and fluid phenomenon, which also created a direct challenge to notions of hierarchy and patriarchy. D'Emilio envisions a radical liberation movement as deinstitutionalizing heterosexuality itself, something that goes far beyond repealing sodomy laws or winning same-sex marriage rights. This would entail the end of privileging propertied and married people over the unmarried and propertyless, which is essentially a socialist vision requiring the overthrow of capitalism. This would not just benefit the LGBTQ community, but straight working class folks as well. D'Emilio is hopeful that these changes will come about due to "movements of the disenfranchised," who will demand a different way of thinking about society and themselves (p. 163).

Visibility is critical, which means all of us coming out of the closet—not just gays and lesbians (Eskridge 2008). It is difficult to achieve critical mass if

At the same time that any movement needs to articulate a clear class position, we have to also support reforms that will assist the working class, such as health care or same-sex marriage. Like Kelly (2006), we understand that real equality is never possible under a capitalist system. We also cannot depend on the legal system to grant equality. But through working toward issues like abortion rights and K–12 comprehensive sex education we can raise awareness to the interconnectedness of sexuality rights and class consciousness. This alludes to Kaplan's (1995) analysis of the three components that make up a "movement": challenging the immorality of laws, questioning blind stateism, and rejecting authority as a natural concept, which asserts that ordinary people and not just rulers or corporations have a right to determine their future. D'Emilio (2003) believes that the most dramatic moves forward happen when people apply radical analysis to social issues and are willing to take a militant stand in order to apply this analysis. Just because the much-awaited "revolution" did not happen in the 1960s doesn't mean that the radical tenets that were part of the movement aren't worth pursuing today.

Sexuality Rights: No Apologies, No Conditions

D'Emilio (2003) describes sexuality rights as happening in alternating waves of rapid change followed by periods of slower momentum. In the 1940s and 1950s, Kinsey's research kicked off the first wave of awareness, along with the founding of the Mattachine Society and the Daughters of Bilitis, just prior to repressive McCarthyism. The second period of rapid change was marked by Stonewall and radical feminism, followed by a period of conservative retrenchment. AIDS and the 1987 march on Washington represented a third move forward, but it was followed by disappointments over Clinton's "Don't Ask, Don't Tell" policy. The recent activism related to same-sex marriage and the gains made in various states on marriage rights could mark a fourth leap ahead. What D'Emilio notices is that these moments of rapid change are not attributable to particular characteristics of the activists involved. Instead, openings are created by social and political factors that motivate people to act en masse. When the leaping forward occurs, it is because people are willing to take a militant stance to make their views known on the streets and in the media. By contrast, the periods between the leaps are not conducive to such militancy. Instead, more traditional avenues are pursued such as drafting laws and policies, setting the stage for more effective moves forward when rapid change eventually hits. The key is an articulated strategy that allows activists to take advantage of the leaps.

When it comes to sexuality liberation, an uncompromising and unconditional stance is critical. Sexuality is a part of life, not a separate facet that can be compartmentalized (Kaplan 1995). Activists must insist that everyone is entitled to the right to determine his or her own sexual decisions, beginning with developmentally appropriate, comprehensive sexuality education that solidarity is risky. The film culminates in a "die-in" commemorating several BYU students who committed suicide after not being accepted as LGBTQ.

Establish an Anticapitalist Stance

In order to end sexual oppression, not only must the working class confront capitalism, it must see that confrontation as being for its own interest.. As McClaren and Farahmandpur (2002) argue, "class for itself action is the only secure and effective means of securing the legal and political apparatuses necessary for controlling the state and its economic hegemony" (p. 50). The ability of capitalism to extract labor power from the working class while using homophobia, ageism, sexism, ableism, and racism to divide the workforce imprisons the people of the world. Working within reformist constraints limits our ability to challenge the imposition of the nuclear family, one of the key means of exploitation (Cloud 2001; Hill, Sanders, and Harkin 2002; Morton 2001). Not only is this family model no longer sustainable for heterosexuals, it is an ideological weapon constantly wielded against LGBTQ people in order to deny key rights and to pit workers against one another. So it only makes sense that conservatives constantly appeal to "the family" as a wedge strategy.

The rise of industrialization in the late 1800s created the space for gays and lesbians to form as a community and an identity (D'Emilio 1983; 1992; 2003; Sears 2005). This fact makes capitalism difficult to challenge with regard to sexuality rights. The LGBTQ community had to invent a revolutionary set of tenets around which to organize in the early 1950s and again in the 1970s since none existed that was relevant or addressed the immediacy of the situation (D'Emilio 2003). However, this should not avert our attention from capitalism as the root cause of oppression, especially as manifested in the government's initial lack of response to the AIDS crisis (Schulman 2008). Much as Marx asserted that capitalism was a necessary stage toward the transformation of society from the limited production potential of feudalism, capitalism is a critical stage in the transformation of the family. For example, capitalism gave us romantic love as a basis of marriage instead of economic dependency or parental permission. It also broke once and for all the procreation imperative. With the ability to now produce enough goods to alleviate material suffering and the enabling of people to freely enter into relationships, human civilization must now take the next step and "build a new socialist society organized not around profit but on meeting people's needs" (Ebert 2001, p. 302). The key contradiction of capitalism is that it has created its own undoing; an organized working class that acts in its own interest once an awareness of its situation takes hold (Hickey 2006). The question asked by D'Emilio (2003) is trenchant: What is left of the sixties that we can build upon in order to bring about a better world and how will this come to fruition?

Freedom to assemble is threatened at Oklahoma Baptist University (OBU) in Shawnee, Oklahoma. An OBU student is afraid to let others know that a small group of gays were allowed to meet at a friend's house to have dinner and watch TV. No one can trust that if they tell someone about a gathering, that person won't tell an administrator that they have suspicions of homosexuality. The presence of Soulforce emboldens a OBU student to speak out against the policy. In return, Soulforce decides to support students who are too afraid to speak out by defying the orders of the administration.

One of the group leaders, however, unilaterally decides to back off and agrees with OBU policy, shutting down the decision to commit civil disobedience. The OBU student is disappointed at the decision and starts to believe that the situation of LGBTQ students will never change. She also fears being isolated and even more of a target in the wake of the publicity of Soulforce's visit and her own speaking out. The fact that the civil disobedience acts were canceled sends a message that in a pinch, Soulforce would rather side with the administration than on the side of justice. This creates a rift in the group. Many members feel guilt at not having stood up and try to console themselves with a small rally held safely off campus, as administrators desired in the first place. The rally makes the more conservative group members happy, but the others express doubts.

Finally, at North Central University, the group takes collective action. They sit in front of the doors of all campus buildings in order to physically remind students that while it is inconvenient to have to push open a door and walk around a person blocking an entrance, imagine the barriers faced by LGBTQ people on a daily basis. Jake makes his speech via bullhorn just outside an entrance about how one North Central student was expelled and had no support, but Soulforce was now here today to support others like that student. This demonstrates to the North Central students that while they might not be able to say anything out loud without consequences, the Soulforce group has more leeway to openly speak. It disrupts the imposed silence and starts the seeds of solidarity, not unlike supporters of military resisters. Though they are eventually arrested, they keep breaking rules and policies, making it difficult for administrators to keep up with the group. This creates momentum as they visit the other campuses.

A Brigham Young (BYU) student expresses the quandary that closeted individuals face who attend right wing colleges: "How do you engage in dialogue with the school, as a student, that the very dialogue that you're trying to bring about makes you guilty of advocacy, of perhaps getting kicked out of school?" While many of the BYU students are more confrontational and homophobic, some of the LGBTQ BYU students are willing to join Soulforce at their rallies, at the risk of being expelled. One student expresses that he is tired of hiding. He shares his story and is thanked by several BYU students who are too afraid to come out. At BYU, even expressing straight

Eventually, the group talks with a student who was expelled from Lee University right before graduation. He was turned in by a "friend" for being gay and was not allowed to graduate. School officials went through his dorm room to spy on him and collect evidence based on testimony from his friend, as directed by the honor code at the school. Once Soulforce breaks the barrier at Lee University, students start to come up to them and tell them that they don't agree with the school's antigay policy. One soon realizes that the police presence on the part of school officials is not accidental; the presence of the activists is beginning to challenge the moral and parental authority of the universities.

At one point in the film, Jake reaches his limit. He is no longer satisfied with leaving campuses only to have another conversation about who is or isn't going to hell. Instead, he wants to advance the dialogue to a point that focuses on civil rights, no matter what, not just who is or isn't a Christian. The groups' response is not surprising. They chastise Jake when he gets a little ornery, telling him to not appear so "angry" so as not to create the "militant activist" (read: atheist) image, thus potentially alienating the university students. Instead, Jake should be happy that their message of Christian tolerance is being spread. They warn him not to make an impassioned speech that he is planning. Pam stands up for Jake, reminding the room that others "hate gay people" and that should be kept in mind. To accept half measures with gratitude all in the name of not breaking the very fragile connection with the right wingers is unacceptable to Pam. She expresses doubt that the desire for dialogue is there at the level some of the members appear to believe it is.

The group reaches a crisis when two competing strategies emerge: one is to work with individual hearts and minds, using the "common ground" approach and converting homophobes one at a time to Christ-like tolerance, and the other, as represented by Jake and Pam, is to recognize and confront homophobia, hurt feelings be damned. Unfortunately, the group decides to avoid arrest at Lee and justify their decision with a symbolic act of superficial solidarity. A few Lee students work to clean antigay graffiti off of the Soulforce bus. This appears to be enough for a majority of the group to feel that they are making a difference, especially when the Lee students join hands with the Soulforce riders in a prayer circle to give thanks for the hate crime that brought them all together.

As the film progresses a growing sense of radicalism emerges. One member says that it sends a bad message to just accept the restraints of the campus administrators as they enter onto each of the remaining campuses. The message is that the LGBTQ students "know their place" and can conform to the status quo. A few group members provide accounts of being victims of hate crimes and the sense of injustice within the group grows. Pam's testimony in particular connects the existence of the homophobic policies of the colleges to the larger climate that creates hate crimes and discrimination.

students do not know the verses behind the "common ground" argument, or run out of versus to volley back, they have an easy out by replying "I guess I have to go back and study my Bible." Absolutely nothing is gained on the civil rights issue. The other side takes their football and goes home, so to speak.

Instead, the challenge should come from a non-negotiation stance. In other words, it doesn't matter what your particular religion says, civil rights exist apart from religion and will not be negotiated, period. When kept in the realm of nuances of interpreting Bible verses and tenets, the group keeps getting drawn into an endless hermeneutic loop. Time is wasted defending themselves against charges of "you're going to hell" and the group leaves Liberty University discussing not future strategy, but how offended they were that the Liberty students had the nerve to declare that Soulforce members were "going to hell," as if that were the worst threat they could muster. At this point in the film, the viewer is wondering when a REAL threat will be asserted by the antigay camp, such as not being able to receive a partner's social security benefit during one's golden years, or being denied health care for your partner; the threats that need to be the targeted focus of direct action.

Prior to a visit to Regent University, one group member, Nate, questions the tactic of getting arrested with the public and media present. He argues that he is "not here to get arrested," effectively saying that it is an insincere act on the part of the group because it sends the message that they are only about making political statements and not about reaching out one-on-one to closeted LGBTQ students. Nate's viewpoint is common within the post-modern understanding of queerness as "performance" and personal connection, not direct activism to challenge homophobia and eventually capitalism. Jake, whose radical consciousness grows throughout the film, reminds the group that it is important to have the media present to help mobilize allies for their cause, which is ending discriminatory policies at these universities. Nate responds that the group still has to hold out hope that they won't get arrested. In other words, he is in the midst of so-called civil disobedience denial syndrome, in which an activist believes that the good intentions and social justice orientation of the group are all that is needed to overcome a line of police and a hostile and violent right wing religious industry.

Pam, another student in the Soulforce group whose radical consciousness is growing, expresses hope that LGBTQ Regent students will see their efforts (including the act of getting arrested) and view their presence as sending a message of solidarity. The act of getting arrested is a dramatic way to show the closeted Regent students that someone is aware of their presence and that they are on their side. One of the Regent students does come to the group off campus to apologize for their unwelcoming behavior. This provides a bridge for the next set of arrests with the media present. Now the university students are aware of Soulforce and attempt to make connections with them as they go from campus to campus, which was one of Soulforce's goals.

many individuals who themselves are gay but who reject that identity (much as famed evangelist Ted Haggard does in the documentary *The Trials of Ted Haggard*).

The group attempts to work closely with the Council of Christian Colleges and Universities (CCCU), an accreditation and policy-making body. Throughout the film, the group is continually betrayed by CCCU representatives who express an initial willingness to have the group appear on campus, and then turn them away when the bus pulls up. Right before the first stop at Liberty University, Jake, responding to the group's nervousness, reminds them that Jerry Falwell is a Christian, so he might not arrest them for trespassing after all. Jake goes on to explain that Liberty University has Christian principles, and that the group needs to approach the situation with that assumption. What would Jesus do? He would welcome the strangers to the campus. Predictably, when Jake enters Liberty property reading a statement regarding gay rights, with the rest of the group following behind, the police first warn and then arrest him. The next group member in line continues reading the statement where Jake left off and he is also arrested, and so forth. Apparently, Jerry wasn't channeling Jesus that day!

This situation illustrates the problems with the "common ground" approach to winning gay rights. Often attempting to argue that the Bible has been "misunderstood" regarding LGBTQ individuals, this strategy utilizes religious-based arguments to overcome the very oppression that happens through religious means; in particular, religion combined with the backing of capitalism. The thought is that the other side will come around to a tolerant way of thinking and that they will find the religious arguments compelling and irresistible on both a spiritual and logical level. In other words, they can change without threatening their religious identity.

At one point in the film, some Liberty students go off campus to talk with the group. They emphatically state that being gay is a sin and that they have the right to not allow gays on campus. Essentially they defend their "religious freedom" over the civil rights of LGBTQ individuals; that their right to discriminate against gays overcomes the civil rights of homosexual students. By LGBTQ groups asserting their civil rights, they are therefore suppressing the religious rights of students at Liberty University. Again, the Soulforce group attempts to appeal with the argument that the policy is discriminatory. But the Liberty students won't budge, even when group members who are Christian and gay tell about their experiences in an effort to reinforce the "common ground" argument.

The problem is that when the Bible is used to defend civil rights, the group is working on the territory of the Christian Right who will continue with an endless barrage of the same old tenets of homophobia (that women and men are meant to reproduce because only they can have babies, that homosexuality is against nature, etc.). Some group members who are skilled at knowing the Bible confront the Liberty students on their turf. When eventually the

the necessity of leadership, and building global coalitions. How we get from point A to point B regarding sexuality in K–12 settings is yet to be determined, but these foundational platforms must be established before we can begin to connect activism to issues of the family, feminism, homophobia, and sexuality rights. Public schools are essential convergence points for people to gather. When sexuality is controversial even at the college and university level, one can begin to understand the constraints facing K–12 educators. If we are to mount an effective resistance, we have to get our act together first.

Case Example: Soulforce Equality Ride

Modeled on the Freedom Riders from the civil rights movement, a group of young LGBTQ adults regularly travel by bus confronting the antigay policies of 20 colleges and universities across the United States. Their effort is called Soulforce Equality Ride. Taking place over the course of two months, the 2007 ride was documented in the film *Equality U* (O'Brien 2008). In many ways, the documentary provides an informative case study in what activists should consider in planning effective resistance to homophobia, as well as strategies to avoid.

First, it should be clear to the viewer that the actions of the young activists represent an enormous amount of bravery. Many of the young adults featured in the film have risked expulsion and the subsequent loss of years of study at conservative Christian universities, and many have little choice since their parents are controlling access to college education via the purse strings. With a college education in danger, future employment opportunities and economic survival are on the line, not to mention the loss of family and friends in a form of shunning that is very hurtful and isolating. As the film presents, the economic climate has also led to the antigay policies of many Christian colleges with powerful donors who continually threaten to withdraw financial support, thus endangering the school's existence. Economic situations, therefore, are determining the existence of antigay policy, one of the many ways material conditions impact sexuality. That many of the activists are themselves religious and are attempting to confront the intolerant tenets of their upbringing demonstrates a willingness to combine spiritual beliefs with social justice in the spirit of Martin Luther King Jr. However, that said, there are several teachable moments within the film that can shed light on paths that activists should try to avoid as they mount a fight against sexual oppression tied to the larger problem of economic exploitation via capitalism.

Jake, one of the leaders of the Ride recounts a conversation with a man he met who was attending a conservative Christian college. Asking about how the student coped with being gay in a homophobic setting, he replied that he could not ever show any indication of being gay or siding with gay causes. When Jake suggested working to overturn the antigay policies of the school, the student defended the policy, saying that being gay was a sin. This provides the first challenge for the group: overcoming the self-censoring of centers began to appear in both urban and remote communities to address the epidemic alongside demonstrations of a size not seen since the 1960s, even in places like Houston and San Diego (D'Emilio 2003; Johnson 2009). Civil disobedience became more frequent, including "zaps" of the FDA in support of access to life-saving drugs by the newly formed ACT UP, and a militant protest of the Supreme Court in the wake of the *Bowers v. Hardwick* decision (D'Emilio and Freedman 1997; Schulman 2008).

What the AIDS crisis also brought to the LGBTQ community was an awareness of the extent of societal and legal exclusion based on homophobic policies that had remained unchanged since Stonewall. Partners were not able to visit loved ones in the hospital, inheritances were challenged, and custody battles played out before a public that was either openly hostile or ignorant of gay and lesbian issues (D'Emilio 1992, 2002; D'Emilio and Freedman 1997; Eisenbach 2006). High-profile stories about individuals such as Rock Hudson and Ryan White finally forced a government response to AIDS in the late 1980s, but by that time 46,000 had died and 82,000 had been diagnosed (Eisenbach 2006, p. 308). In the early years of the epidemic, the majority of casualties were gay men born between 1945 and 1969, revealing a deeper historical problem of ostracism as many from that generation had escaped repressive heteronormative settings attempting to find a likeminded community. Yet this community was not insulated from capitalism, as AIDS revealed that those with the means had greater access to health and end-of-life care.

One of the most dramatic art installations overseen by Harvey Milk intern Cleve Jones encompassed the size and scope of the epidemic in an accessible and moving manner. The Names Project featured memorial quilts that spanned two football fields, using the collective metaphor of piecing together people from all walks of life as one unified whole. The combination of the AIDS quilt and the march of 500,000 on Washington proved too much for the government to ignore (D'Emilio 1992). One thing that the activist response to the AIDS crisis demonstrated at the close of the 1980s was that a space for mobilization was possible even during the most entrenched conservative period. AIDS/HIV prevention is now a common feature of all sex education classes, which was unheard of until this awareness was raised. This historical understanding proves most encouraging as we stare down the barrel of the twenty-first century.

Tactics and Strategies

Comprehending barriers to solidarity and history is one facet of activist work. The other is developing sound tactics and strategies around which to organize. This section begins with a case analysis of sexuality activism at the college level followed by a discussion of key tactical and strategic areas in need of development: establishing an anticapitalist stance, sexuality rights with no apologies or conditions, considering unionization and organizing in churches,

The high-profile elections of Harvey Milk and Mayor George Moscone in San Francisco were cut short by their assassinations (D'Emilio 1992). It is notable that Milk's constituency was not just the LGBTQ community, but senior citizens, the disabled, teachers, and union members. Wolf (2009c) outlines how these broad alliances helped to defeat the Briggs Initiative, which would have enabled the state to fire gay and lesbian teachers or their supporters. These broad coalitions were expressed in large marches to commemorate Milk, such as the first March on Washington for Lesbian and Gay Rights in 1979, which drew close to 75,000 participants (Eskridge 2008). Dan White, who murdered Milk and Moscone, received the lightest possible sentence, which spurred the White Night Riots. These protests centered in the bar culture and its growing militancy against state repression, which was embodied in the police who sided with a killer (White) against the homosexual community (D'Emilio 1992). There was a growing sense that the act of coming out of the closet, though critical, wasn't enough to fight oppression. This, along with the dying vision of a revolution engineered a shift from a radical analysis to an identity politics approach.

Whereas the immediate post-Stonewall activism directly challenged capitalism as the root of LGBTQ oppression via the nuclear family, the late 1970s saw a shift to queer theory and notions of difference, along with subsequent emphasis on "democratic rights to consumption, tailored to differences" (Morton 2001, p. 14). Organizations began to focus on single-issue politics and reform efforts, such as allowing gays in the military, rather than an overthrow of the existing system or cross-group alliances (D'Emilio 1992; Eisenbach 2006). D'Emilio (1992) and Kupfer (2005) describe the proliferation of LGBTQ media images and consumer cultural spaces during this time, creating the illusion that there was no longer a need for Marxist solutions since equality was apparently a done deal. The conservative ascendency and postmodern sensibilities supported each other in muting social class difference as relevant to the LGBTQ community. Activism around these issues slowed down tremendously while organizations worked steadily on legal battles.

The discovery of AIDS and HIV would provide a shocking and abrupt collective realization that such equality was merely illusory. First reported by *The New York Times* in 1981, the disease was labeled gay-related immune deficiency, ostracizing an entire community during a time of extreme conservative backlash (Eisenbach 2006; Schulman 2008). The newly elected Reagan administration had slashed the Centers for Disease Control and Prevention budget by half, leaving those with the disease and their loved ones on their own (p. 270). The virtual silence of the government forced the LGBTQ community to abandon identity politics to begin building coalitions, often using the mass media as a form of protest in order to force national attention to the growing crisis. LGBTQ people of color were acutely aware of the discriminatory aspects of health policy to begin with and were among the first to address the issue (D'Emilio 1992). On a smaller scale, not-for-profit help

The fighting continued for five days with only a handful of arrests and one casualty, a cab driver who had a heart attack during the riot. One of the key lessons that the rioters took from the event was that street fighting wasn't just limited to the stereotypically macho. The media coverage of Stonewall, though often fraught with derogatory language toward the "effeminate" participants, displayed a different side of gay culture to the country. This bunch was out and proud, not closeted and ashamed, marking a significant departure in activist tactics and strategy.

In the wake of the Stonewall uprising, openly Marxist gay and lesbian groups were formed, though many in the leadership were more Maoist than Marxist (Wolf 2009c). The most famous was the Gay Liberation Front (GLF), named in honor of the South Vietnamese resistance and focused not just on ending homophobia, but all of U.S. imperialism (D'Emilio 1983, 1992, 2002; Eggan 2008; Eisenbach 2006; Mecca 2008; Wolf 2009a, 2009c). The later-formed Gay Activists Alliance (GAA) built off of the earlier homophile movement and had more of a reformist orientation, though both groups would engage in collective activism depending on the issue (Eisenbach 2006). Both sought to challenge the psychiatric classification of homosexuality as a mental disorder along with opposing sodomy laws and building a sense of gay pride and sexual freedom, though the GLF tied these issues to capitalism and the GAA did not. As with the earlier societies, such as Mattachine and Daughters of Bilitis, these groups sustained a gay and lesbian press to provide coverage of protests and resistance movements (Eskridge 2008). The National Gay Task Force, Parents and Friends of Gays and Lesbians (PFLAG), and Lambda Legal Defense and Education Fund were other early 1970s groups that reflected more formal organizing concerns such as gaining straight allies (Eisenbach 2006). Members of these post-Stonewall groups often reported that the activism was more potent than the years of therapy they had received (D'Emilio 2003).

At the same time that the GLF and GAA were staging demonstrations and public resistance, lesbian and transgender groups were setting up communally run bookstores, co-ops, self-defense workshops, home repair classes, health clinics, publishing houses, restaurants, housing, and other networks to meet the needs of low-income and street people in addition to the LGBTQ community (D'Emilio 1992; Mecca 2008; Shephard 2008). Lesbian and transgendered individuals felt the need to form these groups since they were often left out of the male gay-centered culture of the GLF and GAA (D'Emilio 1983, 1992; Shepard 2008; Wolf 2009c). With the rise in right wing backlash in the early 1970s, there was a growing pessimism that the revolution would never come, so these groups felt the need to address local concerns on a smaller scale (D'Emilio 1992). This, coupled with the chaotic leadership of the GLF signaled the end of the Marxist phase of sexuality activism (D'Emilio 1992; Eisenbach 2006; Wolf 2009c).

"gay is good" were adopted, inspired by "Black is beautiful," the key slogan of African American militants (Eskridge 2008; Wolf 2009a). For those not able to afford college, gay bars provided a key organizing point for disaffected members of the LGBTQ and straight communities (D'Emilio 1992; Wolf 2009c). D'Emilio (1992) recounts how the candidacy of José Sarria, a celebrity drag queen from San Francisco who ran for city supervisor in 1961, mobilized the gay bar vote. Even though he didn't win, the mere act of running for office caused other gays and lesbians to reconsider the closet and its political implications.

The gay bars were an important site for political organizing and socialization, which made them constant targets for police harassment and violence, as most cops assumed that LGBTQ people would be stereotypically passive (Wolf 2009c). While often portrayed as a spontaneous act that arose out of the blue, the Stonewall bar uprising in New York City was a logical outcome based both on the activism of the times along with who took part in the riot (D'Emilio 1983, 1992, 2002; Mecca 2008; Wolf 2009c). The protestors were predominantly subaltern, working class, and multiracial. They also included a good portion of transgendered people who were the most vulnerable to harassment (Perrotti and Westheimer 2001; Spade 2008). As Morton (2001) explains, these were groups of people for whom waiting politely or assimilating was not a possibility: "they were objectively under more intense economic pressures of class and sexuality and therefore had, compared to other lesbians/gays/queers, less to lose by coming out" (p. 15). Therefore, Stonewall was significant not only for who participated, but also by the results: extensive political organizing and a direct mass challenge to the closet (D'Emilio 2003; Eggan 2008; Mecca 2008; Wolf 2009c).

The Stonewall raid was part of a police action attempting to bust a scheme that involved blackmailing closeted gay patrons, the proceeds of which went to the Mafia (Eisenbach 2006). Gay bar raids had become routine and patrons were used to being accosted at any time (Eskridge 2008). However, as the cops separated staff from customers for questioning, they assumed they could humiliate, intimidate, and physically harm LGBTQ customers, especially the transgendered. In the past, most would talk back a little and agree to be searched, but this June night in 1969 was different. First there was just verbal taunting of the police, then fighting back on the part of gays and transgendered individuals that began to take on a political tone, as Eisenbach (2006) recounts:

> A couple thousand were milling around chanting "gay power, we want freedom now, equality for homosexuals, Christopher Street belongs to the queens, and liberate Christopher Street." The atmosphere around Christopher Park took on the feeling of a political rally, attracting the curious and the angry. As the crowd grew into the thousands it spilled into the streets. (p. 97)

that gays should organize around the cause of civil rights. The call for gays to become politically active was a risky one to make in the McCarthy era (Cory's name was a pseudonym for Edward Sagarin) and the term "homophile" was chosen so as not to arouse suspicion (Eisenbach 2006; Eskridge 2008; Fone 2000). Hay extended the mission of the Mattachine Society to include the formation of a gay identity as a distinct minority group. He brought an openly Marxist outlook to the group in its early years as it sought to build a gay constituency that would grow strong enough to challenge homophobia (D'Emilio 1983; 1992; Eisenbach 2006; Eskridge 2008; Goldstein 2003; Wolf 2009c). Eisenbach (2006) and D'Emilio (1992) discuss the rise of the gay press during this era (*One Magazine, The Ladder, Mattachine Review*), which was a remarkable feat during McCarthyist repression. Lesbians formed the Daughters of Bilitis, a networking organization and support group for women (D'Emilio 1983; 1992; Eisenbach 2006; Wolf 2009c).

Hay's Mattachine Society was secretive and had a cell-like organization that was common to many Marxist organizations. This was done to help avoid detection since being discovered as openly gay could lead to being fired or permanently blacklisted (D'Emilio 1983; 1992). Mattachine members were understandably concerned. The McCarthy hearings began to target "sexual perversion" along with Communism as a threat to national loyalty. While the Mattachine Society attempted to steer discussions about gay oppression to a materialist form of analysis, including emergent plans to openly challenge homophobia, a growing faction of the membership wanted to take on a more assimiliationist stance (D'Emilio 1992). As the pressures of McCarthyism mounted, red baiting occurred within the society and eventually the more radical membership was purged in a quest to make the organization more respectable (D'Emilio 1992; Eisenbach 2006). Even though this destroyed the Mattachine as an activist organization, it set the stage for the next necessary step of challenging "the regime of the closet" in the pre-Stonewall era (D'Emilio 1992, p. 239).

As the civil rights and anti-war movements proliferated in the early 1960s, it became more difficult for gays and lesbians to remain hidden. After hearing about Cuba's expulsion of homosexuals from universities, homophile groups demonstrated in 1965, though the numbers were small (Eisenbach 2006). College campuses also became the site of a growing gay resistance. The Student Homophile League (SHL) was the first group of its kind to be recognized in the United States, providing an important venue for socializing along with building gay-straight alliances in colleges and universities. According to Eskridge (2008), the group's formation at Columbia University in New York helped to counter the image of homosexuals as mentally ill and perverse, mounting a challenge to the medical establishment. As the SHL received media coverage, its activism grew, along with the number of students refusing to remain closeted. It would not be long when the SHL tied its activism to the antiwar and women's movements. Eventually, slogans like of rights for gays and lesbians (Wolf 2009c). These historical examples demonstrate that establishment political figures do not head up or sustain revolutions- they tend to follow suit and only after prolonged struggle as with the decades long civil rights and Vietnam anti-war fights. And even then, there is always the risk that reactionary elements can re-establish dominance (Bageant 2007).

History of Sexuality Activism

An understanding of history is critical when it comes to sexuality activism. Relying only on day-to-day life situations for feedback can often distort one's long view and promote a paralysis of cynicism, as D'Emilio (1992) so eloquently reminds us, "the past vanishes before the image of the next task or crisis" (p. 271). A lack of historical understanding also includes a virtual erasing of the Marxist roots of sexuality activism . This is similar to what happened to labor movement history after the red purges of the 1950s (D'Emilio 1992; Wolf 2009c). Sexuality activism also faces the challenge of being associated with what D'Emilio (2003) terms the "bad '60s," when a rising conservative backlash and harsh repression brought down the countercultural movement in the 1970s. The Stonewall uprising happened in 1969, after the nadir of 1968. Consequent activism is linked to 1973 and beyond, which saw the beginnings of global imperial capitalism and retrenched reactionary conservatism coupled with a declining welfare state.

According to Goldstein (2003), the socialist party of Germany was the first political group to attempt a systematic organization of gay and lesbian people in the early twentieth century. The work of the party included speaking out against sodomy laws, which led to an openly gay community being established in pre-World War II Berlin. The Nazis would eventually target this population as part of its campaign of nationalist "family values"; but at the time, Berlin was one of the leading cultural centers in the world. The situation was far more unorganized in the United States. Other than smaller groups like the Cercle Hermaphroditos, created in 1895 to oppose the targeting of homosexuals, and Chicago's Society for Human Rights, formed in 1925, sexuality activism was essentially nonexistent (Fone 2000). While there was a growing gay and lesbian counterculture movement in film and literature, even large urban areas didn't experience much in the way of an open LGBTQ community until after World War II (D'Emilio 1983, 1992; 2003; Wolf 2009a, 2009c).

In 1951, two watershed moments occurred: the publishing of Donald Cory's (1951) *The Homosexual in America: A Subjective Approach* and the founding of the Mattachine Society in Los Angeles by Harry Hay. Cory made the case that the problems facing homosexual males had more to do with a prejudiced society than with anything of an internally pathological nature, and only fuels reactionary policies toward LGBTQ people as a whole. She uses the example of Ernst Röhm's outing and subsequent mocking as a lynchpin in the rounding up of lesbians and gays for the concentration camps. While Röhm may have been a fascist, the use of homophobic rhetoric in his outing was a source of concern—and should be a cautionary tale for activists today.

Lack of Vision

All of these barriers trace back to the original source: a lack of a coherent and committed vision around which to build the necessary strategies and tactics for sexuality rights and, ultimately, the liberation of the working class. Davis (1981), in discussing the life and work of Lulia Jackson, an African American Marxist, relates how Jackson, while at a political gathering, responded to pacifists who were opposed to the militant language of some leftists in challenging the U.S. involvement in World War I: "Everyone knows the cause of war; it is capitalism. We just can't give those bad capitalists their supper and put them to bed the way we do with our children. We must fight them" (p. 157). Such clarity is needed today when there is a lot of hemming and hawing about the root causes of sexual oppression, which is usually overpersonalized. McClaren and Farahmandpur (2002) suggest that activists remain cautious of anything that will only end up buttressing capitalism in the end, though they acknowledge that some immediate reforms like health care and maternity leave could provide needed relief for the working class. A vision has to go beyond calls for tolerance and changing attitudes to challenging state power outright, as occurred in the civil rights movement (Bourne 2002; Davis 2005). It has to challenge capitalism itself (Cotter 2001; Hickey 2006; McClaren and Farahmandpur 2002; Wang 2001).

The vision has to mature from a platform of reacting to right wing assault to one of an uncompromising commitment to sexual liberation. D'Emilio (2003) views the "fight the right" meme as creating "within the organization a sense of embattlement, being besieged by enemies" (p. 118). Instead, unions need to acknowledge that the conservatism of a portion of their membership should not stop efforts to deliver the clear message that LGBTQ rights are non-negotiable (Dolor 2008). A notion of community needs to be prioritized because capital is hostile to cross-coalition building, especially concerning issues of sexuality (Goldstein 2003). Historical understanding is important in defining a vision in order to effectively critique capitalism and its obfuscations (Cotter 2001; Hickey 2006). Without a vision, protests and actions come to naught because there is no plan for the long haul (Holt 1977; Lynd and Grubacic 2008). A lack of vision also enables more reactionary elements to take hold, as occurred in Stalinist Russia and in Cuba with regard to the retraction specific material conditions into "texts" to be deconstructed (Cloud 2001; Nowlan 2001). Activism becomes another way to survive existing conditions rather than attempting to change them, therefore, why bother; especially since it seems much more risky than assimilation? The failed Soviet experiments have only increased this cynicism (Wolf 2009c). Revolution might still be discussed in some circles, but when it comes to questions about sexuality what does it mean to get from here to there? We want to replace the nuclear family, but with what (D'Emilio 2003)? Still another problem is the assumption that we have to wait for "word from above," while those at the top continue to assume our passivity and blame us for it (Holt 1977). However, Lynd and Grubacic (2008) view this uncertainty as a potential strength as history demonstrates that the least likely groups have successfully pulled off resistance actions when the conditions were right. We also have to remember that activist work happens in a much larger social context and the pattern of two steps forward and one step back is the actual work, not just a detour (Perrotti and Westheimer 2001).

Within-Movement Bickering

Power inequalities are a common complaint within activist organizations (Eisenbach 2006; Kaplan 1995). Those with the most available time and resources tend to be the ones who default to leadership positions. In leftist groups, leadership and membership tend to be mostly white and middle class, and the majority of LGBTQ group members tend to be college-educated gay males (Lynd and Grubacic 2008; Sears 2005). Less powerful members are afraid to speak up for fear of having to take action or propose an alternative in front of the group. There are often different levels of radicalization within any group, as was the case with Jane, the underground abortion service operated by women in Chicago (Kaplan 1995). Some members wanted a more political direction for the group, to connect abortion to larger economic and social issues, while others wanted to focus only on their individual roles within the group. In the early Mattachine Society, the cell-like organizational structure of the group, which was designed to maintain confidentiality, tended to put off newer members (D'Emilio 1992).

Within the feminist movement, many straight women were resentful of lesbians and saw their agendas as being too radical, whereas some lesbian groups were hostile to the presence of male allies or transgendered individuals in "women only" spaces (Wolf 2009c). There has also been intergroup conflict between gays and lesbians, particularly in the wake of Stonewall where some lesbian groups felt the need to splinter off as a result of hostilities from gay males (Eisenbach 2006). Other disagreements concern the practice of "outing" conservative or right wing gays who oversee homophobic policies. While Michelangelo Signorile and Peter Tatchell use the media to out such individuals, Wolf (2009c) argues that this is a regressive strategy that

Such liberal gatekeeping leads to prominent conservative lesbians and gays representing the entire LGBTQ community and derailing activism. Eisenbach (2006) provides the example of the safe sex campaign in the wake of AIDS that faced resistance from the same bathhouse and nightclub owners who funded gay activist groups! Schulman (2008) expresses his regrets that leftist groups failed to tie AIDS to the larger issue of national, single payer health care rather than securing AIDS drug treatments for those with the money to afford them.

Inaccessible language and movement politics also serve as gatekeeping functions (Lynd and Grubacic 2008). When many people fail to act, gatekeepers use it as an example of the futility of expecting the working class to fight for itself. But all gatekeepers and their conservative friends should consider themselves duly warned by Kollontai (Holt 1977):

> The masses will renounce inactivity only after the necessary historic moment, when the proletarian vanguard—the socialist parties of all countries—throw off their paralyzing social reformism and boldly advocate all forms of all methods of struggle which revolutionary creativity suggests. (p. 103)

Uncertainty about How to Build Movements

It is important to understand that for those of us born in the 1970s and later, including myself, activism is a distant historical artifact in this conservative ascendency. The closest many of us have been to marches and demonstrations is through our television screens or YouTube clips. E-mailed petitions to MoveOn.org or being a part of Obama's online mailing list updates may be the extent of our familiarity with organizing. The idea of getting arrested for civil disobedience seems like putting yourself out there with no one to back you up after the fact. A form of cynical paralysis sets in where no one wants to be the first one to take action. Not understanding how to build movements is a critical barrier to change. If we aren't sure where to start, it is even more difficult to reach out to people like those in Bageant's (2007) rural communities, who are often more atomized and at risk of being left in the cold. As he asserts, leftists have lost the ability to undertake grassroots organizing beginning with the most basic skills of communication. In addition, we do not have a full grasp of the radical history of activist movements, particularly in the areas of sexuality, the family, and education, which leaves us without moorings (D'Emilio 1992). We think there is no way out when in reality there has always been resistance. Lack of awareness about our history contributes to our despair.

The conservative political climate has cast a pall over social movement organizing (D'Emilio 2003). The inability to initiate a movement has a lot to do with the postmodern belief that all is discourse and illusion, transforming

Liberal Gatekeepering

A liberal gatekeeper is someone or who seeks to limit the radical agenda of a particular movement. The perception that activism has pushed the limits or a fear of "too much too soon" can derail any movement often more quickly than conservative backlash. In fact, liberal gatekeeping and right wing ideology work hand in hand as gatekeepers allow space for conservatives to privatize education, among other public sector services (Moeller 2002):

> The argument that first we must attend to the rights of the whole and then later to those who are excluded by discrimination from the rights enjoyed by the majority weakens the union…if we go down this path, we would have to forfeit the illusion that we care for the children we teach and ignore the fact that not every child in public education comes from places of privilege (Stewart 2008, pp. 174–175).

In the wake of Harvey Milk's assassination, Diane Feinstein extended a public message to the gay community a few days into her role as mayor. She asked that they "respect the sensibilities and standards of the majority" (D'Emilio 1992, p. 91). The Democratic Party is a classic gatekeeping institution. It has continually hedged on issues such as health care; gay marriage; Don't Ask, Don't Tell; women's reproductive rights, and public control of education (Wolf 2009c). While Democrats actively seek the "multicultural" vote, after an election has passed, they bargain away promises to avoid upsetting conservatives.

In the LGBTQ community, gatekeeping has helped to transform the Marxist vision of 1970s groups like the Gay Liberation Front (GLF), where social needs were primary, into queer theory and assimilation-oriented endeavors in which capitalist desires are prominent (Eisenbach 2006; Kuban and Grinnell 2008; Nowlan 2001). Sears (2005) articulates that reform-based movements are now the norm, including strategies such as shaping nuclear families out of gays and lesbians while leaving capitalism intact. According to this assimilationist and often closeted thinking, which harkens back to the 1950s Daughters of Bilitis and Mattachine Society, only clean-cut lesbians and gays should represent the movement to avoid troubling potential allies (Eisenbach 2006). Straight, liberal allies often turn up their noses at street-level activism and view it as a middle-class phenomenon, leaving no other option but assimilation, "the wages of that illusion is death in small doses. But the alternative is not alienation. It's individuality" (Goldstein 2003, p. 55). The once-Marxist Mattachine Society learned this the hard way when conservative gay gatekeepers identified and expelled more radical members, even threatening to provide names to the FBI (D'Emilio 1992). The result was a fractured and ineffective group that could only play the respectability card in the hopes of relying on the good graces of the larger heterosexual society.

class. However, as Hill, Sanders, and Hankin (2002) state, "the socialist project is not about denying the fact of social difference, but about the need to construct solidarity" (p. 172). Far from suppressing difference, Cole and Hill (2002) assert that "women can [either] struggle as women in localized and general struggle, workers as workers, youth as youth, without silencing other identities or other aspects of their own identity" (p. 94). While isolated marches and protests by different identity groups might achieve some media attention, governments are more likely to respond when they see mass movements across coalitions on the street, as with the 2006 immigrant rights marches. Likewise, identity groups formulated on the narrow basis of victimhood by gender or race (Take Back the Night, Million Man March) are not as effective as broadly based solidarity movements in challenging white supremacy (Superson 2002). Overt personalization of socially derived problems like racism, sexism, or homophobia tends to lead to a superficial level of action, such as tolerance training, street renaming, and sports teams' logo makeovers, without ending oppression in the long run (Bourne 2002). What D'Emilio (2003) calls the "permanent outsider status" of leftist identity groups reduces activists to a constantly reactive state, dependent on the right wing agenda.

There are several examples of identity tribalism hampering what could have been effective forms of solidarity. Eskridge (2008) believes that the setbacks in the overturning of sodomy statutes could have been avoided if gay rights groups had formed connections with feminists, who were busy defending the privacy rights of women seeking services from Planned Parenthood and who likewise could have benefitted from the support of gays and lesbians. Many feminists resented the presence of lesbians in the movement and shut out valuable allies so as not to offend straight women (Wolf 2009c). As part of the conservative backlash of the 1980s, sodomy laws were eventually used to deny visitation and custody rights to mothers who came out as lesbian, illustrating the need for inter-LGBTQ solidarity. Sexual violence against women is connected to homophobic violence against the LGBTQ community—a link that the right wing has long recognized and used to its advantage (D'Emilio 1992). D'Emilio also outlines how many LGBTQ groups in the wake of AIDS were unprepared to confront the rhetoric and prevalence of Christian fundamentalism due to their single-issue orientation. For example, gay men had assumed that reduced harassment by police in the bars and clubs of the late 1970s was a sign of freedom being achieved. The lack of a government response to HIV/AIDS woke up the gay community who was unprepared to take on right-wing attacks on civil rights. Another group that has not been tapped is poor whites, who have often been ignored as part of coalition building. Yet almost 20 million people live at or below the poverty level in the United States and more than 50 percent of all low-income people are white (Bageant 2007, p. 9).

consciousness from school board meetings to the local lunch room. Within the workplace, these right wing views become widely disseminated, but not necessarily because the workers are themselves politically active:

> You are talking about millions of people who are not on political mailing lists, couldn't care less what's on the Internet, and wouldn't know a Blackberry from a garage door opener. They spend eight hours a day listening to talk radio through ear phones as they work. And they are aware of the politics of their supervisors and bosses. It would be wrong to say that supervisors put pressure on workers to vote conservative. They don't have to. They merely let their politics be known, and the desire to curry favor with the boss does the rest. (p. 79)

Bageant reminds his readers that this is a situation of reactionary networking that has deep ties. The elites of the town went to school with the working class and their intimidation factor has been quite strong over the years. Since these elites are also overrepresented on school boards, city government, and the business sector, which employs the working class in the town, they are an almost impenetrable force. The elites also repeat verbatim the sound bites coming from right wing Web sites like NewsMax, FrontPage, or Town Hall. By contrast, any progressive or leftist message sounds isolated and takes on an "outsider" persona since it is so rarely heard or the media is preoccupied with airing "both sides" of an issue.

In addition, Bageant (2007) addresses the problem of functional illiteracy, which affects nearly half of the U.S. population. For a significant portion of the U.S. population, filling out a job application or reading a couple of consecutive paragraphs is a major challenge, and the problem is worse for the poor and illiterate. Functional illiteracy combined with the primacy of television and the church (one of the only working institutions left) further entrenches the reactionary pipeline from elites to workers. The fundamentalist religious outlook of submission to authority permeates the workplace where employees are required to bring in a note from a doctor if a sick day is taken. Few workers resist since the terror of losing a low-wage job is very real while reactionary populism happily steers anger toward minorities or gays. Activists would be wise to consider Bageant's analysis, which is based in class analysis, before attempting to declare such workers as irredeemably lost.

Identity Tribalism

Identity tribalism refers to the isolation that often results from identity politics where the emphasis is on difference, division, and the social markers of such. Often, postmodernists accuse Marxists of downplaying the uniqueness of people in order to focus on the "oppressive" modernist concept of social

Recognizing that a good portion of the working class holds racist, sexist, homophobic, and other self-defeating views is not the same as arguing that these problems originate *within* the working class itself; Thomas Frank's (2004) analysis to the contrary. Activists often lament that low-income conservatives refuse to act or that they "vote against their own self-interest" when the problem originates at a much deeper level (Bageant 2007; Hickey 2006; Holt 1977). At the same time, the mass media and ruling elite are all too happy to declare the working class conservative by nature. When leftists do the same, they run the risk of misinterpreting important political events, such as viewing the defeat of California's "No on 8" campaign as primarily due to right wing backlash rather than a lack of effective outreach to often neglected working class and minority neighborhoods (Johnson 2009). The message from activists needs to be consistent and firm: straight people, especially if they are working class, do not benefit from anti-LGBTQ legislation and policies just as working class whites do not benefit from racism (Wolf 2009c).

Leftist ethnographer Joe Bageant's (2007) *Deer Hunting with Jesus* provides an important case study of reactionary populism written with much wit and insight, since he himself came from a working class and rural background. He begins by stating that his numerous interactions with supposedly "right wing" poor people indicate that they are desperate for options, particularly real health care reform. But in the absence of such options or politicians who aren't wedded to big money (including the Democratic Party), the void that is created allows reactionary ideology to step in. Democrats have long discounted the "red" states altogether, to focus on voters in larger urban areas. This vacuum creates confusion as people are not sure who or what is harming them; they only know that they are being screwed financially and emotionally by poverty itself. This, of course, provides a distinct advantage to elites:

> Getting a lousy education, then spending a lifetime pitted against your fellow workers in the gladiatorial theatre of the free market economy does not make for optimism or open-mindedness, both hallmarks of liberalism. It makes for a kind of bleak coarseness and inner degradation but allows working people to accept the American empire's wars without a blink. (p. 71)

Bageant (2007) gives no quarter to the romantic myth of the provincial small business owner. Not since Marx has there been such a trenchant analysis of the middling bourgeois. He launches a pointed attack at these people who "destroy land use and zoning codes, bust unions, and generally keep wages low, rents high, and white trash down" (p. 41). Not only are the small business owners key players in these communities, it's the *failed* ones who tend to be more hardened in their right wing views as they spread false questioned because "wealth is proof of God's love and bestows the power to mess over anyone who disagrees with you" (p. 86).

Appeals to the nuclear family are used to buttress neoliberal policies and intensify in the midst of economic crisis (Wolf 2009c). In order to protect its interests, the capitalist class has to ensure that divisions based on race, gender, religion, ability, class, and sexual orientation remain in place. Even a classic Enlightenment concept like "personal freedom" is now equated with buying cheap goods from Walmart as infrastructure investment has gone toward unsustainable suburban development at the expense of urban and rural areas (Bageant 2007). In this climate, capitalism has to enforce hierarchies where people have no absolute rights, such as those attempting to seek medical services when they have no money (Kaplan 1995). When political influence is reduced to "consumer rights," those with few means are shut out of the picture, as Sears (2005) points out:

> Those who have gained the most are people living in committed couple relationships with good incomes and jobs, most often white and especially men. At the same time, queer people of color, street youth, people with limited incomes, women, people living with disabilities and transgendered people have gained less... recipients of social assistance now find their eligibility for benefits contingent on their partner's income. Street youth are now hustled out of queer areas by cops acting on behalf of gay or lesbian residents intent on creating comfortable middle-class neighborhoods. (p. 93)

It should be clear that individualized acts of sexual rebellion alone are not up to the task of taking on neoliberalism (Wolf 2009c). As Bageant (2007) writes, "when the deal goes down, all hell is going to break loose in this country. But for now we live in an eerie space purchased through sheer denial" (p. 114).

Reactionary Populism

While the right wing has always been an ever-present feature of U.S. society (for an excellent historical overview of America's elite, see Fraser 2005), today the right is more organized, better funded, and bigger and broader in its goals than in the past (D'Emilio 2002). Faux populism is essentially the ideas of the ruling class representing themselves as "authentic" views of the working class. One of the large-scale goals of the right wing is to secure this collective lack of class consciousness on the part of workers since a divided working class is highly beneficial to the capitalist class (Wolf 2009c). As Hypolitio (2008) explains, a significant portion of union members also ascribe to conservative views, making it difficult to achieve solidarity. However, the fact remains that there is an irresolute contradiction between the owners of capital and the workers, despite reactionary populist rhetoric, and activists could be able to find space within these contradictions (Lynd and Grubacic 2008).

capital and the fluctuation of world currency markets make us subject to powers beyond individual control, the need for dignity, security, and freedom at the level of intimate relationships and the uses of the body have become more important than ever. Solutions to problems in these areas, of course, cannot be divorced from changes in the rules of international economics. (D'Emilio 2003, p. 42)

At the same time, D'Emilio (2003) reminds us that if activists ignore issues of sexuality and the family, they will fail to communicate a meaningful and compelling message for a better world. This message is desperately needed when 33 percent of working Americans earn less than $9 an hour and a quarter of two-earner households fail to reach a combined income of $35,000 annually (Bageant 2007, p. 31). The mounting pressures of credit card debt, high medical bills, home foreclosures, and out-of-reach education costs constantly push down on families, contributing to an entrenched atomization:

It's a class thing. If your high school dropout daddy...never read a book and your momma was a waitress, chances are you're not going to be president of the United States, regardless of what your teacher told you. You are going to be pulling down eight bucks an hour at shift work someplace and praying for overtime to pay the bills. And you are going to be pitted against your fellow workers...to hang onto that job. You are going to draw the inescapable conclusions that it's every man for himself. Solidarity be damned. The much-needed eight bucks comes first. (pp. 10–11)

Within the realm of K–12 education, neoliberalism utilizes the practice of standardized testing in order to shape the curriculum and the populace at large in its image (Weiner 2008). Schools and students that produce higher test scores are often shielded from the harsher policies of No Child Left Behind. Reduced opposition to testing is likely to come from these settings, "because they see the world refracted through the eyes of U.S. capitalism" (p. 259). These policies increase the divide between the lower-income schools and those that serve higher-income communities. While the working class is constantly being lectured by politicians about "personal responsibility," unless one is lucky to be a member of the "deserving poor" and is subsequently hand-picked based on test scores to attend a privatized charter school, most low-income people face the following choices: poverty-wage work, the military, or prison (Bageant 2007). Bageant goes on to describe how small, rural towns are minicartels in a fiefdom run by a cabal of ultra right wing small business owners (also school board members) who seek to keep wages low and the political climate friendly for Walmart and agri-farms. He provides an example of his home town holding a conference on "the future employment needs of our youth" at which the main speaker was the head of a rendering plant (p. 29). The well-to-do residents of these small towns are never

Challenges to Building Solidarity

If oppression and exploitation are not natural features of human society, as many conservatives are happy to argue otherwise, then why is it so difficult to spread the word about alternatives to the existing situation? It would only seem logical that people would gravitate in large numbers toward a message of sexual liberation and freedom connected to a larger movement of worker liberation from capital. Yet this is certainly not the case. Every step of the way is a struggle. While there have been significant gains regarding sexuality rights in the past 50 years, much resistance remains. As soon as one gain is made, as with the growing activism after Proposition 8, another setback emerges, such as Maine's Question 1. It is critical that activists remain clear-eyed about the common challenges they will face and to analyze the root causes of these challenges. Predictably, stronger barriers originate from the well-funded right wing, yet a significant number are tied to shortcomings in leftist organizations. The following challenges will be explored in this chapter: neoliberalism, reactionary populism, identity tribalism, liberal gatekeeping, uncertainty about how to organize and fight back, within-movement bickering, and an overall lack of a coherent vision, which is necessary to sustain an effective movement.

Neoliberalism

The most recent incarnation of capitalist philosophy is neoliberalism, which exerts a profound influence on the material conditions and worldview of the majority of the world's population. Neoliberalism refers to the intensification of markets with globalization, the outsourcing of jobs, and the privatization of formerly public services such as education. Capitalism is making use of the atomization of the majority of people as it enters its imperial phase, encompassed in neoliberalism. Early phases of capitalism were more local in nature with a focus on markets limited to nations with a few trading partners. As local markets reached their profit limits in the 1970s, U.S. capitalists had to go abroad to third world countries in search of cheap labor. Now, that labor—along with harder to find natural resources—is located at the point of a gun in the form of various military engagements all across the globe. The problem is that sexuality is framed as a personal, individual choice, ignoring the larger economic forces at work. Within the United States, the absence of class-based language and overreliance on individualist analysis combined with ruthless competition in a climate of scarcity creates a situation in which few can thrive (Bageant 2007; Cotter 2001). Precarious economic conditions create an urgent need to link sexuality with larger social issues:

> The connections between macro level world economics and the micro reality of personal life are real and substantial. As the movement of global

statement in five classes. Several students reported concerns about the racist and homophobic content. One of the students in this teacher's class was gay. The teacher was reprimanded by administrators and the school board, but was still allowed to teach (Dailing 2009, par. 3). So we have a situation where a teacher promoting a pro-tolerance standpoint is terminated while one who makes an openly racist and homophobic statement is merely reprimanded.

While adults seem too frightened to pull together a coherent, activist response to LGBTQ oppression within K–12 settings, Will Phillips, an Arkansas fifth grader, made a bold move. On October 5, 2009, he refused to stand up for the Pledge of Allegiance in his classroom, declaring he would keep up his protest until gays are allowed to marry, despite the angry response of teachers and classmates as the week progressed (A Boy and His Flag 2009). After a few days, the exasperated substitute teacher threatened to bring Will's mother and grandmother to school to make him stand. Will described the rest in his own words:

> She got a lot more angry and raised her voice and brought my mom and grandma up. I was fuming and was too furious to really pay attention to what she was saying. After a few minutes, I said, 'With all respect ma'am, you can go jump off a bridge." (par. 6)

According to Will, he could no longer bring himself to say the Pledge because it would be a lie as long as gays and lesbians were not allowed to marry. After his "bridge" remark, he was sent to the principal's office to look up information on the flag and what it represents, even though there is no law compelling any student to stand up for the Pledge. As part of the price for his protest, Will has been called homophobic names like "gaywad" and "fag."

This chapter will present analyses of existing resistance movements that work in the areas of promoting comprehensive sexuality education, and the fight for LGBTQ rights that could be applied to K–12 settings. In addition, these cases will be linked to larger social movements that promote workers' rights, connecting the family, partnerships, sexuality, and work to the struggle for liberation from capital. Some of the groups might not identify as Marxist, but could be considered allies for worker liberation. Strategies for a militant insistence that LGBTQ rights are human rights and the concern of all workers, starting with the schools, are presented. This issue must be addressed no differently than gender, race, ethnicity, disability, age, and other categories articulated in class formation and struggles in and for education. The chapter opens with a review of common challenges activists face when attempting to create coalitions for sexuality rights, followed by a brief overview of sexuality activism from the 1950s to the 1980s. The final section looks at tactics and strategies in and out of schools that are important for educator activists to consider. Understanding what happens in K–12 settings requires a connection to the larger world.

Chapter Six

Building Effective Educational and Resistance Movements in K–12 Classrooms

As discussed in previous chapters, the process of addressing sexuality issues in K–12 settings is challenging, especially if one is a teacher committed to activism. There simply hasn't been much organizing around sexuality rights in these settings. D'Emilio (2003) asserts that bringing up issues of sexuality and the family are like salting a mass wound. Hypocrisy and contradictions are the norm. On the one hand we have 24-hour, coast-to-coast presentations of heteronormative, commodified sex in the form of pornography, unrealistic body images, and objectification of women and children. People view sex as one more thing to purchase, making it quite accommodating to capitalist co-optation. Yet K–12 classrooms are reluctant to initiate discussions about sex so as not to offend parents who are most likely attempting to monitor sex-saturated media through Internet filters and cable blockers. Students are not getting nuanced and informed views about sexuality, to the detriment of their health and well-being. Images of the human body are conflated with indecency, as in public high schools that don't allow art classes to produce work featuring nudity (Bowman 1999).

In rural Oklahoma, a teacher wa fired for overseeing a student production of *The Laramie Project*, a well-known play and 2002 movie about Matthew Shepard (Toppo 2009). Even though the teacher had secured permission ahead of time for first showing the movie and then having her students film scenes from the play, the principal unexpectedly halted production after a few weeks. When the students protested this censorship, the teacher held a ceremony off of school grounds to hear their grievances. Not long after, she lost her teaching job. A suburban Chicago consumer education teacher made the following statement as part of a taxes unit on the National Endowment for the Arts: "How would you feel about your tax dollars going to pay some black fag in New York to take pictures of other black fags?" This teacher repeated the harassment, but speech or actions that contribute to a hostile environment—much to the horror of conservatives (Cudd 2002b). Yet when attempts are made to report harassment women are often told to consider sexual harassment as a "personal" or case-by-case phenomenon, not part of the workplace as an institution functioning within capitalism (Cotter 2001). The individualistic approach has the effect of locating the negative behavior within the harasser, and puts the onus on already cash-strapped workplaces to address the situation, as in Cotter's account of her own harassment at the hand of a colleague at a university. The overall effect is one of downplaying collective activism while letting the legal system run its course.

Conclusion

As Coontz (2005) argues, "no one could have survived very long in the Paleolithic world if individual nuclear families would have had to take primary responsibility for all food production, defense, child rearing, and elder care" (p. 38). Indeed, collective solutions to marriage and child care are necessary along the lines of Levine's (2008) social support recommendations of universal paid maternity leave and provisions for household relief and Coontz's (2005) doable proposal of flexible scheduling while workers still receive full benefits. Kollontai's (Holt 1977) proposals included socialization of domestic chores where people would have access to laundries, kitchens, nurseries, and schools all run by paid professionals rather than privately borne by mothers. Hayden (2009) describes Tynggarden, the 1970s Danish housing project where each family gives up 10 percent of their interior square footage to a shared center for 10 to 15 families. These affordable shared dwellings combine much-needed private space with communal space for mailboxes, washers and dryers, kitchens, and meeting rooms. Similar transitional dwellings in Holland have been designed "to sustain a new family form—the single-parent family—in urban society" (p. 52). Leacock's (1981) research of egalitarian societies reveals household collectives and communal child rearing—including no child being seen as illegitimate–as the norm, not an aberration. In these societies, women and men were just as limited by their surroundings as we are today, but they were self-directed, not alienated from their labor. Ever the optimist, Kollontai (Holt 1977) saw egalitarian society as a possibility, not a mere illusion:

> The science of economics and the history of society and the state show that such a society must and will come into being. However hard the rich capitalists, factory-owners, landowners, and men of property fight, the fairytale will come true. The working class all over the world is fighting to make this dream come true. (p. 135)

Chapter 6 explores the possibility of another way of envisioning society, using sexuality as a major point of organization and resistance.

income (p. 240). The issue is urgent as 50 percent of the population in the United States depends 100 percent on the government for assistance as they age (p. 245). The majority of those people are women.

Like other forms of workplace discrimination, sexual harassment "is not just the result of the 'bad' behavior on the part of a few failed individual men but is rather a practice systematically connected to and enabled by the larger structures of sexism within capitalism" (Cotter 2001, p. 210). Sexism functions institutionally in capitalism by creating resentment in working-class men at female targets who are perceived as "taking away" their jobs or horning in on their territory (Maybee 2002; Superson 2002). Much of this rhetoric, as with racism, originates not with the working class, but in the ruling elite, who distribute it in the form of backlash rhetoric via the right wing media (Chamallas 2002; Cudd 2002b; Moeller 2002). The group dynamics of sexual harassment are often, as Cudd (2002b) describes, "preemptive," where the victim is so taken aback at the harassment that they aren't able to immediately respond. At the same time, harassment is considered merely a crime of offending someone's sensibilities. However, harassment goes much deeper, as Cudd outlines:

> Women do not, as a class, complain about sexual banter because it is "off color" or rude, they complain because it both reveals and reinforces the position of women as the subservient sex, as the sexual objects of men. They complain because it is difficult to believe that one's colleagues, professor, or supervisor is taking one seriously as a fellow worker or student when they discuss women as "cunts," "bitches," "whores," or simply sexual conquests, or when their fellow workers' or professor's preferred office decorating scheme portrays women as sexual objects for use by men. Then even if the colleague, professor, or supervisor is otherwise treating her fairly, a woman in this position is burdened with (well founded) anxiety and feelings of inferiority that are neither fair nor conducive to optimal work. (p. 224)

The lower and more precarious employment status of women also contributes to their reluctance to come forward, which, of course, benefits capital by creating a more compliant labor force overall, impacting working-class men (Cotter 2001). At the same time, Cudd (2002b) explains, harassers are given ideological protection in the form of arguments that their "free speech" rights are being infringed upon and that antiharassment policies create a chilling effect. According to Moeller (2002), the irony is that women who come forward to report sexual harassment are the ones often portrayed as fascistic, along with grass roots organizations that insist on taking a collective approach to the problem.

Recognizing the many facets of workplace intimidation, federal law has finally forced workplaces to implement a systematic policy against sexual harassment, using the stricter standard of not only traditional quid pro quo are members of minority groups who were denied legal protections available to married couples, such as qualifying for public housing or health insurance (Kotulski 2004, p. 101). The connection between marriage equality and queer poverty impacts all workers, especially as retirement approaches. Kotulski states that currently there are only six LGBTQ retirement centers in the United States, none of which offer more critical care. Gay seniors are more likely to live alone, having few living family members, which makes them more vulnerable, or they have to hide their identity in order to find lodging in nursing homes hostile to LGBTQ individuals.

The majority of senior citizens are female, making retirement and health a relevant issue for socialist feminists. As more women remain single or do not remarry, the issue of retirement and workplace benefits becomes more prominent and shouldn't be tied to marital status alone. Socialist feminism attempts to address these deficiencies while at the same time attacking the foundation of the problem, which is capitalism. Bageant's (2007) harrowing account of eldercare for the poor is a call to action. While privately owned nursing homes regularly make millions of dollars in profit, many of their clients have to prove they are destitute in order to receive government-funded minimal care in subpar facilities that cost thousands of dollars monthly. Medical expenses are the leading cause of bankruptcy, yet over a third of those filing for bankruptcy have health coverage, indicating that as retirement approaches, medical coverage rapidly decreases (p. 234). Even with Medicaid funding, nursing homes have a limited number of beds available, necessitating a waiting list for families. Because assisted living is considered too expensive to fund, families are steered to nursing homes, as Bageant explains:

> The family had meetings with doctors and with a care center manager, and, after some creative paperwork, it was determined that Ruth had dementia. The family would have to pay a couple hundred dollars a month in addition to what Medicare paid, but everyone breathed a sigh of relief. She was in. The dementia route is often the easiest way in for families desperately needing access to elder care. (p. 243)

Ruth was, of course, treated appropriately for her "condition," and was administered a wide array of drugs from the profit-making pharmaceutical industry. Over time she resembled someone with the symptoms of dementia even though she would have done fine in assisted living before her placement in a nursing home.

Financially speaking, women are hit particularly hard at old age and marriage is no protection. According to Bageant (2007), married women who worked outside the home receive the same Social Security benefits as women who did not pay into the program via payroll taxes (p.239). Older women are the bulk of people drawing Social Security benefits, 24 million of whom depend solely on the monthly check representing poverty-level workforce are buttressed by an overall valuing of jobs that require so-called masculine attributes over feminine ones (Superson 2002). Because women often work in the most vulnerable sectors (part-time, retail, and service industry work that is not far removed from the household drudgery of the past), they are often reluctant to make waves, a preservation strategy that starts at a young age:

> They know that the harm they cause girls will often go unpunished or even unnoticed and that the girls will be deemed "crybabies" if they complain, which only intensifies the gender division.... Boys come away from such discipline believing that they still hold power over girls and may come to resent girls for snitching on them and trying to take away some of their power. (p. 96)

This separatist behavior also extends to males who do not display the appropriate degree of masculine traits; something that has contributed to widespread discrimination against transgendered individuals on the job.

Leacock's (1981) research on egalitarian societies places labor in a different light. While there has always been some degree of gender segregation, the divisions we assume are natural and timeless did not become absolute until the arrival of trade. An individual within a collective was dependent upon the group, not upon someone of a higher class. Authority came not from possessions or brute strength, but from the ability to contribute to the overall functioning of the group. The group was invested in improving everyone's material well-being because this worked to the betterment of the group as a whole. In other words, someone who was more physically able didn't throw a less able person out of a job, as in today's society. Everyone's needs were met because it made no sense to allow someone to go hungry or homeless. Because egalitarian society was organized around these concepts, a woman would not be at a disadvantage because of her ability to bear children and there was no need to segregate labor to the advantage of males, for example.

Kotulski (2004) addresses the important issue of workplace discrimination in the form of denying essential benefits to LGBTQ individuals. This ends up enriching employers and the law's refusal to fully recognize gay and lesbian families provides cover for such discrimination. While civilian federal employees were covered under the Family and Medical Leave Act of 1993, much remains to be done for LGBTQ employees. Even if an employer chooses to provide civil union benefits, there is no guarantee that the benefits will be permanent, as was the case in the Exxon-Mobil merger, in which Mobil employees immediately lost their domestic partner benefits. Heterosexual married couples are not taxed for these benefits while domestic partnerships are.

Contrary to media portrayals of a leisurely, wealthy, gay elite, many LGBTQ people depend on these benefits for survival. In the South, 34 percent of lesbian and gay couples are raising children and many of these families

Willet's (2002) evaluation of academia finds that its increasing productivity demands fall hardest on women who often don't have the same kind of support as male colleagues:

> This rise in productivity did not accompany the entrance into the academy of GIs from World War II, which suggests that women and minorities have been targeted unfairly. At the same time, the standards for excellence in child rearing (from the continual pressure on our children to perform well on standardized tests in the school system to the proliferation of extracurricular activities) are also rising. (p. 122)

The overall corporatization of the university and all other aspects of life has made leisure more and more of a luxury commodity. Even mothers who stay at home feel the pressure to do more to justify not bringing a paycheck into the house.

The Workplace: Wages and Benefits

Over the past 30 years, though the wage gap between men and women is slowly narrowing, the average woman's wage remains stable at 70 percent of the average male wage, even taking into account the issue of part-time labor (Kelly 2002; Yalom 2001). What isn't often explained, however, is that this narrowing of the gap is also due to the overall fall in *male* wages (Smith 2005; Wolf 2009c). While equality in the workplace (including wages) alone won't remedy the situation of working women, focus on this issue is important to the overall liberation of the working class (Engels 1972; Holt 1977; Leacock 1981; Wang 2001). Women often contribute to the reserve army of labor that is routinely used to keep wages low, enabling capitalism to intensify its exploitation (Kelly 2002). The nuclear family structure works hand in hand with capitalism to keep the threat of unpaid labor as a leverage point by dredging up gender divisions in the workplace. This was a point of conflict when women entered the workforce en masse during World War II when it was useful to capitalism. After the war, women were expected to return to unpaid domestic labor, as part of capitalism's strategy to manage employment levels by promoting marriage and college enrollment in the form of the GI Bill. A two-tiered form of assistance was the norm: the enormous degree of federal funding and tax policies that overwhelmingly benefited married couples, and charity-based or lower-level welfare assistance for unmarried women. These policies created an overall incentive for marriage and deincentivized women to seek paid labor (Coontz 2005).

Today, many employers take advantage of women's need for more flexible hours by having women work off the books, not requiring a signed contract, doing piecework, and so forth. This is all part of the "flexible workforce," which is a gendered workforce (Kelly 2006). The global divisions within the bulk of their ideology is anti-woman and pro-nuclear family, leaving the LGBTQ community out of the equation altogether. What appears to bother conservative movements more than the divorce rates, is the fact that fewer women are remarrying as other relationship options, including remaining single, present themselves (Coontz 2005). Men are also rejecting the remote macho "head of household" image, taking a more involved role in child care, from being present in the delivery room to staying home to raise the children (Page 2006).

Coontz (2005) and Yalom's (2001) review of divorce is also enlightening. The Victorians were no better at marriage than we are today; it's just that there were few options for survival in industrialized society other than within the nuclear family, which, for the time being, was relatively sustainable on the male breadwinner model. Alongside the growth of the nuclear family came the simultaneous development of divorce laws, with many countries liberalizing their policies in the late 1800s. Nearly a third of U.S. couples who married in the 1950s obtained divorces, many of them no-fault (Coontz 2005, p. 252). As more and more women found the nuclear family to be untenable, they sought divorce as a relief, despite being put at a material disadvantage by the process. In general, divorce represents an almost 30 percent decline in women's standard of living while men gain 10 percent (Yalom 2001, p. 394). Women who were homemakers for wealthier men are valued more highly in settlements than women who were wives of lower-income men. Even though conservatives have proposed making divorce laws stricter, Coontz (2005) believes that this will not address the problem of the nuclear family, which is far more systemic and tied to economic forces.

Under capitalism, women are exploited within the nuclear family as they are expected to provide unpaid labor in the form of child care and operating a household (Kelly 2006). In today's situation where the majority of mothers work outside of the home, they are devalued further as their wages are viewed as a supplement to a man's income, even if they are a single head of household. The state represents the interests of the ruling elite, who use a combination of social punishment, religious piety, and economic punishment in their administration of a meager and humiliating welfare system (Leacock 1981). Real "family value" policies such as paid leave, universal health care, access to contraception, food subsidies, and child care are opposed across the board by politicians who declare themselves to be "pro-life," insisting that these policies are unnecessary for women who "choose" to work (Page 2006). One result is 10 to 20 percent postpartum depression rates in new U.S. mothers (Levine 2008). Illustrating the importance of context, Fiji, a relatively poor country, has almost no postpartum depression: "For Fiji women after childbirth, there is mandated extended relief from domestic responsibilities such as laundry and cooking for 3–4 months as well as relief from work on the family farm for one year" (p. 28).

percent of all persons marrying by the 1960s (p. 225). In the modern nuclear family of this era, a woman could expect her marriage to last 45 years. In 1850 the average was 29 years, mostly due to lower life expectancies (p. 226).

Coontz (2005) explains that one of the most fundamental shifts to cause us to rethink the nuclear family is the number of people waiting until their mid-twenties and early thirties to marry for the first time, if at all. Between 1970 and 1999, the number of unmarried U.S. households increased seven times over, and the U.S. census revealed 40 percent of cohabiting couples with children under age 18 (p. 266). A good percentage of births to unmarried U.S. women (40 percent) were intentional, and single women making up one-third of adoptive parents in 2001 (p. 270). These are trends that don't appear to be decreasing, indicating a need for alternative family forms and policies where social relief isn't dependent upon one's marriage status. This can be difficult to promote, because while people might be experiencing firsthand being single parents, they still ascribe to the abstract notion of marriage, which only puts pressure on themselves to live up to the image:

> In hunting and gathering bands and egalitarian horticultural communities, unstable marriages did not lead to the impoverishment of women or children as they often do today. Unmarried women participated in the work of the group and were entitled to a fair share, while children and other dependents were protected by strong customs that mandated sharing beyond the nuclear family. This is not the case today, especially in societies such as the United States, where welfare provisions are less extensive than in Western Europe. Today's winner take all global economy may have its strong points, but the practice of pooling resources and sharing with the weak is not one of them. (Coontz 2005, p. 49)

Hardisty (2008) analyzes the traditional marriage movement, which links the nuclear family to poverty prevention and productivity, casting those who don't choose marriage as lazy and prone to dependency. Authors such as Gilder (1981, 1986) argue that marriage channels male aggressions into productive work and regularly point to bachelors as having a lack of direction or purpose. On a national scale, the Heritage Foundation worked with the Bush administration to offer federally funded marriage promotion courses from a decidedly fundamentalist/evangelical Christian outlook. The fatherhood movement also offers a series of conferences and media appearances that focus on divorce as the downfall of society, the unfairness of child custody laws, and the importance of child support, and generally harass single parents and same-sex families. While there are some authors within the fatherhood movement who take a less reactionary and more multicultural approach (i.e., community improvement efforts as the focus), the

Feldt (2004) and Page (2006) believe that we can no longer rely on the courts to protect reproductive freedom. If Roe were to fall, states would be able to determine their own legislation concerning abortion and contraception, and the more conservative states would either outlaw or severely restrict access. The remaining states would be likely to have protected access, but mostly for women with the means to obtain health services or to travel to locations where services are available. With several state laws being vague in their existing language, doctors who provide such services—and the women who seek them—would be open to prosecution, in some cases for murder. Congress could then ban abortion nationwide, reenact sodomy laws, or pass the Federal Marriage Amendment, so no privacy precedents could stand in their way. Activism combined with the proposed Freedom of Choice Act (2009), which would invalidate any restrictions on reproductive decisions or attempts to discriminate based on reproductive choices, seems to be called for.

Marriage and Child Care

Kelly (2002) relates how during the 1984–1985 miners' strike in England, miners' wives found that the nuclear family had no applicability for maintaining the strikers' needs or handling their own households. As a result, the women quickly organized communal means of child care, cooking, and laundry, only enhancing the message of solidarity essential to a successful strike. In egalitarian societies of the past, women publicly produced work in a similar shared manner, but after the arrival of private property, the family became a private economic unit limited to one household (Leacock 1981). Our failure as a society to respect domestic work (child care, laundry, house cleaning) as a significant form of labor "conspires with a larger failure to reflect upon how we as a society define and value work" (Willett 2002, p. 124). Capitalist ideologies about marriage, divorce, and the care of children lead to nothing but dead end solutions for working women and their families. The economic crisis is straining the seams of the already fragile nuclear family model, forcing some things to give way.

In looking at the course of human history through a materialist lens, it is quite revealing that the term "family" referred not to the ultra-privatized household of today, but came from Roman lexicon to refer to those to whom a patrician had supreme rights: the wife, slaves, mistresses, and children (Leacock 1981). As class society became further entrenched, social penalties for sex outside of marriage and illegitimate children intensified (Coontz 2005). The concept of "out of wedlock" births was nothing new in eighteenth-century Europe. In France and Germany, the rate of these births quadrupled between 1740 and 1820, especially among groups which did not hold property (p. 157). Marriage rates peaked at the height of the nuclear family form, the 1950s, where nearly half of women were married by age 19 and 70 percent by 24, eventually reaching the astonishing statistic of 95 conservatives are using these scientific advances to bolster their cause that the fetus is a small human being:

> Opponents of abortion seek to have "potential life" treated the same way in all contexts—and that is as an "unborn child" who has, from the moment of conception, a right not only to medical benefits and the protections of the state, but to life. (p. 341)

The fetus-as-person laws would invariably impact poor and minority women the most, as history and current events demonstrate. From 1989 to 2001, pregnant indigent women who tested positive for drugs at the University of South Carolina Medical School were immediately jailed or were jailed directly after giving birth while still in their hospital gowns. Nearly all were African American and were charged with drug possession and child neglect, facing sentences of 2 to 20 years (Feldt 2004). Florida Governor Jeb Bush even tried to have a guardian appointed for a fetus in the case of a 22-yearold mentally disabled woman who had been raped at a group home (Canedy 2003). While wrangling over fetal guardianship, it became too late for the woman to have a safe, legal abortion and she was forced to have her child without full consent. Taking the personhood concept to extremes, the Nightlight Christian Adoptions Web site features information on adopting frozen embryos, or "snowflake babies," including a link asking "Want to adopt multi-ethnic embryos?" where one can list background preferences of potential families.

In 2009, the North Dakota House of Representatives approved 51–41 a law granting personhood to fertilized eggs. Montana followed suit and other states considering such measures (Colson 2009). Though a ballot measure in Colorado was defeated in the 2008 election, it is a chilling reminder that the decision reached in *Roe v.Wade* (1973) states that if the personhood of the fetus could be established, then Fourteenth Amendment protections would follow suit. Feldt (2004) and Rowland (2004) also point out that while the Supreme Court in *Planned Parenthood v. Casey* (1992) upheld the right of an abortion prior to the viability of the fetus, it makes it harder for obstacle-producing state laws to be overturned since the plaintiff has to show evidence of the law creating undue burden or that it was intended to block abortions. In *Webster v. Reproductive Health Services* (1989), the Court upheld the constitutionality of the words "the life of each human being begins at conception" on a Missouri statute restricting abortion, claiming that there wasn't sufficient evidence at the time that the statute placed undue burden on women and that, if found to be overly restrictive, could be overturned (Rowland 2004). After the decision, anti-abortion groups introduced more than 600 state bills listing restrictions on reproductive health services (Feldt 2004).

People in favor of reproductive rights need to be clear in their opposition to these policies and not seek "common ground" or compromise with lawmakers and others who support restrictions on contraception and abortion.

Smith 2009), one can only conclude that the fetal personhood movement is as much about the hatred and exploitation of women as it is about protecting the concept of the "pre-born." Kaplan (1995) views the elevation of the fetus as a signpost to how society values women as only being worth the children they carry to term, nothing more. Margaret Atwood's novel *The Handmaid's Tale* (1998) reveals a dystopian society where infertility prompts leaders to outlaw sex outside of procreation, done only by the handmaid, one of a handful of women employed specifically to reproduce under strict penalties.

At the same time as "woman as incubator" is upheld, Kaplan (1995) explains, women are portrayed as the potential enemy of children and in need of regulation. This has the added effect of building hostility toward genuine pro-family solutions like health care, paid maternity leave, and child care since the woman is viewed as the source of the problem, especially if she is single and poor. One key example of this is the Children's Health Insurance Program (CHIP), which provides meager health coverage to the children of low-income parents who are not eligible under Medicaid. The Bush administration defined eligibility as conception to age 19 in its definition of "child," but without providing additional funds to cover the expansion to zygotes (Feldt,, 2004; Rowland,, 2004). Here the fetus is literally provided more health coverage than the woman! Feldt (2004) also points out that in the tragic aftermath of Laci Petersen's murder, the Unborn Victims of Violence Act was passed by the U.S. House in 2001. This law allowed for a separate offense if a fetus is killed or injured in the course of a crime, yet no mention of the woman is made. When given the chance to vote on a more sensible bill that would enhance sentencing for the murder of pregnant women (The Motherhood Protection Act), the House voted against it. Currently 34 states have passed unborn victim legislation and fetal protection laws, some of which apply to early stages of pregnancy or prior to documentation of pregnancy (Page 2006, p. 149).

Bader (2008) and Rowland (2004) explain that anti-choice activists view life as beginning at fertilization, not implantation, a view not in league with most religious groups who assisted women in obtaining illegal abortions prior to Roe (Feldt 2004; Kaplan 1995). Taking the fertilization stance means that in addition to abortion, most forms of contraception would be made illegal were the definition of personhood changed to include zygotes. The consequences would be unimaginable, as Colson (2009) outlines: pregnant rape/incest victims would be forced to have a child, extreme birth defects couldn't be interfered with, in vitro fertilization treatments would be outlawed (unless a woman was ready to accept all the harvested eggs as implants), miscarriages would be investigated as possible homicides or considered as part of wrongful death lawsuits. This has particular relevance because the advances of science have made it possible to sustain even a one-pound fetus outside the womb, posing new problems for pro-choice factions still arguing over the finer points of privacy rights (Rowland 2004). Along with unborn victims of violent acts, abortion to breast cancer and population control in the guise of abortion and contraception aimed at minority groups.

After the Helms Amendment passed, a variety of gag rules, local and global, were enforced. Gag rules specify that if any mention of abortion is included in reproductive health care, federal funds will be cut. This targets clinics that offer a wide range of health services, not just abortions. Since clinics are not universally and unconditionally funded as part of health care, they are vulnerable to both ideological and economic forces. Under the guise of preventing abortion, the gag rule is an attack on all forms of family planning and gains momentum from a misinformed public easily moved by anti-choice propaganda and shame about their own sexuality. The ruling in *Rust v. Sullivan* (1991) found that gag rule limits were not unconstitutional since the government had the right to determine the limits of any activity it funded, including requiring physicians to essentially violate their ethical standards by not allowing them to fully inform their patients about their options and fast-approaching malpractice status (Rowland 2004). Feldt (2004) and Smith (2005) describe how gag rules change back and forth depending on who is in office at the time, leaving the future uncertain when it comes to funding. Arizona, North Dakota, and Minnesota have enacted laws that prevent *private* funding for abortion services, even if these services are separate from other family planning options (p. 50). Feldt relates how, like enforced abortion "counseling," gag rules only increase the likelihood and risk of late-term abortion.

The congressional ban on so-called "partial birth" abortion in 2003 has been the strongest hit Roe has taken since the Hyde Amendment in 1976. In the mid-1990s, the anti-abortion movement was in a public relations lull, mostly due to bad PR and incidents like clinic attacks (Rowland 2004). Despite the majority of abortions taking place in the first trimester, abortion opponents decided to focus their energies on the fewer than .05 percent of abortions using intact D&E (Smith 2005, p. 82). Florida Representative Charles Canady, with the help of the National Right to Life Committee, decided to introduce a ban on intact dilation as a way to "refocus the debate" on the fetus, rejecting initial naming efforts ("brain suction abortion" and "partial delivery abortion") and finally settling on partial-birth abortion (Feldt 2004). These abortions are often last resort choices made under the most tragic of medical circumstances, but this didn't stop the right wing from taking political advantage of women facing such situations. The 2003 ban doesn't just impact late-term abortions—its vague language potentially criminalizes other late-term procedures done after 12 weeks (Rowland 2004).

Behind this array of tactics lie efforts to elevate the zygote and fetus to the status of a person. This is done with religious-like fervor and language (there are even songs written from the perspective of a fetus), imbuing a collection of cells with feelings and thought, and immediately convicting the woman who chooses contraception or abortion of murder. Since it is impossible for both a woman and a fetus to simultaneously have human rights (Rowland 2004;

about the right to privacy so as not to alienate independents. Between 1992 and 2004, no national pro-choice marches took place in Washington, D.C., during an election year. This is nothing new. Betty Freidan, one of NARAL's founders, recalled colleagues in the early 1970s stating that women's rights had nothing to do with abortion (Kaplan 1995). Rowland (2004) describes women and clinic staff being grabbed, shoved, pushed, yelled at, threatened, or spit on as they attempted to seek reproductive health care. It wasn't until doctors became the targets of anti-choice violence that the Clinton administration took action.

With such minimal resistance, the Right's incremental approach has been successful. Stan (2009) relates how anti-choice activism uses progressive language to cast Planned Parenthood as a powerful corporation bent on victimizing teenage girls and minority groups through targeted eugenics efforts. The latest face of parent consent laws is Lila Rose, a woman who poses as a young teenager visiting Planned Parenthood clinics with a hidden camera. Viewers are shocked as they hear clinic staff advising Rose on how to work around parental notification laws. When Rose appears in the media, she often quotes Martin Luther King Jr. and Saul Alinski (p. 8). Phill Kline, former attorney general of Kansas, in a declared effort to find child rapists and illegal late-term abortions, subpoenaed the medial records of roughly 90 Planned Parenthood patients. Kline's efforts buttressed the narrative of the underage teenager impregnated by a man several years older being further victimized by ruthless abortion doctors in search of profit. Seigel (2009) describes a recently passed law in Oklahoma that requires doctors to report information about women who had abortions to a Health Department Web site. The medical information includes: (1) date of abortion; (2) county in which abortion is performed; (3) age of mother; (4) marital status of mother; (5) race of mother; (6) years of education of mother; (7) state or foreign country of residence of mother; and (8) total number of previous pregnancies of the mother (par. 2).

Feldt's (2004) review of laws requiring waiting periods is sobering. Often these waiting periods also require "counseling" where women are read misleading scripts about the hazards of abortion, accompanied by graphic images. More than 20 states combine this humiliation with enforced waiting periods of 24 to 48 hours (p. 157). Some states require that women return for a second appointment, only increasing the time and expense needed for an abortion and driving many women to put off getting an abortion until the second or third trimester, when it is more difficult and costly to obtain (Feldt 2004; Page 2006). Unlike pharmacists or cashiers being able to opt out of dispensing birth control, clinic staff in several states is not allowed to forego the reading of pro-life propaganda during counseling sessions (Page 2006). In the media, anti-choice advertising is allowed, as evidenced by the "life, what a beautiful choice" ads from the 1990s while pro-choice advertising is typically rejected for being too controversial (Feldt 2004). Kaplan (1995) and Feldt (2004) discuss other scare tactics such as phony research linking setting up a precarious situation that impacts all women as the economy worsens: "If poor women can be denied sexual and reproductive freedom, so can all women" (p. 59).

Indeed, the cost of an abortion for pregnancies up to 12 weeks ranges from $450 to $600 in the northwest, with costs escalating after 12 weeks gestation (Wilhelm 2009, par. 2). Women who live in Oregon now face unemployment hovering above 12 percent, which drives up demand for community reproductive health care assistance (Wilhelm, par. 10). In Idaho, for example, women can't obtain an abortion after 14 weeks, forcing them to travel to Oregon and garner additional expenses and time off from work. Idaho will also not allow private insurance to cover abortion other than for life endangerment reasons, unless the patient is willing to pay a higher premium. In contrast to this situation, Kaplan (1995) calls for freestanding, outpatient abortion clinics charging no more than $100 and covered by federal funds as part of national health care; 100 percent availability of abortion into the second and third trimesters; paramedics, not just MDs performing the procedure; and counseling done by women at each stage of the process. Most importantly, consent would come from the woman herself, not her doctor, husband, parent, or boyfriend.

After Hyde, the compulsory pregnancy movement continued to chip away at Roe. Their tactics included "abortion counseling" and parental consent laws; the gag rule; targeting minority populations with abortion-as-genocide messages; inventing a nonexistent medical procedure (partial birth abortion); and, most importantly and recently, their efforts to legally define a fetus—even an egg—as a "person." Reviewing each of these tactics in turn reveals how they are linked to distinct aspects of reactionary ideology aimed not only at women, but at the working class. At the heart of these efforts, however, is the elevation of the zygote over the woman, contributing to the creation of what can only be best described as a fetal cult, complete with dedicated followers and even martyrs (Scott Roeder, Paul Hill, Eric Rudolf, and James Kopp):

> On October 23, 1998, Dr. Slepian was killed by a shot through the kitchen window of his Buffalo home as he stood near his wife and two sons. Dr. Slepian worked at a Buffalo-area women's clinic. Prior to the murder, Dr. Slepian's name was posted on an Internet website called the *Nuremburg Files*. The site is where Neal Horsley admitted collecting the names, addresses, and photos of hundreds of doctors across the country who perform abortions. After the murder, Horsley reportedly tapped the strike key on his computer keyboard. It drew a black line through Dr. Slepian's name. (Rowland 2004, p. 277)

In response, the feminist movement has been woefully inadequate at stemming this violent tide, as Smith (2005) describes the leadership of NARAL Pro-Choice America instructing their staff not to frame abortion in terms of women being able to control their bodies. Instead, the message was to be abortion would be ruled as legal. As Kaplan explains, "any reform, no matter how liberal, was incompatible with repeal [of anti-abortion laws] because it did not address the root of the problem: the right of women to full control of their destinies" (p. 81). Even after the revised law passed in New York, few hospitals offered first trimester abortions, and those that did charged prices beyond the reach of low-income women. The back alley abortion remained the only option for many women, using dangerous techniques such as the insertion of catheters, which could puncture the uterus and lead to infection.

Essentially, *Roe v. Wade* (1973) was a conservative ruling. For example, states can still restrict abortion during the first trimester but abortions must be performed by a physician. In the second trimester, the state's interest in the woman's health entitles it to regulate abortion as the only way necessary to protect maternal health. However, during the third trimester, when the fetus is considered viable, the state may regulate or prohibit abortion except, according to medical judgment, to preserve the life or health of the mother (Kaplan 1995; Rowland 2004; Yalom 2001). What Roe did provide was to enforce the precedents for privacy found in the Griswold and Eisenstadt cases and extended this right to abortion, though in a limited manner. Kaplan (1995) recounts how women's liberation activists handed out blank sheets of paper as a form of protest to represent their ideal abortion policy: no restrictions and abortion on demand.

It would not take long for Roe to lose its legal impact. As Feldt (2004) asserts, "rights without access are meaningless" (p. 125). Poor women were the first target as they lost Medicaid funding for abortions under the Hyde Amendment, passed in 1976 and reapproved annually, though the government still happily pays 90 percent of the cost of a sterilization (Kaplan 1995; Rowland 2004; Smith 2005). Prior to Roe, African American women made up half of those who died from illegal abortions (p. 66) and after Roe they were denied Medicaid funding. The passage of Hyde cast 23 million women into pre-Roe situations (Feldt 2004; Rowland 2004). The false 1980s stereotype of the promiscuous welfare queen prompted the Clinton administration to pass the Personal Responsibility and Work Opportunity Reconciliation Act in 1996, which stated that marriage was the foundation of society, further stigmatizing unmarried women and LGBTQ individuals, not to mention heightening social class discrimination.

The most insulting aspects of the 1996 welfare reform were the so-called family-sized caps and incentives provided to the top five states to decrease their ratios of out-of-wedlock births, including funding for abstinence and marriage promotion to the tune of $300 million annually (Feldt 2004, p. 59). Feldt argues that low-income women receive the brunt of this discrimination as every reproductive decision is scrutinized and judged, despite having the same average number of children as the nuclear norm and receiving less than $400 per month in public assistance (p. 57). At the risk of losing this money, women have to answer embarrassing questions about their abortion history,

Historically, abortion was not a controversial concept in the United States, as Smith (2005) outlines:

> The shifting needs of capitalism explain why abortion only became illegal in the mid-nineteenth century. Until that time, early abortions were widespread and legal. Abortions before quickening were socially acceptable for many centuries as a means of controlling fertility. The introduction of antiabortion legislation beginning in the middle of the nineteenth century...was rooted in the needs of a rapidly industrializing economy. (p. 84)

According to Smith, industrial capitalism, which increased its output 14 times between 1870 and 1929, required more labor than ever before. To meet this demand, immigration restrictions were lowered and homemaking was promoted, along with higher family size, despite women's interest in controlling fertility. A similar situation happened in Rome, when abortion was permitted by law until the end of the second century C.E., indicating that the prohibition of abortion reflected social and economic conditions and not only religious concerns (Yalom 2001). The Catholic Church had no sanctions against early abortion or distinctions between fetal development until 1869, when abortion was forbidden except in the cases of ectopic pregnancy or uterine cancer (Feldt 2004). Even today, almost 70 percent of U.S. Catholics believe a woman who has an abortion for reasons other than to save her life could do so and still be "a good Catholic" (p.109).

By 1900, with the assistance of the AMA, abortion was outlawed in every state of the U.S. (Feldt 2004; Rowland 2004; Yalom 2001). Kaplan (1995) explains that regular practitioners, in an attempt to gain control of their profession, used the illegality of abortion to put midwives out of business, so many women lost their only option for terminating a pregnancy. In addition, the AMA took on a moral tone with women in their overt promotion of purity and motherhood, insisting that sexuality not be separated from procreation (Yalom 2001). With abortion driven underground, women had no choice unless they were relatively wealthy. Kaplan's (1995) account of pre-Roe abortion is as harrowing as it is infuriating; women had to pay extortion-level rates, face injury or death, or had to offer sex in return for the abortion. Poor women had the fewest options as a result of a lack of medical care. Women would go to great lengths not to set foot in a hospital because medical personnel would contact the police if they suspected an attempted abortion. By 1965, 17 percent of deaths related to pregnancy and childbirth were a result of back alley abortions. Some hospitals, including in Chicago, had entire wings devoted to after-care from botched abortions (Feldt 2004, p. 122).

In the remaining years leading up to Roe, ten states had revised their abortion legislation to include "therapeutic" abortions for severe health reasons (Kaplan 1995). However, the language of these state laws was so vague that even in cases of rape, incest, or fetal damage, there were no guarantees that an was abstinence. Ironically, Coburn, while pretending to support the cause of HPV prevention and cervical cancer treatment, effectively delayed legislation that would have helped such uninsured women obtain coverage for their treatments (Page 2006). The condom fallout continued. After being threatened with losing funding, the Centers for Disease Control and Prevention removed a fact sheet from its Web site that outlined more realistic policies for condom use and their reliability statistics for preventing HIV/AIDS (Feldt 2004). The updated fact sheet did not include information on how to use condoms, and emphasized their failure rates and trumped up the infallibility of abstinence as sexual policy for teenagers and unmarried adults.

Arising from the women's movement, abortion made public the issue of reproductive freedom and its ties to equality. Though having access to contraception was important, unless a woman had the right to terminate a pregnancy, any other political gains would be without meaning, which is why women's liberation activists didn't frame the matter in terms of "choice," but "freedom" (Feldt 2004; Kaplan 1995). Women, and single mothers in particular, began to demand that they alone, and not clergymen, the government, significant others, or parents, should have the authority to end a pregnancy, regardless of relationship status, because it was they who bore the physical, social, and emotional responsibility of childrearing; in particular (Smith 2005). Anti-choice rhetoric assumes that women's bodies should be under the control of others (usually males), the implication being that abortion is chosen for "convenience" or "frivolity" and that women can't be trusted with decisions related to their reproductive faculties (p. 76).

Most of the women currently of childbearing age have grown up with the federal right to an abortion, yet in 2005 only 51 percent of Americans considered themselves to be pro-choice (Page 2006, p. 151). Younger people aged 18 to 29 are the group most likely to favor strict limitations on abortion access; 72 percent called abortion "morally wrong" and 32 percent of that group believes that it should be illegal to obtain an abortion under any circumstance (Colson 2009, par. 34). Smith's (2005) assessment that abortions are less accessible than they were 30 years ago is quite pertinent. Only 1,800 doctors are available to provide abortion services to nearly 1.3 million women annually, while 600 of those doctors practice in states in danger of making abortion illegal were Roe to be overturned (p. 147). Yet more than 20 percent of pregnancies end in abortion, with half of U.S. women experiencing an unintended pregnancy (Colson 2009, par. 30–32). An astounding 40 percent of U.S. women who have abortions are evangelical Christians or Catholics, and 61 percent of U.S. abortion seekers already have children (Page 2006, p. 58). As with contraception, it is mystifying why abortion is considered "controversial" considering it is a widely used medical procedure across a diverse population of women, many of whom consider themselves conservative.

and censored government Web sites and programs that mentioned condom use and statistics (Feldt 2004). Instead of funding increases for Title X, the Bush administration funneled federal monies into ineffective abstinence-only programs, including for adults up to age 29 (p. 44).

The conservative opposition to the morning-after pill and mifepristone (RU-486) illustrates an important strategy used by the right: conflating birth control with abortion. Bader (2008) explains that movement leaders like Peggy Hammil of Pro-Life Wisconsin deny that there is any such thing as a privacy right, and that such a notion is anti-family. Jodi Wagner of Pharmacists for Life International refused to fill birth control prescriptions, calling them "chemical abortions." According to her logic, the pill doesn't allow the uterine lining to support a baby (p. 4). Feldt (2004) describes how Senator Bill Frist proposed an amendment in 2002 to the Safe Motherhood, Safe Babies Act that would reclassify birth control pills and IUDs as abortifacients and would define pregnancy as happening at fertilization, not implantation. These policy efforts confuse the public into thinking that the morning-after pill, which is simply a super-dose of the regular pill, is the same as mifepristone, the abortion pill. David Hager, a fundamentalist obstetrician and Bush appointee, nearly single-handedly convinced the FDA to reverse its approval of the morning-after pill, using no peer-reviewed research or testimony (Page 2006). Reacting to the outcome, Judy Brown of American Life League commented, "the best thing the FDA can do now for American women and their progeny is to take the next logical step and remove the pills from the market altogether" (p. 119). As a result, a majority of American women are unaware that there are pregnancy prevention options within 120 hours after unprotected sex.

The lowly condom is also under attack by anti-choicers, stemming from initial research in 1993 by Susan Weller, who sought to review existing effectiveness studies of condoms. The fact that Weller's research combined frequent users with those who seldom used condoms didn't trouble conservatives who had struck ideological pay dirt (Page 2006). Feldt (2004) and Page (2006) outline how Oklahoma Senator Tom Coburn and Indiana Representative Mark Souder proposed warning labels for condom packages that would explicitly state that condoms are not effective at preventing HPV. Building on their profound ignorance of type II errors used in medical research, in which a false hypothesis is supported in order to build a stronger case for the effectiveness of a drug along with taking advantage of the public misconception that most sexually active women will contract HPV and cervical cancer is easily prevented by Pap smears, conservatives argued that the "lack of definitive evidence" either way on preventing HPV meant that the government shouldn't promote condom use.

Bush repaid Coburn for all of his hard work by appointing him to co-chair the Presidential Council on HIV/AIDS and spread the word that the only sure prevention for the spread of sexually transmitted infections than 4 in 1900 (Coontz 2005, p. 171). This was despite lobbying by the American Medical Association, long a foe of the working class, for state laws banning abortion and contraception starting in 1857 (Eskridge 2008; Yalom 2001). The National Woman Suffrage Association responded to the backlash by tying demands for fertility independence to their platform of equal rights, made manifest in Stanton's (1898) biting response to the heart of religious patriarchy in her *Woman's Bible*.

Kaplan (1995) and Coontz (2005) describe how women were denied information about sex, leading to rampant misconceptions (literally) and confusion. Coupled with their second-class status, the inability to control reproduction impacted every aspect of a woman's life, including pressure to marry someone if she got pregnant or the threat of being fired for getting pregnant. *Griswold v. Connecticut* (1965) broke the social condom of Comstock by allowing married couples the federal right to purchase birth control (Bader 2008; Eskridge 2008; Feldt 2004; Kaplan 1995). Feldt (2004) describes how leading up to the case, Estelle Griswold, executive director of Planned Parenthood in Connecticut, organized shuttles to transport women out of state to obtain diaphragms. She also challenged the state ban by opening a women's clinic in New Haven that openly handed out birth control, knowing full well she would be arrested (Bader 2008). A few years later William Baird was convicted of distributing contraceptivesto unmarried Boston University students. In 1972, the ruling in *Eisenstadt v. Baird* allowed unmarried people the right to use birth control (Eskridge 2008; Feldt 2004). The significance of these rulings, other than setting important precedents for privacy in *Roe v. Wade* (1973), was to emphasize that marital context did not matter when it came to reproduction. If this were the case, then one's sexuality shouldn't be a factor either, which was the conclusion of the Court in *Lawrence v. Texas* (2003).

The fallout from *Eisenstadt v. Baird* (1972) was immediate. In the United States and Europe, what Coontz (2005) calls "head and master laws" were repealed, sending the message that husband and wife were—equal partners, at least in a legal sense. The backlash to this seismic shift in sexual politics continues today. Whether it is claiming that the human papillomavirus (HPV) vaccine will encourage girls to have premarital sex (Page 2006) or that using birth control is selfish and interferes with God's plan for procreation (Bader 2008), opposition to contraception impacts health policy. For example, Planned Parenthood had to campaign aggressively to require that employers cover contraception as part of prescription drug plans and to fail to do so constituted gender discrimination (Feldt 2004). Cost was obviously not on opponents' minds, as coverage for birth control pills is $360.00 compared to $20,000 for carrying a child to term (p. 230). Page (2006) outlines how the compulsory pregnancy movement has assisted in passing legislation in states like South Dakota and Mississippi that protect not only the pharmacists who refuse to dispense birth control, but store cashiers as well. The Bush administration appointed anti-condom politicians who spread false research about the ineffectiveness of barrier methods is succinctly reflected in a quote by Joseph Scheidler, founder of the Pro-Life Action League: "I think contraception is disgusting–people using each other for pleasure [and] for those who say I can't impose my morality on others, I say just watch me" (Jill 2007, par. 6). With abstinence as their only proposed alternative to pregnancy, Feldt (2004) sees the success of the compulsory pregnancy crowd in making people feel ashamed of sex and ambivalent about the boundaries of reproductive rights on a wide range of issues, even if it means further restricting their own options.

In the 1800s and even into the mid-twentieth century, doctors and psychiatrists assumed that women had no sex drive. It was seen as solely a male attribute that explained their aggressiveness and ability to handle politics and business (Coontz 2005; Yalom 2001). Coontz (2005) and Yalom (2001) identify the segmentation of women by the double standard into "good" and "bad" as a mechanism used by elites to attribute unwanted pregnancy to lower-income women, further dividing working and middle class women from each other. The higher one's social standing as a female, the more pure and virtuous one was supposed to be, and this meant not enjoying sex or admitting one was sexual (necessary when using birth control); virtue was especially important for middle-class women. Fallen women were seen as violating the norm of purity and were often portrayed as abnormal, mostly as a way to deny one's own sexual nature and keep it safely away. The segregation of women via the double standard was a way to ensure that contraceptives would not be available because they had the potential to redistribute power to women; first in the home, and eventually out in the world (Page 2006). Neill (1960), the founder of Summerhill, critiqued the purity standard on the grounds that it led to enforced hypocrisy, promiscuity, and economic hardship:

> Promiscuity is neurotic...if the term "free love" has a sinister meaning, it is because it describes sex as neurotic. Promiscuous sex—the direct result of oppression—is always unhappy and shameful...real freedom and love does not lead to promiscuity...it is lower middle class and the poor who are left, literally, holding the baby. There is no other alternative for them. If a middle-class girl tries hard enough, she can find her a doctor who will perform an abortion for a goodly sum. Her poorer sister either runs the danger of abortion—an unskilled, perhaps unscrupulous abortionist—or she has to have her child. (pp. 236–238)

The history of contraception reveals an arena of struggle and backlash. As women began to assert reproductive rights in their own personal lives, they were often met with hostility from spouses, churches, and government. In the United States, up until the Comstock Law of 1873 that outlawed birth control and abortion, a wide range of contraceptive devices were freely advertised and used, with differing degrees of effectiveness, leading to an overall drop in birth rates for married couples, from more than 7 children in 1800 to fewer recent Commonwealth Fund report, not only are health care costs higher for women due to their biological status as child-bearers women are less likely to receive employer-provided health care due to holding part-time, service-sector jobs. Instead, many women forgo insurance and rely on meager public health insurance or use their spouse's coverage. Using spousal coverage is also risky because according to the report, since the year 2000, health insurance deductables have tripled within the span of eight years, and quadrupled for companies that employ fewer than 200 people (p. 3). According to the March of Dimes' Thompson Report, *The Healthcare Costs of Having a Baby* (2007), the average vaginal birth cost $7,737 and cesarean sections cost more than $10,000 in 2004 (p. 7). If one has private insurance, most of these costs are paid. However, for lower-income women and families who are uninsured, these costs are prohibitive. Nearly 5 million women rely on the services provided by Title X–funded clinics, which prevent an estimated 1,331,100 pregnancies and 632,300 abortions for adult women and 385,800 pregnancies and 183,300 abortions for teenagers, as of 2004 (Feldt 2004, p. 43).

Page (2006) estimates that 42 million American women are sexually active and do not wish to become pregnant and only 5 percent of women between the ages of 15 and 44 do not use any contraceptive method during intercourse. From this small percentage originates nearly half of U.S. abortions (p. 6). The opposition to managing reproduction is puzzling considering that 90 percent of Americans are in support of the concept with nearly two-thirds believing the government should fund it for women in financial need (Feldt 2004, p. 33). Even a majority of women who call themselves pro-life want to control their fertility, and 96 percent of Roman Catholic women have used non–church sanctioned contraception at least once (Bader 2008, pp. 4–5). In 1969, in response to opposition to the proposed Title X, then Congressman George H.W. Bush stated: "We need to take the sensationalism out of this topic so it can no longer be used by militants who have no real knowledge of the voluntary nature of the program...if family planning is anything, it is a public health matter (Ross 2009, par. 8). This would be the last time conservatives could publically and emphatically speak out in support of reproductive rights and still remain politically viable.

The opposition to contraceptives rests in the sexist double standard and sexual repression, especially aimed at women who are portrayed as the ones to blame for unwanted pregnancies (Coontz 2005; Yalom 2001). Page (2006) argues that most major anti-abortion groups also oppose contraceptives, meaning that we can no longer assume that abortion and contraception are separate issues: "To allow birth control would mean tolerating a lifestyle that allows people to enjoy sex outside of marriage and parenthood," which is intolerable for many conservatives (p. 5). Presenting the argument that contraceptives would prevent abortions is useless with many anti-choicers because they are opposed to sex apart from procreation to begin with, let alone the notion of distributing birth control to unmarried individuals. This

Reproductive Rights

Socialist feminists and many mainstream feminists understand that a woman's ability to control her fertility is central to women's equality as a whole (Feldt 2004; Kaplan 1995; Kelly 2006; Page 2006; Smith 2005). Reproduction is tied directly to marriage decisions and economics, particularly the availability of paid labor, as Coontz's (2005) analysis of birthrates prior to widespread contraception demonstrates. During the Middle Ages, when demand for high birthrates predominated, different approaches to the control of reproduction were also viewed as being the responsibility of a particular gender as in abortion being a female sin and withdrawal being a male sin (Yalom 2001). Kaplan's (1995) history of Jane, the underground abortion collective that operated prior to *Roe v. Wade* (1973), begins with the assertion that reproductive freedom is an essential component of women's liberation:

> They could not control their lives without having control over their bodies. If they did not have the right to regulate this basic biological function, their reproductive ability, what chance do women have to control any aspect of their lives? It was women who got pregnant, bore children and raised them, but men—husbands, fathers, clergymen, legislators, and doctors—had the power to judge women and punish them. (p. 20)

Recent right wing victories, such as the so-called "partial-birth" abortion ban, are part of a highly organized and well-funded backlash against reproductive rights, which Feldt (2004) argues boils down to opposing women's recent ability to separate sex from pregnancy. This backlash hits not only working-class women, but working-class *families*, as Sargent's (2008) summary of policy proposals emanating from social conservatives indicates: eliminate federal funding for Titles IX and X, offer tax breaks for married women to have children, fully fund marriage and motherhood classes as career options or for welfare recipients, repeal laws protecting victims of domestic violence or sexual harassment, de-fund legal assistance for women seeking counseling for divorce or abortions, end child support, revoke equal pay laws, completely ban abortion even in the case of the woman's impending death, censor contraceptive information until marriage, or ban it altogether (p. 8). Yet access to reproductive options also benefits capitalism, indicating a significant rift within conservative thought (Smith 2005). The family-values wing seeks complete control over women's reproductive decision making while the neoliberal wing sees value in the timing of managed family size (at least for some people).

The health care costs associated with reproduction are one of many ways the capitalist class privatizes the overall cost of raising the future workforce. They prefer that workers bear these expenses in the form of high-priced private health insurance and unpaid domestic labor mostly performed by mothers (Lapon 2009). Women are especially vulnerable when it comes to health care access. According to Rustgi, Doty, and Collins (2009) in a

At the same time that socialist feminists advocate for LGBTQ marriage rights in order to provide material relief to many, they recognize that winning these rights alone will not end the exploitation inherent in the nuclear family, which is an unsustainable institution. Kollontai (Holt 1977) understood this when she argued that simply engaging in "free love" and open relationships did nothing to remedy the effects of capitalism, especially if one had little access to material necessities.

As Willett (2002) explains, "capitalism increasingly reduces our social life to the family, and severs the family from the socioeconomic sphere" (p. 124). This shrinking of social circles and the reduction in the number of adults children can depend upon goes hand in hand with the increasing costs that the private family unit has to bear: health care, education, and child care (Smith 2005). In many ways, it is the basic fight for essentials such as health care and reproductive freedom that is more radical than what many postmodern feminists propose in the way of individualism and analysis on a psychological level. The ability to transform the nature of work apart from the alienation of capitalism to a more family-friendly concept would be a good starting point, as Willett (2002) proposes building on the 36-hour work week used in Europe. Here, leisure is seen as a human right, tied to family and sustaining communities, yet not excluding those who are single or childless. In contrast, the U.S. workweek automatically discriminates against those who have to provide child or elder care; that is, the majority of American workers. Another potential solution would be along the lines of the Garden Cities, where feminist concepts of cooperative housework were taken into account by architects and designers who featured shared courtyards and kitchens in their designs (Hayden 2009). The Garden Cities approach was also used to build apartments for singles and the elderly in the early 1900s.

Cloud (2001) and Coontz (2005) argue that the majority of workers have never fully identified with the nuclear family ideal, as evidenced by the sheer variety of family forms prior to, and in spite of, capitalism. The transformation from band societies to individual nuclear families represented a large-scale "speed-up" imposed on the working class as collectivity was replaced by privatization of the domestic sphere. Women and children were directly impacted by this shift as Leacock (1981) describes with the introduction of corporal punishment in Montagnais-Naskapi households after the arrival of the fur trade and Christianity. The European family model of patriarchy and female submission, along with ending the right to divorce, immediately placed women at a material disadvantage. Even if a woman is able to amass property and wealth within capitalist society, she can usually do so "only by maneuvering within the family's reproductive system...rather than by resisting male dominance within marriage...she ends up strengthening the patriarchal family" (Coontz 2005, p. 131). Smith (2005) further links homophobia to the enshrinement of the nuclear family, which is by design a heterosexual institution in support of capitalism.

the United States during settling of the western frontier when homesteaders couldn't comprehend Native Americans' "usufruct" (common use) and Native Americans were puzzled by individual land ownership (p. 130).

Egalitarian societies organized around entirely different structures, with important implications for women's rights. Because the sharing of resources was essential for survival, band societies viewed values such as generosity, keeping one's word, and putting in a fair share of labor above personal property holdings (Coontz 2005; Engels 1972; Holt 1977). Hoarding supplies in individual households was unheard of so resources were readily distributed and consumed, based on the needs of the group as a whole. Those with the most knowledge of a task or topic were considered leaders as new situations arose, so women were viewed as authority figures, not just in charge of domestic chores (Leacock 1981). After the fur trade imposed by white settlers entered the picture, the Montagnais-Naskapi began performing more isolated forms of labor that took them away from group endeavors, such as hunting, placing them in a more dependent role (men being dependent on trading post goods and women on the men):

> When men became trappers, the sexual definition of functions and spheres of interest became sharper, for the wife and children began to be set apart as a family who were provided for, as compared to the men who were the providers. At the same time, there was the breaking up of the family bands into smaller units. (p. 37)

Leacock finds it ironic that preclass societies are often characterized as consisting of drudgery, slavery, hard labor, rigid gender divisions, and a focus on status, all of which are hallmarks of capitalism today. Rather than a product of evolutionary biology, it was only 10,000 years ago that the notion of formalized subordinate roles for women began to appear in the archeological record, yet these roles are portrayed as ever present in human history, communicating the message that nothing much can change (p. 288).

Socialist feminists locate the oppression of women within the basic structure of the nuclear family, which is the primary social instrument for privatizing the domestic labor that capitalism needs (Cloud 2001; Engels 1972; Holt 1977; Leacock 1981; Smith 2005; Willett 2002). However, the nuclear family is one result, not the ultimate cause, of the oppression of women (Smith 2005). Its abolishment alone would not be tantamount to achieving a socialist society (Holt 1977; Leacock 1981), but its destruction would certainly be part of the process. Cloud (2001) elaborates:

> In so far as defending gay and lesbian families can be a form of resistance to gay bashing and homophobia, it should be defended. However...we must recognize that the assimilationist approach upholds the idea of "the family" even as it attempts to redefine who counts as a family. (p. 73)

power-feminist logic, one can be a stay-at-home mom or a famous politician without having to focus so heavily on being a woman or drawing negative attention to the exploitation of women worldwide.

Socialist Feminism

The socialist feminist approach is different from the various forms of mainstream or traditional feminism, particularly in its centrality of class analysis. Recognizing that capitalism has benefited from the divisions between different identity groups, Kelly (2002) sees the fusion of feminism and socialism as life-changing, since it deals with many of the day-to-day issues impacting not only women, but working people in general. It also doesn't require activists to be college educated in order to participate. Socialist feminists examine both patriarchy, commonly understood as male power, and capitalism, which exploits all workers, male and female (Smith 2005). Finding culture and religion to be inadequate frames for understanding women's oppression, Wang (2001) views the exploitation of labor in rural India and industrialized nations as resulting from private property, or "freedom from property for the working class and the poor" (p. 222). By situating analysis in dialectical materialism, religion and culture then become offshoots of the superstructure rather than the sole causes of oppression. During the Russian Revolution, when many elite feminist groups were attempting to reign in female workers' demands for change, Kollontai (Holt 1977) resisted, asserting that "a sound class instinct and a deep distrust of 'ladies' saved the working woman from being diverted into feminism and from any long and permanent connection with the bourgeois feminists" (p. 44). Indeed, Russian women actively utilized historical materialist analysis by picking up forks, rakes, and brooms in 1905 to drive soldiers out of their villages.

Marxist analysis is not about focusing on genetic or biological differences, or whether there is some mysterious scientific rationale for discrimination against women. Rather, it examines how ideologies about gender are situated in class-based societies and used to maintain divisions (Ebert 2001; Kelly 2006). If we can study, for example, egalitarian societies, we might be able to see that our existing situations are not inevitable, as Leacock's (1981) work demonstrates. In her thesis, women's elevated status in more egalitarian social arrangements is part and parcel of the social and economic structure of those societies, not a separate matter or an aspect of societal temperament. Likewise, socialism is linked with the liberation of women from unpaid household labor as one cannot happen without the other (Holt, 1977; Kelly, 2002). It was the society itself that supported more equal relations by not being based on private property ownership in Leacock's (1981) assessment. Without a market system, traditional concepts of "group leader" or "dominant male" are not applicable for analysis because items necessary for survival were produced and distributed by all able-bodied adults in society. This created friction in in more controls over the movement of women. Feminism has assisted in the exploitation of women, all egalitarian rhetoric about helping Afghan girls to attend school aside:

> Feminism has been used by colonial powers, and, of course, the real aim of the colonizers has not been the improvement of women's rights, but rather the control of the natives by labeling them barbaric and primitive...for instance, the Euro-American media has promoted a growing number of Iranian and Muslim memoirs, such as the neo-orientalist book *Reading Lolita in Tehran*. (p. 39)

While Hillary Clinton represents the no-nonsense side of free-market and national security feminism, no politician best embodies the current direction of power feminism than 2008 vice-presidential candidate Sarah Palin. According to Scher (2008), Palin managed to take rather tired conservative talking points from moribund think tanks and transformed them via the equity feminist message that is appealing on a populist level. It is quite telling that many prominent women balk at using the "f-word" (feminism) while Palin regularly calls herself one. Her political career "serves to reveal how much mainstream feminism has become a benign politics of difference, rather than a revolution to overthrow a system of patriarchy" (Sargent 2008, p. 8). In the eyes of the Christian media, Palin is able to overcome the contradiction of being in a higher position of authority than her husband as she is compared to the Biblical figures Esther and Deborah, who were chosen by God to lead in a uniquely female way (Scher 2008):

> To these activists, Palin is normal....She wears makeup. She is pretty. She is an evangelical Christian. She is anti-abortion. She is also white. That is normal within the sphere of these conservatives. But "traditional" for these young people is no longer a woman who stays home with the children while the husband works, or who submits to her husband. Todd Palin's active domestic role is not so unusual. (p. 22)

Because younger people have adapted their notion of "traditional" to more closely fit reality, Palin remains safe from the kinds of critiques that activist feminists have successfully used against women who assimilated rather than resisting exploitation. Evangelical Christians experience the same degree of economic hardship as the rest of society, they often "talk right and lean left," meaning that they eventually recognize that certain conditions of the nuclear family are no longer sustainable yet still have to deal with the dissonance this realization creates with conservative Christian beliefs. Figures such as Palin represent the false illusion that women can achieve anything if they try hard enough, a bedrock belief within the power feminist front. One can still hold on to the individualism inherent in conservative Christian ideology. Per who questioned her ability to make sound judgments on the bench (Khan and Tapper 2009). Echoing nineteenth-century thinking, radio talk show host and felon G. Gordon Liddy further speculated that in addition to Sotomayor speaking "illegal alien," her menstrual cycle would contribute to bad decisions on the bench (Somoza 2009). The larger significance of Sotomayor, though many of her decisions have benefited corporations over people, is that her presence is an unpleasant reminder to elites that there is little "justice" in the U.S. criminal justice system for minorities and the poor. The fact that she has spoken openly about her background only intensifies the fears of mostly white elites.

In the case of domestic violence, the message aimed at women is to "gather their strength" and end abusive relationships, similar to the power feminist recommendation for confronting sexual harassment. In a 2009 episode of *The Oprah Winfrey Show*, former model Tyra Banks guest starred to provide inspiration for women to summon their self-esteem and leave a bad situation. As Schulte (2009c) relates, "nowhere in this discussion was there any recognition of how difficult it is—financially and emotionally—for women to get out of battering relationships, much less a real answer to why battering takes place" (par. 10). The fact that women have less and less access to emergency shelters due to state budget cuts, or that many women lack the credit and work histories needed to "start over," seemed to be lost on Winfrey and Banks as they relentlessly promoted the equity feminist line to audience applause. Schulte questions the use of the term "cycle of violence" as it psychologizes what is essentially a public, social issue; that is, the second-class status of women, which benefits capitalism and provides the very conditions for domestic abuse. Instead of counseling, Schulte recommends "alternative therapies" such as free health care and child care; living wages; and free college tuition so that women can leave abusive relationships; as well as policies that would promote rather than work against families. At this point, there seems to be little indication that mainstream feminism will back these solutions.

Third world countries are often the targets of equity feminist projects. The women who live in these countries are portrayed as being "held back" by backward ways, even though, as Leacock (1981) reminds us, prior to colonialism, women and men shared authority. It was only with the advent of class society that practices such as the submission of women became common. Today, Islamic women are held up as the latest noble "cause" for American feminists to rally around. Khanlarzadeh (2009) pinpoints Western imperialism in the form of economic sanctions and not "internal factors" (i.e., Islam, patriarchy, the Mullahs, etc.) the American media constantly highlights as barriers to women's equality. For example, economic sanctions directly impede Iranian and Muslim women from taking part in public life since they lead to job scarcity and a lack of basic necessities. The constant threats launched against Iran by the United States have only served to militarize the country, resulting

Women who are not content to receive an apology from their harassers and who come out publicly against sexual harassment are viewed with particular hostility by power feminists who accuse them of manipulating the system for revenge or power. The woman who speaks out against sexual harassment and rape is transgressing what Cotter (2001) calls "the valorization of personal experience" in which violations of women's rights "can only be understood on a case-by-case basis as an individual issue" (p. 207). The very group that tends to deny social context suddenly wants to see it invoked when it comes to women making their exploitation a very public issue, connected to labor (Maybee 2002). Pointing out how even Anita Hill joined the campaign to smear Paula Jones, Smith (2005) views free-market feminists as reinvoking the plausibility test for women who talk publicly about their harassment. For example, Jones's appearance was brought up to suggest that there could be no way Clinton would be attracted to her, thus rendering her accusations false. Stan (2009) describes conservatives' divide-and-conquer strategy in the 1980s over the issue of pornography when Attorney General Edwin Meese proudly placed Andrea Dworkin on a panel with right wingers, making it appear that they were allies.

What lies behind free-market feminism is an overall hostility toward framing issues in terms of race, class, or gender. Chamallas's (2002) analysis is useful here. She posits that conservatives reject the use of identity, even when it is needed to redress social ills, such as the policy of affirmative action. According to equity feminists, affirmative action is just as discriminatory as barring opportunities from women just because they are female. This makes it difficult to challenge existing power relations, mostly because capitalism seeks to appear neutral at all costs even as it uses identity as a divisive tool. The irony, of course, is that race *was* a major factor in determining citizenship in the United States and used constantly against all ethnic groups. Only after *Brown v. Board of Education* (1954) was the use of race considered a negative thing, contributing to discrimination against whites. According to this logic, the Plessy decision wasn't "judicial activism," but Brown was. Chamallas (2002) continues: "What feminists regard as discrimination, conservative critics ascribe to women's choice" (p. 81). The LGBTQ community, then, doesn't require any "extra rights" because they are making a "lifestyle choice" and should face the consequences. This puts the obligation onto the individual person to assimilate, not on the unjust society to change.

Maybee (2002) identifies one exception to the use of identity being frowned upon: when an African American criticizes another African American, she or he is lauded by conservative groups as being ground breaking, neutral, and objective. Inversely, when Supreme Court nominee Sonia Sotomayor positively referenced her own background as a Latina and a woman, one would have thought society would collapse from the shock. She was immediately referred to as "racist" by Twittering media figures such as Newt Gingrich

Kelly 2002; Moeller 2002). This results in this tiny minority of high-profile women being seen as all that is needed to "make up" for the exploitation of the working class, "without a transformation of the conditions which give rise to these damages" (Cotter 2001, p. 186).

Power feminism is nothing new. Chamallas (2002) views it as an offshoot of evolutionary psychology and evolutionary biology, fields that locate barriers within the individual, not in society. Those who advocate the evolutionary approach lean toward social Darwinism, the only difference being their invoking of "it's in the genes" rather than "it's God's will" to explain exploitation and poverty (p. 74). Leacock (1981) critiques the enterprise of traditional social science that strips poor people from the context of their material situations, promoting a deficit perspective in which they are condemned to suffer due to a "culture of poverty" and the like. Coontz (2005) and Scher (2008) tie conservative feminist thought to nineteenth-century Christian evangelism and the early suffragette movement. These early feminists embraced patriarchal ideologies and promoted fitting into, rather than resisting, industrial capitalism and its attendant nuclear family. In all of these perspectives, conservative feminism is not only ahistoric, it is anti-history in its refusal to consider labor at the center of people's lives. The writings of Alexandra Kollontai in the early twentieth century (Holt 1977) lay out in clear language the differing class aims of people who advance women's liberation, identifying the problem of conservative feminism early on:

> For the feminists, the battle to obtain equal rights with men within the limits of the capitalist world is a sufficient aim in itself...the feminists consider that men, who have unjustly all the rights and privileges for themselves and left women in prisoners' chains and with a thousand obligations, are the main enemy....The women of the proletariat see the situation very differently...as long as a woman has to sell her labor power and suffer capitalist slavery, she will not be a free and independent person. (pp. 50–51)

Discussing Roiphe (1994), Chamallas (2002) points how out many postfeminists seek to "empower" women by encouraging them to, for example, verbally insult sexual harassers in the workplace or to physically fight back if attacked. These solutions are suggested rather than enforcement of the law, because, according to the postfeminist camp, sexual harassment legislation casts women as victims before harassment even occurs. This view is suspicious of solidarity in that women should fight back one on one instead of banding together to combat sexism through the use of the law. Women who, for various reasons, are not able to fight back are viewed as the weak links and in some cases are seen as the reason for the continuance of sexual harassment and rape! By conflating endorsement of legal channels with women's weakness, authors such as Roiphe (1994) and Sommers (2005) downplay the use of the courts as a solution to exploitation, a common right wing trope.

scarcity. By failing to locate the source of women's oppression in capitalism, feminism enables its functioning, and in some cases celebrates it.

Patriarchy theory, Leacock (1981) concludes, hinders women from organizing against exploitation in two important ways. First, contemporary feminism with its embracing of relativism and rejection of any kind of totalizing discourse, makes it difficult to arrive at common goals to effectively meet the needs of women in poverty who have the most desperate situations requiring immediate relief. Second, patriarchy theory divides women from men and other identity categories (LGBTQ, racial) at the very moment when such alliances are critical. The focus on women's roles takes our eye off of the nuclear family and its relationship to capitalism, which exploits men as well as women. Recognizing this fact involves the inevitable task of making important choices:

> Yes, one can take one's pick among conflicting generalizations. This has been true since the times of John Locke and Thomas Hobbes. Locke, defender of democratic forms, stressed human cooperative interests, and cited as an example the generosity of Native Americans who were still free, living without rulers apart from centers of colonial conflict; while Hobbes, defender of a strong monarchy, argued that the competitiveness of his times was innate. What we understand about ourselves is crucial. Today the humanistic goal of a peaceful and cooperative world has become an urgent need if we are to survive as a species. Generalizations about women are, in effect, generalizations about men and about human society in general. It is important to pick right (pp. 203–204).

Conservative Feminist Thought

Beginning in the 1990s, more prominent feminist authors began changing course from public demands for child care and reproductive freedom to a more privatized, internal direction by questioning the degree to which women experienced oppression (Leacock 1981). According to Smith (2005), Naomi Wolf created the term "power feminism" to counteract what she saw as a rash of "victim talk" coming from women. Power feminists are also called equity or free-market feminists for their focus on rock-bottom essential rights to opportunity, not equality (Scher 2008). The power feminist philosophy advocates women embracing individualism and capitalism, though some, such as Wolf, admit that this can cause exploitation for most and rule by the few. But in presenting no alternative way to equality other than women joining the patriarchy, power feminists take the short view; that is, if they can achieve higher ranks in society (e.g., CEOs, top military brass, etc.), then humanity will be all the better for it. These few prominent women who often espouse free market rhetoric in leadership positions are regularly paraded in front of the media as if they represent the majority of working-class women, when in actuality they represent at best a politics of containment (Cotter 2001;

until the arrival of class-based society. In some earlier cultural mythologies, such as the Arapaho and Aztec, males and females exchanged roles, with some males giving birth. The moon is spoken of as both male and female by indigenous groups and some instances of androgyny and even bisexuality emerged. What has happened, Leacock argues, is that historians and anthropologists have emphasized the male deities and downplayed the female ones, or imposed separation of the sexes on the beliefs of ancient cultures.

In critiquing traditional anthropology, Leacock (1981) emphasizes that for many social scientists and anthropologists, survival of the fittest implies warlike domination. Except for a few instances in Australia, outward expressions of male aggression, such as war, a concern with tracing a child's paternity, and rigid gender roles, are not found in hunter-gatherer societies, leading us to question the vision that social Darwinists uphold. Instead, Leacock suggests that we need to reexamine the notion of "fittest" to imply adaptation to many and new circumstances that often don't involve fighting or war (such as figuring out the best way to distribute goods to the most people). The irony is that following the conservative reading of "survival of the fittest" to its conclusion means mutual destruction, not survival! Likewise, advocating matriarchy over patriarchy does nothing to unseat exploitation. As Leacock reminds us, the alternative to patriarchy isn't matriarchy, but egalitarianism.

Leacock (1981) asserts that our analysis needs to go back much further than the data that was collected mostly in the twentieth century. She focuses much of her own analysis on the arrival of the fur trade and its impact on the Montagnais-Naskapi tribe going back to their initial contact with Jesuits, whose diaries reveal a vexation with their egalitarian and communal ways. The intersection of religion with trade put pressure on the tribe to move from an egalitarian way of operating to a more stratified social functioning. By examining how all of these material factors contributed to the roles within tribal families, Leacock locates the exploitation of women within the rise of capitalism and the nuclear family, replicating Engels' (1972) analysis in *The Origin of the Family, Private Property, and the State*. This analysis is examined more closely in chapter 3.

Leacock (1981) views the feminist focus on public and private spaces as being inadequate for understanding societies that were not organized by class hierarchies. The public/private dichotomy helps to disguise class and make differences between groups of women more personal and cultural, as in Leacock's critique of Margaret Mead's work distinguishing between professional and working-class women's experiences. The fight for better working conditions is not a personal matter, but one requiring professional and working-class women to band together to recognize the common source of their exploitation. Likewise, Leacock viewed the pitting of racism against class exploitation as "a sophomoric sociological enterprise...not Marxist analysis" (p. 15). For Leacock and other socialist feminists, racism and sexism are part of the same system of capitalism that seeks to divide workers by introducing

Wang (2001) reminds us that private property, not male-headed households, was what made possible the formation of the nuclear family and the resulting exploitation of female labor.

Eleanor Leacock's (1981) *Myths of Male Dominance: Collected Articles on Women Cross-Culturally* provides a pivotal and important challenge to both patriarchy theory in feminism and protective/provider ideas in traditional anthropology. Unfortunately, her works are rejected in both camps because she employs a dialectical, materialist analysis of the origins of women's oppression along with seeing the choices before us in today's world as either survival or destruction of the human species. Leacock asserts that the practice of viewing women's oppression or social stratification as an inevitable and unchanging part of "human nature" provides an important ideological tool for the ruling elite. To Leacock, as with other Marxian thinkers, the oppression of women goes hand in hand with the oppression of all workers and is part of the same system brought on by the particular arrangement of society around private ownership. In other words, to examine how women were viewed in society, one only has to examine the rise of social class. This unseats tenuous theories like the "exchanging of women as property" and other explanations for the secondary status of women.

While both conservatives and liberal feminists view patriarchy as an essential component to understanding the role of women, our concepts of male rule are based on artifacts of Western culture rather than some sociobiological tendency of women to be dominated by men. Leacock says as much when she reminds us that it was the effects of colonization or the imposed social rankings after trade entered the picture (the spread of both assisted by Christianity) that resulted in the exploitation of women, not the brute force of men. For example, Western feminists might see Native American women being coerced into child care as proof of a secondary status when in egalitarian societies women took on far more roles, or their roles were elevated. Others claim that there is a universal dimension to male dominance when anthropologists are actually applying Western concepts of government and authority to tribal governance (and finding the tribal ways lacking), making differences between men and women seen "not as ethnocentrism, but as common sense" (p. 135). Yet egalitarian societies featured complementary rather than gender-dominant roles. Sometimes males performed "female" tasks such as child care and vice versa, which is a much different vision than that portrayed by traditional anthropology and contemporary feminism. In egalitarian societies, categories of "male" and "female" have different meanings.

Leacock (1981) also takes on the notion that women, by virtue of their biological functions of childbirth and breastfeeding, were always viewed as inferior to males. This type of essentialist feminism is often featured in postmodern writings, locating the oppression of women in patriarchy and sexism stemming from women being viewed as closer to nature and thus despised. Anthropological evidence suggests that women were not viewed as inferior

What is notable is a profound misunderstanding of Marxism in general and Marx in particular as his ideas relate to feminism. As discussed in chapter 4, postmodernists take for granted that we live in a "post-industrial" world filled with nothing but signs and signifiers. Taking social action is presented as an outdated form of arrogance, because to take action means clearly defining who has or doesn't have your interests at heart; in other words, taking a definite stance and articulating truths about how the world works. Postmodern ideology is a barrier to organizing as a social class because the one form of unity critical to social change—the revolution of the working class—is sacrificed to textualizing and endless readings of identity. Movements for social change exist within a capitalist system. The fact that the same examples of patriarchy and racism one sees in the larger society (which are affected by capitalism's use of divisive tactics) are also reflected in social justice movements is conveniently overlooked. The symptom, in this case patriarchy within liberation movements, is confused with the cause of oppression, ultimately capitalism, which results in Marx's dismissal.

Patriarchy as the Source of Women's Oppression

Contemporary feminist theories share variations on the common theme of viewing the source of women's oppression as arising from male domination, or patriarchy. Men's power and control over women is also linked to the common source of oppression for LGBTQ individuals as a form of heterosexism, or majority straight society. In the attempt to locate oppression's source, social class is seen as a minor issue in this postmodern clash of discourses, being only one textual identity among many (Cloud 2001). Having rejected Marxian analysis of the family and labor as arising from capitalism and being the sites where women experience exploitation firsthand, feminists have created "an ahistorical construct suggesting that sex, not class is the basic division in society" (Cotter 2001, p. 205). Men, according to patriarchy theory, control women's bodies and their labor and have done so throughout history and across all cultures.

In a sense, the development of patriarchy theory was a form of academic resistance to the longstanding scientific-sounding inscription of sexist ideology in anthropology. Coontz (2005) relates how feminists began constructing an argument against protective marriage, the perennial idea in anthropological studies that marriage arose to keep women safe from hostile men who were bent on kidnapping or raping them (of course these same anthropologists could not account for spousal abuse). Feminists also challenged the male provider theory in which men were the natural hunters and their authority came from their position of physical strength. While there are still defenders of these ideas today, mostly in evolutionary biology, "most paleontologists reject the notion that early human societies were organized around dominant male hunters providing for their nuclear families" (p. 37).

health issues never get to see the light of day. Yet the media have provided air time for spokesmen for the compulsory pregnancy movement via sensationalistic interviews. For example, news coverage of murderer Paul Hill contributed to his image as a martyr, thereby legitimizing his beliefs. And little more than a decade later another doctor was murdered—George Tiller of Kansas—by an assassin most likely influenced by media figures like Bill O'Reilly who regularly referred to the physician as "Tiller the baby killer" (Filipovic 2009).

Limitations of Feminism

There are two overarching aspects of mainstream feminist theory that support the ongoing function of capitalism: the assertion of patriarchy as the inevitable source of women's oppression and conservative or power feminism. Both strands of feminist thought buttress, rather than oppose, right wing ideology in a variety of ways. Certain forms of nature-based or essentialist theories lead nicely into biological determinism, while the independence of women is something that just about any conservative gal can endorse even as she denies the material conditions of the working class as a whole. Both result in feminism being of service to capitalism rather than in opposition to it. Both also ignore social context and class difference in favor of trying to unite women as an oppressed identity, implying that, underneath it all, Queen Elizabeth shares commonalities with Liz who works for $7.00 an hour at the local Starbucks, superseding the need for Marxist analysis in a "post-industrial" society. For example, Smith's (2005) important reminder that mainstream feminism speaks for a class of women who already have access to child care in the form of nannies (who themselves are usually underpaid women of color) highlights the many ways that women do not share experiences that make them a "class."

Marxism has been dismissed by feminists as not being relevant to the particular needs of women as an oppressed group. In its place has been put what Ebert (2005) calls "micropolitics," which focuses on the extreme local where feminism is tailored to ultra-specific needs of different cultures of women. In this view, class analysis centers on "lifestyle and consumption" (p. 49). Employing the notion that Marx was a product of nineteenth-century thinking and is therefore no longer relevant to postmodern times, Nicholson (1992) takes aim:

> It was Marxism's very failure to appreciate the rootedness of many of its own explanatory categories—such as the categories of production, labor, economy, and class—within the hegemonic value system and belief structure of its times that also made it politically oppressive for feminists. Moreover, similar arguments could be developed to describe the inadequacies of Marxism in relation to other social movements, such as movements against racism or movements of gays and lesbians. (p. 56)

child would make an already dire situation more desperate (p. 148). Yet these are the very women being denied health care access.

The assault on reproductive freedom has had a measurable impact. A 2009 Gallup poll released prior to Obama's controversial Notre Dame appearance revealed that for the first time since Roe a majority of adults in the United States identified themselves as pro-life (51 percent, which is seven points higher than in 2008). While 53 percent believed that abortion should be available for special situations (to save the mother's life, etc.), the fact that support for the choice position once stood at 73 percent in 1973 should make feminists seriously reevaluate their strategies (Smith 2009, par. 13–14). Feldt (2004) addresses part of the problem to the fact that women guests are a distinct minority when it comes to the prominent Sunday talk show circuit (only 9 percent in 2001). When a reproductive rights advocate does manage to secure a spot, the station feels compelled to bring on an extremist, giving viewers the impression that both sides are equally viable. So for women to be heard at all, they have to appear beside right wingers who represent a slim but vocal minority viewpoint, totally divorced from the realities of the working family. The way that "choice" is promoted makes it appear as if all women have the same access to health services and education, prompting even self-described feminists to not fully support women and their children who were being cast from the welfare rolls in 1996 (Smith 2005).

Even though he had a decisive electoral victory, President Obama has up to this point treaded lightly when it comes to defending reproductive rights. Yet feminists continue to utilize the electoral strategy of relying on the election of democrats to protect access to abortion. Smith (2009) points out that while the threat of pro-life demonstrators descending on the campus of Notre Dame during Obama's commencement speech never materialized, there were no pro-choice demonstrations either. This resulted in a de facto reliance on Obama to represent reproductive choice, a stance he did not forcefully argue. Calling once again for "common ground" with the pro-life movement, Obama mentioned the need to reduce unintended pregnancies and rely on adoption instead of using abortion. Because Obama took on a conciliatory stance in his speech, anti-abortion figure Randall Terry "was able to transform Notre Dame into ground zero for the most maniacal wing of the anti-choice movement without ever being forced to answer a coherent defense of the right to choose" (par. 12).

Feldt (2004) states that though Planned Parenthood has a high approval rating from the public (over 70 percent) or that 85 percent of parents support comprehensive sex education and 90 percent of people in the United States use some form of contraception, those who echo these sentiments on television are presented as the ones with "controversial" perspectives representing only a minority viewpoint. Worse still, if a conservative guest decides to ignore an issue by canceling a television appearance the station will pull the segment so they won't be accused of being unbalanced. Crucial women's males and females were needed in order to invest in the future workforce. Yet the conservative backlash against feminism also hit the schools with a rash of articles and books expressing concern about male students "falling behind" due to an over-feminization of the classroom coupled with women receiving "special treatment" at the expense of males (Kelly 2006; Martin 2006; Spring 2008). Gender-segregated education alludes to the split within conservative ideology: on the one hand capital needs workers from all backgrounds, especially to maintain a reserve army of labor to keep wages low. Women and minorities, particularly undocumented immigrants, fill this role nicely. But the other segment of the right wants women to serve in traditional roles as laid out in the Bible, with men in charge and the modern, heterosexual nuclear family as the norm.

The success of girls in school is a threat to this group because it throws "God's will" into question, subverting the social order with female students doing as well as or even better than males in the educational setting of the classroom. With the scarcity imposed via capital, even in the "prosperous West," women are viewed as taking things away from men (jobs, fathers' rights, etc.). Of course, this division among workers is a boon to capital, which is why the religio-conservative wing will never fully disappear from the political landscape (Hypolito 2008). Their philosophies and ideologies are too handy for periods of intense scarcity for elites to discard. To do away with them would mean defenders of capital having to explain themselves without religion's cloak and the message would be too hard for the general public to swallow.

In the arena of reproductive rights, stunning rollbacks have happened since the *Roe v. Wade* decision of 1973. Different states within the U.S. have passed more than 300 anti-choice laws, with 44 states restricting access to abortions for women under 21 (Feldt 2004, p. 130). Though officially "legal," abortions are not available in 87 percent of counties in the United States and for those counties that do offer services, no federal funding is available. Of the 50 states, only 17 offer funding for abortions that would save the mother's health (p. 130). In the global arena, Feldt describes how 350 million couples have no access to family planning, resulting in 80 million unintended pregnancies per annum. It is estimated that 78,000 women are killed annually during unsafe abortions (p. 202). Almost half of people who have HIV/AIDS are women and they are more susceptible due to their lower status: women are dependent on men for survival, creating conditions where violence and coercion are a way of life. The abstinence-only philosophy endorsed by the Bush administration lured monogamous married women into thinking that they were spared from HIV/AIDS, despite husbands who had extramarital relationships. Indeed, Page (2006) is correct when she assesses that young, minority, and poor women suffer the most from the virtual outlawing of abortion services. Roughly 57 percent of women who get abortions survive on incomes 200 percent below the poverty level, indicating that having a women, but there is also a sense of hope as the working class ranks grow larger:

> The exclusion of women from social production produced in earlier stages of capitalism has set up enabling conditions for an extremely cheap labor force that can be easily manipulated depending on what will make the most profit for the capitalist. Now, in a time of economic crisis, and a subsequent increase in ruthless and reckless global inter-capitalist competition, more women are being proletarianized on a global scale than ever before. (p. 169)

Indeed, Leacock (1981) reminds us that opposition, in the form of Marxist ideas, never stands still, leading to both changed relationships and a transformed society as working class women begin to see that they experience certain things in common, beyond the realm of the personal (Kaplan, 1995). It is most likely that a major part of any future working class revolt will include a large proportion of women, and not just in countries that have been ravaged by war. Wolf (2009b) describes a growing army of baristas, fast food servers, chain bookstore stockers, and cashiers among the mass of minimum wage workers in the United States.

Wang (2001) and Smith (2005) locate the oppression of women generally within the system of capitalism, its specific site of incubation being the heterosexual nuclear family. As discussed in chapter 3, the nuclear family creates a privatized system of unpaid labor geared toward the rearing of the future workforce and provides an important service to the owners of capital. As testament to the services that women provide (often silently), Kelly (2006) reminisces about a 1970 conference where women demanded four things: equal pay, child care on demand, free contraception, and immediate access to abortion. Later demands included the right to choose one's sexual identity and free education. Looking at the first three, "none of these has been achieved...the highest percentage of women are part-time workers, 48 percent is to be found amongst those with dependent children...the high cost of nursery provision means that part-time work is often the only option for women" (p. 7). The problem is that traditional notions of equality do not take into account the special reproductive status of women. Kelly mentions that leading members of the Bolshevik Party viewed women as needing protections from certain kinds of work and during pregnancy, and initial laws after the 1917 Russian Revolution reflected the impetus to protect reproductive rights and motherhood. This outlook would give pause to those feminists who filed a friend of the court brief in *California Savings and Loan v. Guerra* (1987), which sided with the banks. The ruling declared that maternity leave for female employees was not considered discriminatory against men, yet in this case high-profile feminists viewed the pregnant employee as unfairly receiving "special rights" (Smith 2005).

Within schools, the co-education of girls and boys arrived on the scene along with the growth of the human capital argument in education: that both contraception and abortion); child care and family (gender roles and responsibilities, marriage and divorce laws); and the workplace (sexual harassment, pay equity, health care, and retirement). These areas of struggle overlap, so it is difficult to maintain a purity of categorical analysis. In many ways this chapter builds on the analysis in previous chapters. A focus on reproductive rights is prominent as it is a foundational issue for women's sexuality and thus workplace equality.

What is important to understand in this analysis is that the socialist feminist perspective views the entire working class as oppressed, not just women as a separate identity group (Ebert 2005). This includes a point of view in which the granting of full civil and human rights to LGBTQ individuals is essential to the liberation of the working class. While same-sex couples are systematically being denied these rights, heterosexual couples are also suffering from a lack of access to health care, education, and child care (Kotulski 2004). Both groups can benefit from seeing their lot in common rather than engaging in endless dialogues about gender and identity. Leacock (1981) stresses that we have to begin uncovering the harmful writings targeted at women—from both conservative and liberal sources—which mislead them into thinking that being female is the same as being part of a "class." This only serves to turn women against each other and against men rather than aiming for the real source of women's oppression, which is capitalism. As Smith (2005) argues, "working-class men have no objective interest in maintaining the role of the nuclear family as it exists under capitalism, for it places the entire financial burden of reproduction on the shoulders of working-class men and women" (p. 168).

The State of Women Today

In early 2009, when 21-year-old recording artist Rihanna revealed that she had been abused by her boyfriend, a survey of males and females between the ages of 12 and 19 found that half thought she was responsible for being beaten (Schulte 2009c, par. 6). The survey also found that one-fifth of female high school students reported being physically and/or sexually abused by someone they were dating. Forty percent of girls between the ages of 14 and 17 said they knew of someone who had been struck by a boyfriend. This correlates with the reported statistic of 4.8 million U.S. women being physically assaulted or raped. Women facing economic crisis are in an even more vulnerable position when it comes to domestic violence than those with the financial means to escape, like Rihanna.

Women bear the brunt of the more predatory features of neoliberalism, including growing unemployment, lower pay, stratified educational opportunities, lack of access to health care and reproductive services, and holding multiple part-time and temporary jobs (Levine 2008; Weiner 2008). This is what Cotter (2001) describes as being part of the overall violence toward

Chapter Five

The Socialist Feminist Message: Sexuality, Work, and Liberation

This chapter examines important developments within socialist feminism with a focus on the writings of Coontz (1997, 2005), Feldt (2004), Kaplan (1995), Kollontai (Holt, 1977), Leacock (1981), Page (2006), and Smith (2005, 2009). While not explicitly identified as socialist, these writers address critical implications for a movement toward worker liberation, providing a promising alternative to postmodern identity politics. More so than the other chapters, this one will look outside of K–12 education to broader social trends impacting not only women, but the working class as a whole. This can be difficult to articulate among a dizzying array of media coverage proclaiming that women have reached a "post-feminist" era and no longer need "special protections" like sexual harassment laws (Roiphe 1994; Sommers 2005). Sites such as Abolish Alimony.org use feminist rhetoric as a cover to avoid paying child support or alimony to parasitic spouses who should be financially independent since women's rights have been successfully achieved; at least according to the "fathers' rights" movement. Race and class intersect with power feminism as high-status women like Oprah Winfrey and Condoleezza Rice are held up as proof that anyone can make it. However, as Maybee (2002) reminds us, "the economic reality of the wage gap leads to the reproduction of women's oppression, even if no one involved expresses any negative attitudes about the ones who are oppressed by the structure (p. 137).

This chapter begins with an overview of the state of women in the world today. This is followed by a critique of the limitations of both essentialist/nature-based and conservative/power feminisms, including authors' readings on the origins of the oppression of women and the theory of patriarchy. Socialist feminism is defined and explained as it fuses Marxist analysis with the unique ways in which women experience oppression under the system of capitalism. This socialist feminist perspective more clearly articulates the areas of struggle most important to women: reproductive freedom (access to

Taken with the overview of materialist readings of the family in chapter 3, this moves us toward a total comprehension of sexuality's role in the overall exploitation of the working class. Chapter 6 will carry these understandings forward to look at the formation of activist tactics and strategies in K–12 settings and beyond.

> liberating man from the domination of all these phrases.... Nor shall we explain to them that it is possible to achieve real liberation only in the real world and by real means...people cannot be liberated as long as they are unable to obtain food and drink, housing and clothing in adequate quality and quantity. Liberation is a historical and not a mental act, and it is brought about by historical conditions. (p. 44)

This is similar to Anyon's (1994) conclusion that "useful theory will have as a primary goal not the refinement of concepts, but successful political activity" (p. 129).

Because of the distrust of metanarratives as a whole, it is only logical that postmodernists accuse Marxists of being reductive or objective in their application of dialectical materialism. The fact that there are distinct differences between the owning and the working classes is regarded as an oppressive viewpoint that overlooks the "spaces" between what is seen as a fundamentally exploitative relationship between both classes (Hill, Sanders, and Hankin 2002; Wolf 2009c). Similarly, postmodernists point to sexual harassment as a simplistic legal construct that overlooks different levels of women's consent to sexual acts (Cudd 2002a; Kelly 2002). This makes it difficult to confront the growing backlash against women, because the backlash itself is constantly fictionalized and made uncertain, and downplays the seriousness of exploitation and the forces behind it and feeds into rightist ideology.

The postmodern rejection of consensus around issues of oppression and exploitation also makes it difficult to mobilize movements for sexuality rights, as with a response to AIDS, violence against women, or reproductive rights (Nowlan 2001; Wolf 2009c). Kelly (2002), in her discussion of postmodern feminism, argues that its theoretical position "underlines the impossibility of the oppressed empathizing with each other and acting together with other oppressed and exploited groups" (p. 222). If we are not able to identify who is or isn't oppressed, group people together, or to embrace concepts like freedom or equality, then we lack important ideas around which to gather, introducing a form of inertia (Bourne 2002; Kelly 2002; Spade 2008). Ebert (2001) calls this "indeterminancy or overdetermination," in which the act of not making a concrete decision "simply leaves the status quo unchallenged as it oscillates between the indecisive logic of 'on the one hand' and 'on the other hand'" (p. 287).

Conclusion

Luckily, there are alternatives to the deconstructive nihilism and consumer-saturated positioning of postmodern theories. Socialist feminism offers insights that address the contradictions between the working and owning classes, and sheds light on the importance of the domestic sphere.

not inherently oppressive or exploitative. Labeling science as oppressive can feed into right wing populism (i.e., global warming deniers) as well as superstition (Nowlan 2001). Mann and Huffman (2005) explain that while Marxist feminists are critical about the philosophy of positivism, they are not antiempirical. In fact, empirical analysis is precisely what is needed in order to better understand what is happening in order to make more effective plans for social change. In summarizing Marx's dialectical ontology, Gimenez (2005) asserts that we build knowledge when we move on from what appears to be common sense or localized knowledge. The reason that cultural concepts are not trustworthy on their own is that they are often put forth as if they exist readymade, with no historical context or analysis of the social relations that created a particular concept to begin with, leading to a false sense of neutrality (Anyon 1994; Cotter 2001; Ebert 2005). Instead of looking at "women" as a concept, we should be analyzing concrete aspects such as domestic labor and gender roles, because these ideas have more to do with social relations, as "things are what they are because of their relationships with other things, which are not always visible to immediate perception" (Gimenez 2005, p. 26).

For postmodernists, distinctions between theory and practice are blurred. It becomes difficult to distinguish from whence each emerges, which leads to inherent contradictions. For Marxists, theory and practice are integrated but in a way that facilitates social change through an analysis of purposeful human activity (Anyon 1994; Ebert 2005). The postmodern view of truth or reality is always a matter of human perception, which is presented as too constantly changing to pin down (Wolf 2009c). One of the main contradictions of postmodernism that Anyon (1994) observes is its outward eschewing of binary categories, which are seen as an oppressive aspect of modern science, yet postmodernists continually employ binaries such as structural/poststructural and truth/no truth. Anyon also locates a postmodern "metanarrative" of "indeterminency," of being certain about not being certain (p. 122). Wolf (2009c) sees a similar reluctance within postmodernism to use precise definitions—yet just because our concepts of people are shaped by capitalism doesn't mean the categories we use for analysis are automatically oppressive. Binaries, like other aspects of human categorizing and language, are not oppressive entities in themselves; the system of capitalism imposes many of the most exploitative binaries, as Ebert (2005) points out.

Unlike the postmodern rejection of empiricism, Marx (1998) defends the importance of the scientific method, though he does acknowledge that science can be pressed into the service of capital. Without applying historical materialism, everything becomes an abstraction, which makes it impossible to begin to craft strategies for change. The constant act of deconstructing social categories replaces what is needed for revolution:

> The liberation of man is not advanced a single step by reducing philosophy, theology, substance and all the rubbish to "self-consciousness" and by

critique women's oppression in the workplace are portrayed as making victims out of women, especially if they favor safe workplace policies. Postmodern and conservative theorists see the acknowledgement of a hostile work environment as overly sheltering women, even though this policy applies to all individuals and not just women or heterosexuals. According to Cotter, this "reproduces the reactionary ruling class fantasy that people only need to look within themselves and recognize their power—that is, pull themselves up by their own bootstraps—in order to be powerful" (p. 179). Soon, human needs are made into "special rights."

In a similar vein, conservative gay author Bruce Bawer (1994) argues that the LGBTQ rights movement, gay studies programs on college campuses, and so on, are politically correct and that the more overt examples of gender identity displayed by some within the LGBTQ community are giving a "bad name" to respectable homosexuals and harming the movement's goals of equality (defined as, of course, equality of opportunity, not equality of human rights). Using the terms "subculture-oriented gays" and "mainstream gays," Bawer endorses the latter as being more respectable (p. 35). This is clearly conveyed in his outlook and explanations. Following classic backlash logic, he describes a 1993 gay pride march where, in his view, the subculture-oriented gays were creating "image problems" for the movement (p. 222). In contrast, he praises King's 1963 march on Washington, D.C., because "he hadn't called for revolution or denounced American democracy or shared the podium with stand-up comics" (p. 221). This, of course, is a conservative tactic: ignoring King's radicalism and presenting a more palatable portrait of the civil rights leader. Goldstein's (2003) analysis of the homocons finds that their salience has much to do with reflection of the postmodern nature of our time: reject labels, get the government off your back, eschew solidarity, be your own person, and so forth (Wolf, 2009c).

Bawer (1994) also asserts that gay studies programs reinforce victimhood, though he does make a salient point about many within gay studies who cannot envision coming together for reasons other than gay identity. But correcting this problem with the solution of worker solidarity is not Bawer's goal; in fact, he goes on to say "many subculture-oriented gays don't really have very much pride in themselves as individuals; for it would never occur to an individual with pride in himself to feel a need for group-oriented pride" (p. 183). Here we have objectivism applied to sexuality activism; that is, solidarity indicates low self-esteem! D'Emilio (2002) offers another possibility, where "going it alone" should be rejected in favor of building alliances and working together. It is individualism that is pathological, not solidarity.

The Rejection of Reality

The postmodern view of empiricism as an oppressive master narrative confuses capitalist appropriations of science with the scientific method itself, which is endless differences (a form of individualism) and skepticism about equality, postmodernism isn't much different than racist logic itself. When Anders (2008) states that "tolerance is conditional and subject to revocation; freedom of choice is an absolute," he is embodying the postmodern resistance to concepts of structural economic forces as explanations for racism (p. 91). Instead, for Mann and Huffman (2005), people are neither free to create history nor does freedom mean doing what one wants when one wants to. All the while, there are real barriers acting upon people that no amount of individualism or possibilities of difference can overcome.

When sexuality was finally divorced from procreation in the late twentieth century, it led, along with a rapid expansion of the middle class, to the false conclusion that one could fashion an identity that existed apart from material reality, as D'Emilio (1992) explains. In actuality, capitalism has continued to shape sexual practices even if it might appear that sexuality is only a personal issue. Similarly, individualism is a key ideology employed by capitalism in order to facilitate the cooperation of the workforce. By isolating workers from each other, capitalists can offer up a hollow substitute for solidarity, replacing it with workplace loyalty, freedom of choice, and other aspects of false consciousness (Ebert 2005; Marx 1977). Marx (1998) observed that a "relapse into idealism" seemed to occur when the conditions were right for global upheaval (p. 47). Working in the service of capitalism, postmodernism offers the idealism of continual deconstruction and the hope of rootless ideology, when all the while "not criticism, but revolution is the driving force of history" (p. 61). Within Marxism, no social issue or person is autonomous, everything and everyone is connected to larger economic forces and free choice/free will is therefore a moot point and an illusion until capitalism is abolished (D'Emilio 1992; Nowlan 2001; Wang 2001; Wolf 2009c). As Goldstein (2003) puts it, "all of us inhabit a halfway house that calls itself freedom" (p. 4).

It doesn't take much for this individualism to blend into economic objectivism and outright libertarianism, against a background of real attacks on the working class (Chamallas 2002; D'Emilio 1992; Wolf 2009c). Cloud (2001) finds many similarities between queer theory and family values rhetoric, especially in their privatization of the social. Both postmodernists and conservatives are hostile to the idea of mobilized responses to societal needs and turn instead to "ludic textualist strategies" (p. 72). Conservatives yearn for the idealized family, but they look inward as part of a message of personal responsibility rather than seeking universal child care, heath care, or education as pro-family. The queer theory equivalent is individualized performance as part of a strategy of disruption of the nuclear family construct. In the end, both offer ineffective solutions and support personal desires over human needs. Likewise, Cotter (2001) taps into the disdain that many on the right and postmodern left have against antisexual harassment policies that go beyond quid pro quo to establish a safe working environment. Those who do with a complex series of social interactions stemming from who owns or doesn't own property, waged versus unwaged labor, and pay differentials in the workplace. The fact that women are often relegated to the unproductive sphere of domestic labor has to be considered a major factor behind sexism, wage differentials, and sexual harassment.

Anti-Solidarity

Neoliberalism derives from traditional liberalism, an early capitalist philosophy that stressed the notion of society as a collection of individuals seeking their own self interests amid inevitable scarcity (Robertson 2008). Opposed to collective solutions, liberalism stresses that each person is in charge of his or her destiny and that, presented with an array of market-based choices, he or she will make rational choices with little interference needed. Neoliberalism ramps up the privatization and individualism of traditional liberalism to a global scale, with vast social resources being devoted to the protection of a small number of ruling elites, usually with increased militarism. In the aftermath of Hurricane Katrina in 2005, police were sent first to protect Walmart and the oil fields rather than distributing food and water sent. This sent home the message that we are all finally alone and can no longer rely on government to meet human needs (Berkowitz 2008; Eggan 2008). Within education, the state continues to divest from K–12 schools and universities, stressing the development of "local control," charters, parent involvement, and so on (Kuehn 2008). The language of neoliberalism has filtered down to schools where business-oriented teacher unions no longer resemble a force that can effectively fight for smaller class size, supplies, benefits, basic facilities, or other essential needs (Murphy 2008).

Indeed, Marx (1977) explains that the unity of workers is a necessary ingredient to combat capitalism. In fact, it was mass industrialization that created the potential conditions for capitalism's undoing in the first place:

> As the number of the co-operating workers increases, so too does their resistance to the domination of capital, and, necessarily, the pressure put on by capital to overcome this resistance. The control exercised by the capitalist is...a function of the exploitation of a social labor process, and is consequently conditioned by the unavoidable antagonism between the exploiter and the raw material of his exploitation (p. 449).

Contemporary culture, assisted by postmodernism, tends to recast structural problems requiring collective solutions into private ones solved only on an individual basis (D'Emilio 2003; Goldstein 2003; Wolf 2009c). Larger problems like racism are separated from their historical and institutional foundations and translated into existing "within" people, composing the simplistic equation of racism = power + prejudice (Bourne 2002). By both focusing on imagination, because, "it is not a question of the practical abolition of the practical collision, but only of renouncing the idea of this collision" (p. 304). Recent academic efforts to transform stripping, prostitution, sadomasochism, and sex tourism into "empowering" acts have echoes of this strategy. Instead of confronting capitalism, empowerment becomes a matter of getting people to see their oppression as a "choice" or a matter of linguistics:

> Their often exclusive preference...for arguments that are disembodied, for arguments that ignore social realities, for arguments that dwell completely in the rarified air of the general and universal, reflects the social position of White males who have spread out their vision and made it the dominant one, whose ignorance of the social positions of others has allowed them to be convinced that they really do represent the universal point of view. (Maybe 2002, p. 143)

The instability of the postmodern subject and the floating signifier of power leaves no alternative but to locate meanings within the individual, rather than to external, structural entities, or what is meant by the phrase "the personal is political" (Bourne 2002; Kelly 2002, 2006). This especially creates problems when considering the situation of many working class women and LGBTQ individuals. Asserting a particular identity is no longer a means to activism, it becomes the activism itself as in Morton's (2001) analysis of queer theorists reverting back to the closet and this time framing it as a "choice." The abandoning of the revolutionary project means that deconstruction is all that remains of political action, where postmodernists find themselves constantly interrogating binaries and confusing this with social transformation (Anyon 1994; Gimenez 2005; Mann and Huffman 2005; Wang 2001). Lynd and Grubacic (2008) take on critical legal theorists when they pointedly ask: "How will your critique of all legal discourse and activity assist 10,000 Palestinians indefinitely detained in Israeli jails without criminal charges? What is your plan for bringing an end to indefinite confinement at Guantanamo" (p. 217)?

Deconstruction also leads to a severe underestimation of the force behind capitalist ideology. Because structural forms of analysis are eschewed, problems of racism or homophobia become conceptualized as existing within individuals. Since there is no stable self and power is not located in any one spot (it's everywhere and nowhere), it becomes difficult to pin down where these racist or homophobic attitudes originate. Fighting intolerance then, means changing "bad attitudes," not confronting how capitalism uses race, gender, and sexual orientation as a means of preventing solidarity among workers (Wolf 2009c). One only ends up dealing with social problems on a case-by-case basis while leaving capitalism intact. As a counter to postmodern deconstruction, Gimenez (2005) asserts that inequality and sexism are not simply reducible to biology, intention, development, or power. Instead, sexism has to

Sutherland's (2005) examination of the limitations of postmodern feminism: "Women need to be extricated to provide social change, yet women cannot be extricated without social change. No one is outside the system, but no one can provoke change from within it" (p. 128).

Wolf (2009c) finds much hostility from postmodern and conservative gay circles in reaction to the notion that the working class is capable of leading revolutionary change, particularly regarding sexuality issues. This was absolutely the case in how the aftermath of Proposition 8 in California was presented as a black versus gay issue, sending home the false message that the LGBTQ community could not rely on minority support. Within dystopian thought, nothing can be trusted, including our existing concepts of oppression, which may be illusory. Just as scarcity is taken for granted as the status quo, so too is the inherent conservatism of the working class, which is often pointed to as the reason for the failure of utopian experiments. When postmodernists assert that Marxists do not put forth any clear-cut visions of an alternative society, what they mean to say is that Marxists refuse to put forth compromise-based (i. e., pragmatic) alternatives that leave capitalism and assumed scarcity intact (Cole, 2009).

Deconstruction and Instability

According to Anyon (1994), deconstruction involves "identifying the rhetorical operations that ground an argument and then demonstrating that the terms being used are contradictory or philosophically unstable" (p. 119). Postmodernists view language as shaping human experience while doing away with modernist concepts of change arising from human agency. It is the discourse, therefore, which shapes the subject. While there is much merit in deconstruction as part of a larger process of dialectics, the transformation of activism into acts of deconstructive performance in an effort to fight oppression is ineffective (Anyon 1994; McClaren and Farahmandpur 2002). In *The German Ideology*, Marx (1998) spends a great deal of time critiquing Max Stirner, whom he dubs "Saint Max" and "Sancho," referring to Stirner's preoccupation with ideology that can be simply put out of one's mind. Today's postmodernists view ideology as something that can be deconstructed out of existence, or made less threatening by applying critique, as in categories of gender being viewed as the problem rather than capitalism using gender to promote sexism (Wolf 2009c). Just as gender has been socially constructed, so the logic goes, it can be equally deconstructed, and the act of deconstructing will end the oppression around gender by calling into question issues of power. Likewise, if homophobia is viewed as a "masculinist" or patriarchal power identity to deconstruct rather than a phenomenon resulting from economic forces, the thought is that confronting masculinity will end LGBTQ oppression.

Marx (1998) was highly suspicious of any philosophical attempt that sought to undo the effects of oppression by making them figments of the socially nonconformist than it is to make sure that our lavender seniors can pay their rents and aren't homeless. (p. 45)

Activism around marriage rights does not "take away" from other LGBTQ issues or inhibit the ability of people to engage in alternative relationships as many postmodernists maintain. In fact, the whole antimarriage discussion amongst queer theorists is a moot point, as it "really doesn't matter what LGBT people think of marriage, because right now they're barred from truly making a choice" until federal marriage rights are guaranteed (p. 99).

Individualism and Rootlessness

Marx (1973) traces individualism to the early Enlightenment era of the 1700s. During this time—which coincided with the rise of capitalism—it became fashionable to re-imagine society as consisting of a collection of lone individuals looking after their own self interests. This, of course, is contrary to anthropological evidence, but in Marx's time, as today, academics were busy at work imposing their own interpretations in the service of capital onto the past. "Production by an isolated individual outside society," Marx explains, "is as much of an absurdity as is the development of language without individuals living together and talking to each other" (p. 84). Postmodernism, along with conservatism, has resurrected the specter of the individual but in different ways. While conservatives valorize the lone wolf who heroically exists without making any demands on his or her betters, including demands for a social safety net, postmodernism denies the notion of self but creates the fiction that all forms of solidarity are suspect and will inevitably end up reproducing oppression. Taken together, the kind of rootless individualism that eschews empiricism and historical materialism is a dangerous development.

The Dystopian Project

The Giver, Lowry's (1993) young adult novel featured on just about any district middle school reading list, presents the typical postmodern warning about how a society's attempts to meet material needs is, beneath the surface, a corrupt endeavor. Far from being a progressive work, *The Giver* reinforces the worst aspects of libertarianism and objectivism. In the novel, the possibility of socialized needs (child care, education, health care, a fair criminal justice system) seems irreconcilable with concepts of freedom, diversity, and love. The message of inevitable failure of any kind of communal solution is reinforced at the end of the novel by the main character only listening to the voice of the giver, making his way in the "uncertain" world as he possibly starves and freezes to death. Yet the uncertainty is all worthwhile because true freedom of choice merits the sacrifice of comfort, and scarcity is accepted as a fact of life (Goldstein 2003). This dystopian conundrum is explained by lies in viewing marriage as the problem while putting forth the fiction that under a capitalist system one can fully unleash desires as part of a vision of liberation (Cotter 2001).

Kollontai (Holt 1977) fully recognized the limitations of attacking marriage as the problem and questioned whether "free love" could exist as long as there was capitalism. In challenging the feminist movement of her time, she put forth her socialist critique as follows:

> In opposing the legal and sacred church marriage contract, the feminists are fighting a fetish. The proletarian women, on the other hand, are waging a war against the factors behind the modern form of marriage and family...the feminists and social reformers from the camp of the bourgeois, naively believing in the possibility of creating new forms of family and new types of marital relations against the dismal background of the contemporary class society, tie themselves in knots in their search for these new forms. (p. 68)

Wolf (2009c), Coontz (2005), and Leacock (1981) echo that just as there is nothing inherently repressive about monogamy, there is likewise nothing about alternative sexual arrangements that is by nature revolutionary. For Kotulski (2004), some heterosexual marriages can reflect sexist values but that doesn't necessarily mean that the same will always happen for gay and lesbian couples. Jindal (2008) admits that aspects of queer theory's enacting of desire end up reinforcing racist notions of the sexual availability of people of color. Cloud (2001) analyzes how features of postmodernism and queer theory often legitimate features of conservative ideology, including the privatization of desire and remaining locked within the family values discourse.

The fight for marriage equality does not have to be an endorsement of the nuclear family and capitalism. Wolf (2009c) and Kotulski (2004) maintain that removing the legal barriers to marriage would create important openings for future struggles and the move toward socialism, while "domestic partnerships" only enhance the separate but equal status of lesbians and gays. Most importantly, federal marriage equality would provide needed material relief for the working class as a whole in the midst of continuous right wing assault. As Kotulski (2004) reminds us, Social Security, something that most seniors depend upon for survival, is denied to lesbian and gay survivors, costing this segment of the elderly population close to $125 million in unavailable benefits (p. 44). Medicaid offers protections for the assets of married couples should one partner enter a long-term care facility. However, these benefits are denied to same-sex partners:

> But hey, it's just money. And besides, marriage is between a man and a woman, for hets only. Why should queer people want to participate in something so patriarchal? After all, it's more important to be cool and

into meaning doing what one wants when one wants, with no assurances that exploited groups will be protected, while Marxists are labeled as narrow-minded (Cotter 2001; Sears 2005). Ebert (2001) sharply points out how the postmodern discursive relationship with women's breasts as a site of performativity and subversion mean something entirely different when it comes to impoverished women worldwide who cannot breastfeed due to malnutrition or being able to afford to buy formula: "The affectivity of breasts for these women is not the erotic pleasures of nursing.... The objective reality of these women's working day under capitalism means they must sell their labor power as commodities in order to earn the money to feed themselves and their children" (p. 292). (Perhaps postmodern scholars can rectify this situation by writing an essay on how poverty isn't oppressive by looking at the inability to breastfeed as creating new "spaces" for "mental nourishment" that go beyond the limiting binaries of hunger/satiation!)

A privileging of desire over need can also contribute to advocating counter-revolutionary strands of individualism such as promoting deliberately unsafe sex, pornography based on violence or rape, vigilantism, sexual harassment, or sexual objectification packaged as libertarian acts (Cotter 2001; Goldstein 2003; Mann and Huffman 2005; Sears 2005). In discussing the work of Camille Paglia, Goldstein (2003) finds her privileging of male aggression disturbing, as it does nothing but reinforce sexism. But since Paglia does not account for sexism as anything other than existing in the minds of women who "choose" to be victimized, this does not appear to be problematic: "Like any pornographer, Paglia must raise the rod while calming the fears that lurk in the shadows of sex. She does the former by stroking the male ego, and the latter by attacking its perceived enemies" (p. 59). Similarly, efforts to "reclaim" slurs such as *slut* and *cunt* mean something entirely different for poor women and women of color due to their historical and stereotypical connotations (Jindal 2008). Jindal also explains that sex parties sponsored by wealthy white males are often located in low-income and minority neighborhoods safely removed from gentrification efforts.

The fight for marriage equality is the latest in a long line of struggles around desire versus need. The postmodern left locates oppression within the institution of marriage and therefore rejects marriage equality as playing into the hands of racism, sexism, and homophobia, as Jindal (2008), Schroeder (2008), and Bailey, et al. (2008) argue. Jindal (2008) believes that because many LGBTQ individuals along with subaltern people cannot receive social security benefits, health care, inheritances, and so forth, then marriage is not a relevant issue. Schroeder (2008) asserts that lesbians and gays who have children are only contributing to the overall conservatization of the LGBTQ community. Bailey and colleagues (2008) explain that granting marriage rights to blacks during the 1800s did not minimize discrimination or oppression. Instead, the social benefits that come with marital status should be available to all—something that Marxists would not argue against. The problem the planet. However, the illusion of desire overshadows the more imperative needs of the many, contributing to a philosophy that the elite are entitled to this existing situation. As Nemec and SalMonella (2008) point out, "A lot of gay activism focuses on forcing the capitalist system to create comfort for a particular oppressed group. What about queer activism that attacks capitalism and focuses on using people power, community power, to create comfort for everybody (p. 173)?" Until capitalism is overthrown, it is impossible to base a strategy for liberation on desire alone without at the same time maintaining a system of oppression and exploitation, a limitation that D'Emilio (2002) notes in the following question: "When we make our sexual desires the heart of our lives and our political movement, is this an act of resistance, or does it signify the acting out of the terms of oppression" (p. 70)?

Queer theorists view the sexual revolution and its aftermath as once and for all freeing desire from the chokehold of material need (Morton 2001; Sears 2005). Queen (2008) expresses this viewpoint when he sees acts of sexual rebellion as influential enough to change laws and policies while being able to refashion relationships relatively free of exploitation. Certainly there have been seismic cultural shifts in the family and the formation of personal relationships, but the fact remains that true sexual freedom is an elusive goal under capitalism. Here, the promotion of desire is about nothing more than protecting the rights of a small segment of the population and is itself a form of false consciousness (Ebert 2001). When injustices cannot be effectively overcome without threatening the profit structure, the solution for postmodernism is to simply set aside need and suggest instead that it is desire that cannot be avoided:

> The ruling class's pataphysicians are willing to pay this price because at a time like the present when the logic and effects of exploitation are becoming starkly obvious in the chasm between rich and poor within the U.S. itself, the pathologizing of the social helps to heighten the suspicion of reason. (p. 45)

Dialectical materialism, with its emphasis on historical understanding and rational logic, is held suspect in postmodern circles for a reason. Morton (2001) and Cotter (2001) claim that a sound society would be organized around meeting everyone's needs and that these needs change depending on the particular situation at hand. Class relations, not irrational and addictive desire, are what shape humans (Cotter 2001; D'Emilio 2003; Ebert 2001).

While the right wing targets the LGBTQ community by using "promiscuity" as a moral barometer, the postmodern left, in calling for ludic desire, "ignores the material constraints need inevitably places on desire" (Nowlan 2001, p. 137). Desire being privileged over need is nothing more than an embracing of the self as consumer with individual choice and free will, certainly not a threatening concept to the ruling elite. Creativity itself is reshaped revolutionary goals and more about creating alternative "spaces" for exploring desires, individualism, expression, and the like. Ebert (2005) classifies this as the difference between a "sexuality free from material constraints" and the "exercise of desire free from moral constraints" (pp. 52–53). Postmodern desire postulates that sexuality is a natural phenomenon existing apart from any limitations whatsoever, placing it above material need as a priority (Sutherland 2005). Any limitations are a matter of language and power, which are not tied in any meaningful way to social class. Contrary to assumptions, Marxists are not antisex. Instead, sex is understood to be an important need, but not in the same manner as food, shelter, health care, and the like (Wolf 2009c). Sexuality is a form of social relations that is mediated by capitalism; in other words, it isn't Marxism that is antidesire, it is the economic realities that make it possible for only a few to experience the type of pleasure postmodernists advocate that is the problem.

Far from being natural, human sexual behaviors are shaped by economic forces and exacerbated by capitalist property relations (D'Emilio 1983, 1992, 2002; Ebert 2001, 2005; Morton 2001; Wolf 2009c). As Marx (1998) astutely points out, "the fixation of interests through the division of labor and class relations is far more obvious than the fixation of 'desires' and 'thoughts'" (p. 277). Notions of desire are dependent on the ability of those who are better off financially to enact their interests, often without consent of the objects of their desires (Cotter 2001; Ebert 2001). With the rapid transfer of wealth from the working class to the ruling elite over the past 20 years, it has become more difficult for postmodernists to avoid dealing with the rising importance of need (Morton 2001). In Nowlan's (2001) discussion of the aftermath of AIDS, he views concepts like the liberation of desire as "an ideological maneuver to lower the working masses' expectations of the health care system, a move quite convenient to capitalists who do not wish to be taxed to support public health" (p. 123). Advances in AIDS treatments, which were derived from privatized research efforts, only impact those with the money to pay for expensive therapies. Unfortunately, the queer left's refusal to acknowledge the material realities of AIDS, which has gone far beyond the gay community, provides an opening for the right wing to legislate morality on an international level (banning condom distribution) along with provisions for super-soaring pharmaceutical profits at the expense of suffering populations worldwide.

While the message might be an unleashing of desire for everyone, uncontrolled profits for a very few remain the norm, a "free play of desire for those who 'have,' and continued oppression for those who 'have not'" (Cotter 2001, p. 197). When sexuality is divorced from a materialist understanding, it becomes difficult to work toward a world where pleasure doesn't have to be at the expense of the oppressed (D'Emilio 2003; Goldstein 2003). Ebert (2005) explains that our ability to overproduce means that we have enough resources to provide health care, food, and shelter for everyone on to be able to pass our apartments to each other, rather than for tenants' rights or low-income housing" (Spade 2008, p. 51). As much as LGBTQ marketers are selling a particular lifestyle to those with the money to buy it, so too are pharmaceutical and insurance companies selling health care to the highest bidders (Nowlan 2001). Wolf (2009c) views this as a scaling down of activism to include fights for access to the marketplace and "lifestyle" along with an overall vanishing of class struggle.

A common pitfall within postmodernism is to view social problems and their solutions through the lens of popular culture. This is reflected in Peterson's (2009) interview of hip-hop scholar Tricia Rose. Rose argues that "the beast" of corporate-dominated hip-hop has to be confronted in order for alternative voices to be heard. Yet, Rose continues, "the way to slay the beast is not through some communist, you know, reformation, but by developing a critical form of engagement with the market" (p. 36). In this view, postmodernism does not seek to challenge the market beyond its own terms, only to "problematize" its representations of people. It fails to see that representations of people, such as women being viewed as sex objects in music videos, are shaped by capitalist social relations, going way beyond popular media. As another example, President Barak Obama represents the ultimate in postmodern branding, according to Pilger (2009): "He is a marketing dream. Like Calvin Klein or Benetton, he is a brand that promises something special—something exciting, almost risqué, as if he might be a radical, as if he might enact change.... He's a postmodern man with no political baggage" (p. 26).

The danger of consumerism as activism is that it automatically deflates genuine activism, much as the lead-up to the election of Obama all but killed what was left of the antiwar movement (Moeller 2002; Pilger 2009). Nowlan (2001) argues that a "pragmatic" approach to a cure for AIDS like trendy awareness campaigns will not challenge the fundamental issues of access to health care, housing, education, and food for all people. Ebert (2001) views postmodernism as a form of false consciousness, particularly the way the sex trade is often refashioned into a form of empowerment:

> We need to critique both the false consciousness that blinds us to the coercion involved in all these exchanges and break down the distinctions not only between voluntary and coerced sex work, but between sexual labor and all other forms of wage labor...we need to critique the dichotomy between paid sex work and seemingly "free" sexual intimacy in our daily interpersonal relations (pp. 300–301).

Desire Versus Need

A key difference between postmodernism and Marxism is how each situates desire as opposed to need. Wolf (2009c) outlines how activism deriving from postmodern philosophy is less focused on ending exploitation or on

However, Marx (1977) was clear that capitalism, as a totalizing discourse, ensures no worker is able to escape its ideology. This implies that there is never inherent rebellion in individual acts of consumption:

> Individual consumption provides, on the one hand, the means for the workers' maintenance and reproduction; on the other hand, by the constant annihilation of the means of subsistence, it provides for their continued reappearance on the labor-market. The Roman slave was held by chains; the wage-laborer is bound to his owner by invisible threads. The appearance of independence is maintained by a constant change in the person of the individual employer, and by the legal fiction of a contract. (p. 719)

Indeed, capitalism has been quite adept at containing rebellion by co-optation and rampant marketing; especially when it comes to sexuality and gay liberation (Cole and Hill 2002; Ebert 2001; Eggan 2008; Mecca 2008; Nowlan 2001; Sears 2005; Sutherland 2005; Wolf 2009c). By translating activism into consumerism, those with the most means tend to get the most hearing, as Sears (2005) and Sutherland (2005) point out regarding legal matters affecting LGBTQ and feminist groups. According to Sears (2005), "commodified queer space appears, in part, as a set of market niches, in which people live their sexuality through the purchase of specific goods and services" (p. 108). To be visible within the gay community, then, means one has to be white, male, attractive, physically able, and wealthy—and not necessarily "out." For example, Sears indicates that there is a lack of research on LGBTQ low-wage workers, many of whom occupy the lower rungs of the working class because they are unable to assimilate as easily as more well-to-do white gay males. O'Brien (2008) reminds us that it is often the transgendered who are the most vulnerable to the economic restructuring that has happened worldwide as a result of the IMF and World Bank.

The intensification of marketing the "gay brand" as a chic lifestyle option began in the 1990s as AIDS activism gave way to niche advertising as the primary form of LGBTQ awareness (Wolf 2009c). The market was seen as a way for lesbians and gays to achieve both respectability and equality, provided they weren't too flamboyant. Conservatives also latched onto the brand approach by building on media stereotypes of gays and lesbians as affluent, thereby not in need of "special rights." By pitting African Americans (who are overly represented in all markers of lower income) against so-called "privileged" lesbians and gays, the right wing inserted a wedge between these groups. Mecca's (2008) and Sycamore's (2008) separate accounts of lower-income and transgendered LGBTQ people being displaced from the Castro neighborhood in San Francisco to make way for more prosperous gay home owners is another instance of how marketing destabilizes solidarity. Similar problems have happened in the neighborhoods surrounding the site of the Stonewall rebellion, where "mainstream LGBfakeT organizations choose to fight for homeowners reminds us, "activism is not about identity politics. Activism doesn't wait to be told. Activism doesn't rely on the opiate of hope.... Real activism has little time for identity politics, a distraction that confuses and suckers good people everywhere" (p. 26).

Consumerism as Activism

Even if not all of its proponents intentionally do so, postmodernism "serves the interests of capital's current hegemonic project" (Cole and Hill 2002, p. 92). By downplaying historical materialist strategies, postmodernists are unable to confront the totality of capitalism, which has only grown more rapacious in the wake of neoliberal globalization (O'Brien 2008). In fact, consumerism itself is celebrated as both a cultural artifact and a form of activism, leading to the erasure of class difference, as McClaren and Farahmandpur (2002) explain:

> The postmodernist and post-socialist assumption that culture has suddenly found new ways of winning independence from economic forces, and that somehow the new globalized capitalism has decapitated culture from the body of class exploitation by constructing new desires and remaking old ones, in ways that are currently unmappable and unfactorable within the theoretical optics of political economy, has not only contributed to the crisis of Western Marxism, but has effectively secured a long-term monopoly for capitalist market ideology. (p. 46)

In addition, since social class is taken out of the picture, "it begins to seem as though the products control the producers rather than the reverse, as if decisions about production flow naturally from the value of commodities rather than being made by people" (Sutherland 2005, p. 118). By turning the effects of capitalism into "texts," the "real" is erased, providing the illusion that a shift in language is all that is needed to address exploitation (Cotter 2001). The seemingly progressive language of postmodernism often masks its allegiance with both neoliberalism and assumptions of conservatism (Weiner 2008).

Mann and Huffman (2005) apply dialectical materialism to the development of postmodernism and poststructuralism as academically ideological reactions to the globalization of capitalism, which began en masse during the 1970s. The increase in service industry, low-wage, and temporary work brought awareness of the increase in minority and female workers who were (and still are) filling these jobs: "Hence, it is not surprising that the theoretical discourses during this period placed less emphasis on social class and more on other forms of difference" (p. 80). The temporary and diverse nature of work and "lifelong learning" combined with aggressive marketing of credit cards and bank lending created a deceptive situation where "choice" and "flexibility" made it appear as if social class no longer matters (Kuehn 2008).

age factor (Spade 2008). A study out of Georgia also discovered that African Americans, while they might be more likely due to religiosity than whites to view homosexuality as a sin, are more likely to *oppose* laws that interfere with LGBTQ rights (Wolf 2009c). Wolf's point is that identity politics operating within a capitalist society will take postmodern intentions and spin them to suit the goals of the ruling elite, which is to divide the workforce. The results become an "oppression Olympics" where conservatives counter-pose LGBTQ rights against the history of African-American struggles—as if they cared about either group.

Another problem with the foreclosure of cross-group alliances is that it presumes that all people of the same race, gender, or sexuality share common interests. This might hold true to an extent until class enters the picture. Then, one cannot argue—at least from a Marxist stance—that pro-business New Orleans Mayor Ray Nagin shared the same experiences as the poor and working class African Americans who were systematically denied crucial aid and left behind to rot under his watch after Hurricane Katrina. Working class whites and blacks gain nothing by seeing their interests connected to the capitalist class (Bageant 2007). D'Emilio (1992), Sycamore (2008), Mecca (2008), and Shephard (2008) would be skeptical of working-class LGBTQ individuals considering their interests in common with ruling class gays and lesbians who support laws and policies to ensure that undesirable LGBTQ people are kept out of gentrified neighborhoods. According to Burton-Rose (2008), "If power were all of the sudden handed to a gay ruling class in America, the exploitative relationships would continue. There would still be racism, class oppression, women's oppression…the only thing that would change is there would be less homophobia" (p. 27). D'Emilio (2002) asserts that the problem isn't privilege, which is a neutral concept, but the fact that only a few people have it at the expense of the majority. If leaders who happen to be of a different background than those with which they seek solidarity emerge within activist circles, the real issue is in which direction they work, not whether they are labeled "privileged."

Identity politics is limited in that its activist potential is reduced to individuals, not groups (Cloud 2001). A kind of fortress thinking can materialize along with energies spent fighting a monster with many heads rather than a single one: capitalism (D'Emilio 2003). When males cannot join with women or heterosexuals with LGBTQ people to advocate for universal human liberation, then experience of oppression is all we have (Wolf 2009c). This plays into the hands of budget-conscious capitalists who often benefit from different minority groups competing for small-scale grant funding, as when women's studies competes with LGBTQ studies at colleges and universities (D'Emilio 1992). The good news is that sexuality is socially constructed, meaning that if it was established by people under certain historical conditions, then aspects of this construct can be changed through the efforts of the working class confronting capital (Forrest and Ellis 2006). As Pilger (2009) groups gone (i.e., capitalism as the common oppressor), irreconcilable differences between minority groups was the inevitable result. Combined with the fragmented self put forth by postmodernists, differences between groups made alliances seem an insurmountable problem (Forrest and Ellis 2006; Jindal 2008; Mann and Huffman 2005; McClaren and Farahmandpur 2002; Rikowski and McClaren 2002; Wolf 2009c). As Chamallas (2002) explains, when difference is valorized, along with specificity of context, it then becomes improper to draw analogies between various forms of discrimination. This is no different than anti-gay conservatives claiming that the LGBTQ community is presumptuous to assume that their discrimination is on par with African-Americans'.

While it can be argued as a whole that males and whites benefit from the current cultural setup, for the working class any so-called privileges are compensatory at best (Cole 2009; Hickey 2006). When a focus on difference makes alliances impossible for activism, it tends to place minority groups outside the reach of class (Murphy 2008). When attempting to understand history, such as with labor organizing, this can have an important impact on teachers unions. Murphy explains that educational unions are presented as belonging under the heading of educational history, not labor history. In addition, no-allies thinking makes it difficult to work across sectors of the labor market, as with teachers unions striking to support service industry workers. When the Gay Liberation Front folded, enormous rifts were created between gay and lesbian groups with lesbians splitting off to form their own women-only coalitions (D'Emilio 1992). As patriarchy was viewed as the problem rather than offshoots of capitalism, many social problems such as rape and sexual violence became a matter of achieving consciousness about women's identity, which was changeable, while heterosexism and male domination were viewed as intractable features of society. D'Emilio explains that "this consciousness, once attained, was precarious. It had to be protected and defended...for many, separatism became not a tactic...but a way of life" (p. 255). As Goldstein (2003) puts it, "we may see a gay movement run by an assimilated elite and the gender movement representing the unassimilable" (p. 103).

The focus on within-group identity politics can have dangerous conclusions. Wolf (2009c) presents the example of the outcome of California's Proposition 8, which successfully overturned gay marriage in 2008. Immediately, African Americans were pointed to as the reason the measure passed, to the benefit of conservative policy makers who apparently found the ultimate minority wedge group in their perpetual fight against LGBTQ rights. In reality, age was a bigger predictor of voting outcomes on Proposition 8, with younger people supporting marriage rights across the board than those older than 65 (p. 256). California's Proposition 21, calling for laws to try juveniles as adults, was supported by more well-to-do gays and lesbians while LGBTQ youth opposed both Proposition 21 and 22, which was an antigay marriage law also on the ballot, illustrating the impact of the phenomenon and the categories with which we attempt to characterize it in their historical context means first, to elucidate its conditions of possibility and supports within a given mode of production (e.g., capitalism), and second, to investigate the historical processes leading to its capitalist form. (Gimenez 2005, pp.16–17)

This enterprise is distinct from looking for constructs within gender or differences between sexual categories. The context of relations of production comes first.

There are serious dangers in overlooking the larger picture to celebrate the local. First of all, the overreliance on technology as a social equalizer bypasses labor in favor of relying on consumer products to shape identity. Kuehn (2008) reminds us that all the social networking software in the world cannot overcome the fact that people live within real, live communities. While commodified identities can cross borders, most workers cannot. Secondly, a microscope on the local avoids dealing with subterranean assimilation. Sycamore (2008) and Jindal (2008) express concern about the response to 9/11 where many gays and lesbians donned the stars and stripes and endorsed by proxy American imperialism across the globe. Spade (2008) elaborates:

> Many U.S. activists still fail to make connections between their domestic oppression...and the manifestations of the consolidation of global capital in poor countries...if we understand ourselves as a population struggling against increasingly vicious forms of capitalism, which do their harm through vectors of race, gender, immigration status, ability, and age, how can we begin to envision a strategy that actually engages our struggle? (p. 51)

Indeed, Kelly (2006) questions how postmodernism can explain the current state of affairs of a majority of people in the world, let alone fully comprehend oppression in the first place.

No Alliances Are Possible

Gimenez (2005) and D'Emilio (1992) explain that when postmodernists were able to successfully divorce identity from the relations of production, other causes for sexism and homophobia had to be found to fill the void. For example, within feminism, patriarchy became the central causal factor of women's oppression, foreclosing any alliances women could have with men. Within queer theory, heterosexuality became the major oppressor along with marriage and the family attendant forces of exploitation for both women and LGBTQ individuals. With no economic explanation provided, an intense focus on outlining identities became the hallmark of post-al theories. Not surprisingly, with the major source of commonalities across exploited between men and women, dialectical analysis would examine the impact of declining overall wages in the face of escalating profits that are tied to the family and problems such as domestic violence and sexual harassment. Traditional accounts of women's experiences of sexual violence tend to stop at patriarchy or male privilege, ignoring economic factors.

Small-Scale Outlook

Part of the localizing effect involved with deconstruction involves dissecting social categories perceived to be essentialist, effectively cutting out the role of labor in the process. When labor is discussed, it's usually in the context of increasing the social status of a particular minority group, not equality overall (Bailey et al. 2008; Ebert 2005). Nowlan (2001) and Wolf (2009c) provide the example of AIDS, in which Marxist understandings of the impact of the disease on health care, wages, and civil rights were replaced by smaller reform efforts to ensure that more well-to-do gay men would have access to expensive AIDS therapies. In their reluctance to categorize groups for the purpose of articulating oppression and organizing for future action, postmodernists remain mired in details (Bourne 2002). For example, instead of calling for militant and organized responses to homophobia, the answer for postmodernists is to celebrate performance and self-expression in order to make spaces for difference. In focusing on such differences while ignoring the significance of the relations of production, postmodernists have effectively severed people from key organizing points (Ebert 2005; Gimenez 2005; Wang 2001). This is, in effect, a defeatist strategy along the lines of Bawer's (1994) thesis in *A Place at the Table* where all one can hope for is the space to be one's self. While some segments of the ruling class might be put off at open displays of homosexuality, this does nothing to disrupt the system at large.

Part of the focus on smaller scale issues and concerns has to do with an overall suspicion of solidarity and collective social solutions (Mann and Huffman 2005). Marxism is viewed as inadequate for accounting for sexual desire, racial oppression, or gender inequality (Ebert 2005). However, the capitalist economic relations of production are precisely where homophobia, racism, and sexism become salient and vicious, and not "simply a matter of asymmetrical power relations" (p. 40). As Cole (2009) explains, "One of the great strengths of Marxism is that it allows us to move beyond appearances and to look beneath the surface *and* to move forward collectively locally, nationally, and internationally" (p. 67). To build coalitions means understanding that people can indeed share experiences of exploitation across identities of race and gender but not across the categories of the working and owning classes. Confronting oppression means comprehending how capitalism shapes social relations:

> Marx's methodological insights suggest that we need to look at the inequality between men and women in their historical context...to place a

being addressed within higher education because of traditionally white and male researchers overlooking the private, domestic sphere (D'Emilio 1992; Mann and Huffman 2005). However, Mann and Huffman (2005) distinguish between the vision of early feminists in the 1970s academy who applied materialist analysis to the private sphere in order to link research to street activism and later identity politics feminism that currently focuses on "specializations and difference" within a more narrow range (p. 85). Personal experience narratives were viewed as the answer to the crisis of legitimacy of voice in qualitative research in that such accounts could "speak for themselves," putting forth the authenticity of the emic perspective over the etic. These narratives are also featured prominently in ethnic and gay and lesbian studies, a holdover from an era when much work of an intellectual nature occurred apart from higher education due to its outsider status (D'Emilio 1992, 2002).

Lynd and Grubacic (2008) assert that "if organizing is personal expression alone, more often than not it will fail to give rise to mass action" (p. 109). This is due to postmodernism being yet another social reform movement in its insistence on viewing larger social problems as smaller, local ones filtered through the lens of experience (Cotter 2001; Mann and Huffman 2005). While oppression is acknowledged in these experience narratives, the dialectical means of analysis is lacking. For example, racism has been placed "firmly in the subconscious where it cannot be reached by reasoned argument or rational analysis" (Bourne 2002, p. 207). When racism is viewed as an institutionalized feature of capitalism, it can be challenged and fought. But when it is personalized as part of "discourse," it can then be "negotiated" (Ebert 2005). Ending racism, sexism, and homophobia becomes a matter of making spaces where people can express their desires, not a matter of challenging capitalism per se (Wolf 2009c). Queer theory and identity movements become, then, a form of "performance politics" (Mann and Huffman 2005, p.72).

Among the limitations to placing primacy on personal experience, Ebert (2005) sees postmodernists ending up with a collection of "ethnographies of power" (p. 40) rather than any distinct analyses from which to advance a better society. For postmodernists, power is a floating signifier, but for Marxists, power is always connected to the point of production. Personal experience as presented by postmodernists is rootless when it comes to material conditions, being tied instead to the amorphous construct of power:

> Micropolitics is the politics of bypassing class and putting in its place lifestyle and consumption. It is a politics that erases any examination of the structures of exploitation, substituting instead ethnographical studies of the behavior of the subject in its multiple consuming relations. (p. 49)

As Gimenez (2005) explains, focusing on economic relations transforms identity categories into deeper theoretical constructions that do not stop at the intentions or expressions of people involved. In the case of differences

Other problems with minority status that D'Emilio isolates include relying on establishing "inborn" or "genetic" homosexuality as a tactic to winning civil rights or being considered oppressed. German scientists who argued in the 1930s that homosexuality was inborn did nothing to prevent—and in fact enhanced—Nazi efforts to locate and exterminate LGBTQ people. Sexuality is a fluid phenomenon, which is a likely combination of genetics and situational choice, making it difficult to define LGBTQ people as a distinct minority. To continue to make biology a requirement for membership in a minority means that the situational choices and the larger economic forces at work are ignored in favor of "born that way." For D'Emilio, to accept minority status as solely defined by genetics means that the entire concept of sexuality as a fixed entity has to be accepted as well as its attendant oppressions. Instead, there needs to be an insistence on the right of sexual determination with no contingencies based on biology or choice, much as is done with religious identity, which doesn't have to prove itself as genetically based to gain rights. However, to do this requires moving beyond the local to more global and universal notions of identity as it is fused by economic forces.

For postmodernists, the overthrowing of markers of identity is the path to liberation, as Mann and Huffman (2005) explain:

> The central idea is that identity is simply a construct of language, discourse, and cultural practices. The goal is to dismantle these fictions and, thereby, to undermine hegemonic regimes of discourse. (p. 63)

Identity politics, while aiding in obtaining important civil rights gains, has not ensured that these gains have been distributed equally (D'Emilio 2003). Those with more economic clout tend to be able to get ahead while others are left behind. Indeed, Pilger (2009) and McClaren and Farahmandpur (2002) point out that the most diverse administration in terms of cabinet appointments and Supreme Court justices was the Bush regime. Yet figures like Colin Powell and Condoleezza Rice elevated the ethos of empire to a fine art. There are three components of postmodernism and identity that create barriers to a revolutionary project: the centrality of personal experience, a small-scale outlook, and the impossibility of forming alliances across identity groups.

Centrality of Personal Experience

Beginning in the 1970s, the focus on personal experience emerged throughout higher education as part of an overall revolt against the detached, authoritative voice of the researcher, as with the Annales school's approach to re-envisioning history by featuring the accounts of ordinary people (Burke 1990). Though not tied directly to postmodernism, the use of personal narratives and discourses has been embraced by post-al theory, particularly from the vantage point of oppressed groups. Feminist researchers saw that many of their issues were not limits" (p. 13). Indeed, Gimenez (2005) concludes that despite the continual opposition from postmodern circles:

> Marx's intellectual power and vitality remain undiminished, as demonstrated in the extent to which even scholars who reject it must grapple with his work's challenge, so much so that their theories are shaped by the very process of negating it. (p. 12)

Localized and Micro Identity Politics

For postmodernists, the concept of identity is not a fixed or universal entity but consists of multiple personas, dependent upon specific, local, situations or consumer choices that are freely used to create the self (Anyon 1994; Bourne 2002; McClaren and Farahmandpur 2002). Of course, an acknowledgement of smaller, localized choices necessarily involves by default admitting larger social forces at work, but making generalizations about these forces is viewed with skepticism, particularly when it comes to class (Anyon 1994; Cloud 2001). In the cases of race and gender, Cloud (2001) argues that both of these factors are likely to be experienced outside of social class in some situations, but exploitation and oppression based on race and gender are always linked to material factors. Yet whenever Marxists attempt to go beyond celebrating local differences to analyze the economic conditions that create inequality, they are declared dogmatic and modernist, as Cotter (2001), a professor, found when female colleagues were uncomfortable at her openly connecting her sexual harassment to institutional and not personal factors. As soon as local struggles are transformed into larger global ones, postmodernists cry foul (McClaren and Farahmandpur 2002).

D'Emilio (1992) explores the quandaries involved in establishing exactly what goes into forming LGBTQ identity. Minority group status tends to be a problematic label when applied wholesale to LGBTQ people because historically speaking, homosexuality as an identity (as opposed to "homosexual" behaviors) is a modern invention. Arguing against a notion of a historically enduring LGBTQ identity/community, D'Emilio posits that capitalism itself, through "wage labor and commodity production" created the conditions that allowed people to form relationships outside of heterosexual norms, beginning in earnest in the late 1800s (p. 3). Yet the intersecting factors of race, class, and gender are used to the advantage of capital in creating and sustaining divisions and oppressions within the working class. LGBTQ people have not escaped this situation despite the post–World War II concept of identity politics being tied to the apparent uncertainty of consumer culture and purchasing power. During the Cold War era, for example, gays and lesbians were targeted at the precise moment they began to advance a collective identity with threats of disclosure and black listing; threats tied to economic sanctions.

language to critique a right wing social democrat, von Schwietzer (Wolf 2009c). Wolf does not excuse the language in the letter, but she points out that in referencing von Schwietzer as a pederast, Marx and Engels were not far off; he had indeed been convicted for seeking sex with a boy under the age of 14. Similarly, Emma Goldman's complaint about lesbianism within the movement is pointed to when she expressed a critique of movement politics rather than sexuality per se. Instead of the homophobia that postmodernists charge, Goldman's critique could be read as an attack on early identity politics rather than speaking out against lesbianism. The other source of postmodern opposition is Stalinist oppression, which is continually linked to Marxism despite the two being different philosophies of social organizing. Lenin in particular is linked to Stalin, even though his writings express a polar opposite expression of democratic values (and the fact that he was targeted by Stalin seems to belie the two working in tandem). As Wolf points out, the failure of the Russian Revolution to meet its goals had more to do with post-revolutionary difficulties surviving under internal and external duress than with the revolutionary project itself.

Pinar, one of the leading postmodern writers in education, eschews Marxism in favor of queer theory and opposes the notion of the centrality of the working class, as he explains in response to the early AIDS crisis (1998):

> At the time I was angry with the "macho Marxists" in our field, at their (apparently unconscious) insistence that power (i.e., the phallus) not gender (i.e., [heterosexual] women let alone queers), not race, construed especially as "class" (that same reactionary American working class, that, in a final act of self-hatred and self-immolation, had just helped elect Reagan president the year before) was the primary category of curriculum theory. (p. 23)

Pinar goes on to explain that at one point he felt the need to critique such leftists, but is more resigned now since "their tired, self-serving pseudo-Marxism has long been eclipsed by feminist theory and racial theory, joined now by queer theory" (p. 23). The assumed death of Marx underlies much postmodern research, as Cole (2009) outlines in his critique of critical race theory. With the "old man" conveniently out of the way, the post-Fordists can finally let it all hang out in their dismissal of the working class and class oppression in general.

In removing class from the picture, postmodern feminists and queer theorists have become nothing more than apologists for capitalism. Instead, one solution might be Ebert's (2005) proposition of embracing a "red feminism" that would address issues pertinent to the working class, including the domestic sphere and the family. Robertson (2008) envisions more salient theoretical work around the connections between educational practices and capitalism, which will enable us "to confront, overcome, and transcend those one of the few groups to unequivocally challenge capitalism. This is threatening to both conservative reactionaries and left-liberal apologists alike (Cole, 2009; Ebert 2001; Nowlan 2001).

Marx's (1998) critique of the young Hegelians of his day could be similarly applied to postmodernists, who,

> In spite of their allegedly "world shattering" phrases, are the staunchest conservatives. The most recent of them have found the correct expression for their activity when they declare they are only fighting against "phrases.'" They forget, however, that they themselves are opposing nothing but phrases to these phrases, and that they are in no way combating the real existing world when they are combating solely the phrases of this world. (p. 36)

Whether it is post-Marxist Haraway (1990) looking to the power of machines to reconfigure human society by reducing our need for community, or Tierney (1997) wanting to find space for queer identities somewhere between "anyone can make it" objectivism and "the dour straightjacket of orthodox Marxism or Freudianism that believes individuals and groups have little possibility for controlling their lives" (p. 24), the perception that Marxism is outdated and over-deterministic persists.

For example, consider the following excerpt from Pinar, Reynolds, Slattery, and Taubman's (1995) comprehensive postmodern anthology *Understanding Curriculum*. In declaring Peter McClaren's abandoning of postmodernism for Marxism a "crisis" for curriculum theorizing, the authors declare:

> It is the problem—indeed the crisis—of contemporary political curriculum theory: how to reconcile a view of politics, that, finally, has strikes and street barricades in mind, with a more complex view in which what we think and what we do, i.e. the realm of the symbolic, in a semiotic way, represent the location of political action, not the streets. To assert its own authoritativeness and moral and political self-righteousness, Marxism requires its proponents to dictate that *they* understand history, *they* embody morality and ethics; *they* dictate strategy. (p. 310)

Note that street action is viewed as not only less intellectual, but is dismissed altogether. This is followed by the common characterization of Marxists as narrow-minded and dogmatic. In their review of curriculum theory, Pinar and colleagues conclude that with the departure of Marxism as a viable theoretical frame, the endeavor of political analysis applied to curriculum will divide and disappear altogether, replaced instead by identity frames of race, gender, and sexuality.

Much of the stated opposition to Marxism stems from a letter between Marx and Engels in the late 1800s where both used overtly homophobic

D'Emilio (2002) goes on to explain his skepticism at the notion of being "born gay," asserting that the reasons someone is LGBTQ shouldn't matter:

> The gay and lesbian movement offers interesting interpretive challenges because of the way sexual identity takes shape. One is neither born nor socialized into it at an early age. The identity has to be discovered and accepted. One can choose to not accept it, or one can choose to hide it. At the very least, this suggests that the history of lesbian and gay politics is as much a history of the creation and elaboration of the self-conscious community and culture as it is a story of a social movement...the structure of the community marks a train on which the movement can operate, and the actions of the movement are continually reshaping the life of the community. (p. 237)

Definitions of sexual identity have themselves gone through an evolution with medical explanations dominating in the late 1800s to the 1920s, to "choice" in the 1950s through the 1970s (argued by both conservatives and leftists), back to biological rationales in the 1980s. This illustrates that identity categories—including those with scientific rationales—are dependent upon historical factors, tied to economic conditions. Queer theorists would be wise to pay attention to the shifts that occur with identity categories, rather than focusing only on deconstructing them. Wolf (2009c) concurs with D'Emilio's (2002) analysis that the reason scientific rationales to support "born gay" are sought by many on the left is that "it spares us from the horrifying idea that our sufferings could end if only we exerted enough will or had made a different choice" (p. 160). D'Emilio argues that the conditions that created identity politics are now radically altered. We are living in cynical times with an overall retraction of the social safety net, impacting the majority of the globe, requiring more complete and dramatic solutions than the deconstruction of identity can offer (Sears 2005). As Wolf (2009c) puts it, "despite the radical intentions and proclivities of many exponents and adherents of postmodernism and its political offshoots—identity politics and queer theory—these ideas do not arm people with a worldview that can overthrow the oppression that LGBT people face" (p. 168).

Postmodernism and Its (Dis)Contents

When confronting postmodernism, it is essential to understand that beneath the "pragmatism" and "openness" of poststructural theory rests a profound disillusionment with the perceived and actual failures of 1960s activist movements, as Wolf (2009c) explains. One only has to look at the laundry list of "former Marxists," including right wing media personalities such as Michael Savage and David Horowitz, to see that enhanced cynicism is part of the package, no matter how progressive and left-liberal one claims to be. Within and outside higher education, Marxists are targeted precisely because they are flourish. At the same time, the medical establishment began to carve out diagnoses of different sexual practices, including the ground-breaking release of Kinsey's (1998) research. After WWII, which threw people into same-sex living quarters away from parental supervision, a lesbian and gay subculture grew, though still largely closeted due to Cold War era witch hunts and the threat of economic persecution. It wouldn't be until the early 1960s and the growth of student resistance movements related to antiwar organizing that groups were formed to openly challenge homophobic policies, culminating in the Stonewall rebellion and subsequent activism of a Marxian bent.

Ebert (2005) views identity politics as "the latest formation of the subject under capitalism" (p. 38). Indeed, the development of globalization and neo-liberalism has resulted in the emergence of multiple identities and a focus on identity politics (both conservative and liberal) linked to consumerism (Mann and Huffman 2005). The struggles within sexual identity politics revolve around notions of essentialism (that all members of X group share similar experiences); social construction (that difference is primary and sometimes insurmountable); and identity as the key to liberation verses identity categories as oppressive (D'Emilio 2003; Mann and Huffman 2005). In particular, these struggles played out in feminist theorizing, with the eventual outcome of an overall moving away from Marxist and socialist feminist understandings of oppression to identity politics itself as a way of conceptualizing the exploitation of women and LGBTQ groups. This was partly due to issues raised by feminists of color whose race and social class concerns were overlooked by a largely white, middle-class feminist movement (Mann and Huffman 2005; Smith 2005). The answer was to focus on particular identities with growing assertions that alliances between identity groups were not possible (Cole 2009; Wolf 2009c).

D'Emilio (1983, 1992) concludes that we have to consider distinctions between homosexual *behaviors* and homosexual *identities*. Confusion over these differences has led some identity scholars to conclude that there have always been LGBTQ people throughout history. While at first the notion of the "eternal homosexual" led to potentials for liberation, D'Emilio (2002) feels that this has now constrained the movement for LGBTQ rights because it removes class from the historical equation. Instead, universities have focused on identity politics, stemming from the growth of black and feminist studies departments along with queer theory. The notion of a fluid sexuality might be embraced by the postmodern left, but this fluidity is not tied to historical conditions and is as free-floating as Foucault's notions of power. D'Emilio expresses concern that the issue of gay identity "is in danger of devolving into the search for the first homosexual" (p. 109). He goes on to challenge both identity and queer theorists to re-envision how LGBTQ history is framed. For example, instead of situating the historical account of gay males in theories of sexuality, D'Emilio suggests approaching it from gender or race against a backdrop of class, illustrating commonalities between groups.

obtain minority group status, which was particularly important in the wake of AIDS. The latest phase (1990s queer theory) views the closet as "sublime" and not such a bad place to be. As Morton describes, the closet was once seen as being a matter of economic and political relations, but is now promoted within queer theory as a "textual" matter wrapped up with personal desires (p. 9). A Marxist reading of the closet would instead place its significance related to historical developments, its location Morton refers to as "an annex of the system of wage labor; a place in which inequalities in the exchange of labor power for wages are justified by pathologising resistant sexualities" (p. 1). For Morton, "freedom" refers to universal freedom from need, not freedom for some to enact desires at the expense of others.

While once coming out was a political first step that signaled one's solidarity with others and choosing vulnerability over the protection of the closet (D'Emilio 1992), coming out is currently envisioned by queer theorists who "ultimately seek only to find a way to preserve a space to remain as queerly different as possible within contemporary late capitalism" (Nowlan 2001, p. 147). The sublimity of the closet also depends upon one having the means to sustain a divergent identity, including having access to the correct consumer goods needed to carve out one's space (Sears 2005). Morton (2001) elaborates:

> Coming out of the closet became a practice of the privileged whose ideological goal was to demonstrate under late capitalism there need be no connection between economic exploitation and sexual oppression, that it was indeed possible to 'make it' and be lesbian/gay/queer at the same time. (p. 18)

As a result, the LGBTQ working class is supposed to be grateful they are successfully employed and go no further than to frame their struggle as a culture war and never as a class war. Indeed, accounts of K–12 LGBTQ teachers express the inability to confront economic exploitation while in the closet (Beach 2005; Gold 2005; Ognibene 2005; Ray 2005). As Ray (2005) puts it, "teaching from the closet is a ridiculous notion; way too much of the best stuff gets buried in there" (p. 231).

D'Emilio's (1983, 1992, 2002) overview of the historical conditions leading to LGBTQ identity politics has been addressed in earlier chapters but is helpful to reiterate here. According to D'Emilio, there was no space for a homosexual identity as we think of it today in Colonial America, because economic conditions compelled people to form heterosexual family units. As soon as procreation was no longer a required outcome for survival, capitalism's requirements, or sexual relations, this radically altered the family as well as providing the possibility for a homosexual identity. Capitalism also shaped the movement of labor from agricultural settings to the city, contributing to the growth of urban areas where alternative sexual relationships could

Ellis 2006; Wolf 2009c). In the parallel second phase, sexuality was linked to Marxian concepts of liberation (D'Emilio 1983, 1992, 2002; Forrest and Ellis 2006). However, as the third phase emerged, postmodernists discarded dialectical materialism, instead calling for the deconstruction of heterosexuality to end oppression (Wolf 2009c). In both CRT and sexuality theories, the elimination of the centrality of class has been the most devastating outcome, leading to the narrow fight against "white supremacy" and "heterosexuality" rather than capitalism. One could also compose a similar three-phase evolution for female oppression, beginning with materialist analysis and ending with patriarchy being the target, as discussed in chapter five of this volume (Cotter 2001; Hickey 2006).

Stemming from the postmodernist third phase, queer theory was a result of the overt rejection of Marxist theory and tactics applied to sexual oppression (Nowlan 2001; Sears 2005; Wolf 2009c). Like other postmodern philosophical movements, queer theory is not monolithic, but a collection of approaches related to sexual identity with a focus on deconstructing sexual identity categories. As Sears (2005) describes, queer theory "is not a sexual politics, but an account of contemporary culture that begins with sexuality" (p. 100). Instead, Marxists view sexuality as emerging from economic forces. The medicalization of sexuality (arising from the needs of industry to categorize and train the workforce) beginning in the late 1800s and the advancement of social changes since WWII made it easier to shape and publicly present an LGBTQ identity. Yet the conformity and censorship in the culture at large made it hard to fully come out and thus to resist oppression (D'Emilio 1983). So while the medical model facilitated oppression, it also made it possible for individuals to forge gay and lesbian identities. This occurred alongside the medical model that was connected to the immediate needs of capital. Queer theorists seek to abolish medical and other definitions of homosexuality through continual deconstruction, largely viewing class as irrelevant. Instead of material analysis, queer theorists seek to disrupt heterosexuals (not capitalism) through performances highlighting alternative sexualities, with the thought that subversive sexual acts will alone end sexual and gender oppression (Morton 2001; Wolf 2009c). Cloud (2001) likens these utopian beliefs to conservative familialism where the traditional nuclear family is instead the queerly constructed self; both postmodernism and conservative family values being assimilationist bourgeoisie constructs in service to capitalism.

In addition to identity categories and identity politics, a major source of contention for queer theorists is the closet (Wolf 2009c). Morton (2001) summarizes the various ideological shifts that have occurred within queer theory regarding the significance of the closet. Beginning in 1969 and through the 1970s, coming out was viewed as an essential part of gay liberation as the closet had been a place of enforced hiding. The second phase involved making one's self visible and leaving the closet in order to achieve civil rights and valorizing consumerism, and promoting a libertarian-style rootlessness and individualism in a world apparently (if one takes seriously postmodern ideology) framed by constant "choice."

Basic Concepts: Postmodernism, Queer Theory, and Identity Politics

As Wolf (2009c) explains, postmodernism emerged in the mid-twentieth century as a varied collection of philosophies in reaction to modernist notions of objective knowledge, including abandoning the use of binary categories, structural analyses of exploitation, collective solutions, and metanarratives of any kind, (which includes Marxism in particular) to focus instead on the strategy of deconstruction and troubling (Mann and Huffman 2005; Nowlan 2001). Along with localized concepts of identities as mutually exclusive to any kind of social solution, the problem postmodernists address is power, which is not fixed within the economic system of capital, as Marxists would argue, but is free-floating; it is everywhere and nowhere (Wolf 2009c). According to McClaren and Farahmandpur (2002), the political dimension of postmodernism has endeavored to do the following in its locating of power: (a) treat culture as a factor apart from ideology; (b) remove social class as a central feature of cultural analysis; (c) support features of neoliberalism; (d) communicate that socialism is a dead-end strategy; and (e) assert that we are now living in a "post" world where differences are irreconcilable. While postmodernism has enhanced awareness of the many nuances of identity, the use of post-al theories applied to sexuality has the effect of limiting application to a dialectical materialist solution. In the case of education—the only common organizing point remaining in society today—postmodernism has not ensured that the state will guarantee nondiscrimination within and total access to K–adult settings (Kuehn 2008).

The evolution of critical race theory (CRT) in the wake of postmodernism illuminates what has occurred with theories of sexuality. Cole (2009) outlines three phases in the development of CRT, beginning with the late 1800s, when some authors theorized that while whiteness contained certain advantages, it was but one of many identity forms. In phase two, which escalated in the 1980s and 1990s, whiteness was transformed into a political identity linked to privilege and open to critique as a construct. The third phase contained a postmodern interrogation of whiteness along with calls for abolishing whiteness as a social construct, thus ending white supremacy. Cole's analysis is remarkably similar to what Wolf (2009c) identifies as happening with postmodernist theories of sexuality. In a parallel to Cole's (2009) phase one, early scholars of sexual identity viewed it as a combination of medical and psychological factors, creating the categories of heterosexual and homosexual in the late 1800s, along with identifying emerging advantages to being straight (D'Emilio 1983, 1992, 2002; Eisenbach 2006; Eskridge 2008; Forrest and of knowledge, by the paradox of knowing and not knowing, by the relations of the known and the unknown, the explicit and the inexplicit around homo/heterosexual definition" (p. 389). Claiming that today's classrooms are anti-erotic, Cavanagh argues that applying contemporary educational ethical standards of child abuse reporting to analyze teacher/student sexual encounters is inappropriate. The assumption is that at the time of the "relationship" the ninth-grade student was a willing participant. Yet Cavanagh fails to explain how, years later, the victim reads the incident as a negative one and labels it as harassment and abuse when conducting media interviews. Predictably, she compares the situation of the student and teacher to the 1962 film *The Children's Hour*, overlooking the fact that in this film the "scandal" was applied to the suspected relationship between two adult females who were school teachers, not a teacher and an underage student.

The reader gets the sense that because this is a same-sex scandal it should be overlooked, and that sexual harassment was not possible because women, after all, are on the receiving end of patriarchy and are not privileged in matters of sex. The assumption is that erotic behavior aimed at students is not inappropriate, especially if the student appears to reciprocate. Reacting to the testimony of the former student who reported feeling ill at hearing a certain song from the same time she was abused, Cavanagh (2008) responds:

> But this sickness is a hysterical reaction to contagious knowledge about female homosexuality: it cannot be about me, and if you say that it is, I will throw up. Heterosexual mainstreams refuse to acknowledge that maybe the teenage girls involved in the lesbian teacher sexual improprieties were less than innocent and naive... The sex might be abusive only in retrospect. (p. 394)

Summoning power feminism, Cavanaugh charges that the adult-onset memories of such events has to do with imposed victimhood rather than reality, which, in retrospect, may not have been all that bad. She concludes her analysis by positioning children not as innocent, but automatically queer beings, and expresses outrage at the suggestion that lesbians might be as capable of inappropriate sexual contact with a student as a heterosexual males are. She conflates outrage at this kind of sexual abuse with automatically supporting the stereotype of gays and lesbians as "recruiters."

This chapter will foreground a Marxist response to postmodernism and its offshoots of identity politics and queer theory. A discussion about such theories will begin the chapter, reviewing aspects of postmodern thought, and the historical developments leading up to the oppositional disciplines of queer theory and identity politics. Next, the rejection of Marxism by postmodernism will be explored, along with reasons behind such resistance. This will be followed by problematic aspects of postmodern theories, including a privileging of localized identities, a prioritizing of desire over material need,

Chapter Four

Identity Politics: Limits of Postmodernism and Queer Theory

In 1993, a 31-year-old Canadian female disclosed that she had once been sexually abused as a ninth grader by one of her female high school teachers. Cavanagh (2008) deconstructs the details of the situation and other related cases in her article, "Sex in the Lesbian Teacher's Closet: The Hybrid Proliferation of Queers in School." Instead of being troubled by the unequal dimensions of power and authority behind teacher and student relationships, which contributes to the flourishing of conditions for sexual harassment, Cavanagh targets "child protectionist" discourse as the source of the problem:

> Queer time does not respect chronological age and generational differences that organize thought about the "child," the "teen," and the "adult." Part of what it means to be queer is to upset heterosexual life-stages. The coming-of-age stories we tell about who we are authenticate our gender identities and naturalize our sexual orientations. Queer tales of seduction in school upset the nature and linearity of heterosexual life-stories; they are out of time and place, rendered perverse, abusive, and criminal. (p. 389)

In "complicating" the relationship between a teacher and a student, Cavanagh insists that to view the teacher's actions as abusive is to interfere with the queering of students and is tantamount to censorship. While certainly much of the reaction and hype centered around sex scandals have to do with the fact that they involve sex and should be critiqued on that account (as with outrage over the exposure of Janet Jackson's breast during a half-time football show supported by erectile dysfunction drug ads), Cavanagh's (2008) defense of the teacher's actions are puzzling, unless one understands the logic behind postmodern and queer theories.

Channeling the rhetorical style of Donald Rumsfeld, Cavanagh asserts that "the sex scandals involving lesbian teachers are saturated by the problem established and trade enters the picture, "the family ceases to be necessary for the functioning of society and thus loses its strength and vital capacity" (p. 225). Thus we have the family working within an enforced capitalist system and alienated from its own labor. This immediately resulted in the secondary status of women and the privatizing of child care and domestic work via the role of the housewife. With no property of her own and thus no power, women were reduced to marriage for reasons other than love or free choice, or had to turn to prostitution. The legitimacy of children also became of paramount concern with marriage ensuring that the rearing of the future workforce would be conveniently handled by isolated and individual families.

Kollontai (Holt 1977) was one of the few Marxist writers to speculate on what families would look like under communism. Reminding her readers that the family has always changed forms in response to social conditions, we can expect no less under socialism. Suggesting a fusing of the best features of the old and the much needed new, Kollontai stressed that a critical starting place would be the socializing of domestic labor, as with collective housekeeping (communal kitchens and laundries with professional house cleaners), a view shared by Trotsky (1970). Child care and education would become a state matter, no longer privatized, a view shared by Andrews (1889). We do have historical examples of Works Progress Administration nursery schools and on-site, 24-hour child care facilities that existed during World War II when women had to take factory work (Yalom 2001). Though these factory child care centers were corporate sponsored, their existence demonstrates that the structural organizing of 24-hour care is possible.

Croll's (1986) historical analysis of post-revolutionary Russia in 1917 indicates massive changes that impacted the family, including laws to protect pregnant women, access to reproductive health care, no-fault divorce, and abolishing of laws discriminating against homosexuals. One of the first barriers to this vision, however, was the unevenness of distributing socialized services; only a small portion of rural areas were equipped with nurseries, for example. Another barrier was the inability to foresee just how expensive it would be to fully implement the socialization of domestic labor, and initial efforts were often poor in quality due to the lack of material resources that Russia had at its disposal. Today, we have the resources and materials; it's a matter of wresting them from the system of profit.

In chapter 4, we move from an analysis of the family to postmodern readings of identity, sexuality, and society, with an eye toward K–12 education. Since the late 1980s, postmodernism has been the dominant framework for understanding sexual identity. However, Marxism is beginning to take on these issues while challenging the notion that postmodern performances of self can exist apart from an economic context.

into the workforce replacing the need for full time housekeeping. Gilman envisioned communal kitchens and child care centers that would facilitate the obsolescence of privatized domestic work.

In one of Gilman's (1994) most provocative passages, she addresses what she sees as the false consciousness surrounding the nuclear family:

> That the masses of the people do not notice existing conditions, and that they are not pleased with those who do. This is the one strong reason why the sexuo-economic relation passes unobserved among us, and why any statement of it will be offensive to many...it is easier to personalize than to generalize." (p. 81)

Echoing Engels (1972), Gilman targets the nuclear family form as disrupting solidarity since it places people into little isolated groups concerned only with their wealth and paternal children. Within these small groups, the interaction of sexuality with economics furthers along the effects of alienation. On the one hand there is alienation between men and women within the nuclear family, while on the other there is the family itself being alienated from other families. The political manifestation is a distrust of socialist ideas, since "the popular mind is so ready to associate socialistic theories with injury to marriage" (p. 115). Since marriage disrupts collectivity, then people conclude that collective solutions, diverse family forms, and so on will automatically threaten marriage.

Trotsky (1970) noted that one of the challenges of establishing class consciousness was the conservative nature of the nuclear family form: "Inertia and blind habit, unfortunately, constitute a great force. And nowhere does blind, dumb, habit hold sway with such force as in the dark and secluded inner life of the family" (p. 35). Part of the difficulty concerned this cellular/atomized structure which created barriers to collectivity. Trotsky actually viewed the task of dealing with the family as "infinitely more arduous" than the larger aspects of revolution (p. 24). Another problem was the lack of examples of egalitarian relationships for married couples to follow, making it difficult to envision both the transformation of the family alongside the overthrow of the system of profit. Motherhood in particular was targeted as a special role in need of protection at the same time that equality of women was being sought. As Trotsky maintained, "Motherhood is the hub of all problems...each practical step in economic and social construction must also be checked against the question of how it will affect the family" (p. 55).

Kollontai's essay, *Theses on Communist Morality in the Sphere of Marital Relations* reveals similar diagnoses (Holt 1977). Prior to the ownership of private property, the family performed the bulk of the economic functions necessary for survival and the formation of civilization. Patriarchy was nonexistent and women were not economically dependent upon men; everyone was part of the labor structure needed for survival. When social classes become for common use, placing it in the hands of the people. Native American and other tribal cultures were also surviving examples of earlier family forms and living proof that private property ownership wasn't a requisite for a functioning civilization and that there is always the promise of things changing:

> The society which organizes production anew on the basis of free and equal association of the producers will put the whole state machinery where it will then belong: into the museum of antiquities, next to the spinning wheel and the bronze ax. (p. 232)

Bebel's (1910) dissection of marriage and the family in *Woman and Socialism* follows Engel's (1972) outline of the development of private property and its impact on the household, from what he considers Matriarchate (pre-class society) to Patriarchate forms beginning in the ancient era. His frankness about not only marriage but also prostitution, abortion, reproduction, and sexuality made him at risk for censorship, even by fellow socialists. Arguing that women and men should be viewed as equally sexual, he made the connection between such freedom and economic constraints:

> Of course, it is impossible to assert the free choice of love in bourgeois society, as we have shown by our entire line of argument, but if the community were placed under similar social conditions as are enjoyed today only by the few who are materially and intellectually favored, all would have the possibility of a similar freedom. (Ch. 28, par. 8)

Only until people could be free of capitalism, could they be able to freely enter into a union with another. This is a different line of thinking than expressed by the libertarian anarchists, who believed that free love could prefigure a better society.

Gilman (1994) grasps at the roots of female oppression by pointing out that humans are the only species where the female has to depend on the male for a constant source of food and other necessities. Although writing during the late 1800s, her analysis is applicabile today. Locating the source of female oppression in "sexuo-economic" relations (p. 23), Gilman also puts forth the notion that there is a social class divide among women, with more well-to-do women doing less work than poorer ones. Here she separates mother work from housework, positing that "it is not motherhood that keeps the housewife on her feet from dawn till dark; it is house service, not child service" (p. 20). Because the family involves both sexual as well as economic relations, both forces have combined to create what Gilman terms the abnormality of sex divisions in the human race. In Yalom's (2001) discussion of Gilman's measured idealism, it was industrialization that prompted important changes to the household. Just as the machine taking over farming freed people from intense labor, so too was the movement of women nuclear family form wish to interpret. To be considered promiscuous, there would have to be some sort of inhibition or prohibition, and during this era, no such sexual restrictions existed.

The second development, the Punaluan family, began to inscribe incest taboos on children of the same mother and also not allowing marriages between cousins. Because it was easier to establish maternity, emphasis was placed on maternal descent, not paternal. Engels speculates that this family form existed among all peoples and across all cultures, including early ancient societies. In both the Consanguine and Punaluan families, there were no gender divisions or socially significant roles attached to either sex. In other words, even if males hunted, it didn't mean that men were "protectors" and women were "weak" or "gatherers." The notion of hunting extended to all of the phases of the activity, from setting up camp to scoping out animals, to killing, skinning, preparing, eating, and celebrating.

Eventually these earlier forms blended into the Pairing family, which included the primary husband and wife, along with secondary wives. Missionaries would misinterpret Pairing families as adulterous when all involved were considered part of the family. As civilization developed, it became more difficult to maintain the ancient family forms, as Engels (1972) explains:

> The continuous exclusion, first of nearer, then of more and more remote relatives, and at last even of relatives by marriage, ends by making any kind of group marriage practically impossible. Finally, there remains only the single, still loosely linked pair, the molecule with whose dissolution marriage itself ceases. (p. 112)

Thus, we have the "nucleus" of the nuclear family! The Pairing family became the Monogamous family, based on the essential components of patriarchy: strict division of male/female roles, devaluation of female roles, obsession with legitimacy of heirs, the double standard (women are to be monogamous, but not men), and the financial dependence of women on men. Bell (1986) provides a more contemporary example of this occurring with colonization in Australia, as Aboriginal cultures took on the sexist practices of Europeans. Similarly, Shashahani (1986) maintains that economic factors such as the displacement of smaller businesses and farms by outside global investments have led to an intensification of patriarchal household relations. Typically, poverty and female oppression are blamed on some amorphous concept of backward cultural traditions rather than the impact of capitalism.

Engels (1972) continues by reviewing features of ancient civilizations, such as Greece, and how class society and the establishment of private property further entrenches the nuclear form. Only a brief respite in the form of so-called barbarian societies represented a departure from class-based society. For example, Engels relates how the German barbarians seized Roman lands

(Ruiz 2006). Antonio (2000) posits that figures like Buchanan are able to appear at such venues with little resistance because cultural politics have been placed over class as a form of social analysis, shifting "politically oriented collective action and social criticism from a primarily modern, material, universal plane to a postmodern, discursive, local one" (p. 51).

It doesn't take long to draw connections between this form of right wing anarchism and homophobic beliefs; indeed, Troy Southgate, one of the key ideologues of the National Anarchists, suggests the establishment of "queer villages" for homosexuals since they engage in "unnatural" sexual practices that quickly devolve into pedophilia and other social ills (Sunshine 2008). Southgate claims to be more open-minded than many religious groups in that he doesn't call for the extermination of LGBTQ individuals, only that they live *separately* from the rest of society. In attempting to create a better society, activists must be mindful of the company they keep.

Materialist Readings

For materialist writers, the nuclear family is not a static entity, weathering historic change. Nor is it just one of many coincidental family forms with as much a chance of appearing as any other. Instead, as Wang (2001) explains, the nuclear family "needs to be situated in the capitalist mode of production and analyzed accordingly" (p. 235). Part of what makes the nuclear family form essential to capitalism is its function in reproducing the future workforce, while privatizing all the accompanying expenses. Labor always plays the essential role (Engels 1972). The nuclear family also promotes and entrenches the system of private property, by the legal and social function of marriage, which serves to legitimize some people at the expense of others, thus benefiting the ruling elite by ensuring its continued dominance (Beneria and Sen 1986; Hickey 2006). By understanding the significance of the nuclear family, we can begin to broaden the scope of our struggle to include multiple fronts, not remain mired in the ineffectiveness of identity politics (Bailey et al. 2008).

Led by Morgan's (1877) *Ancient Society*, a ground-breaking materialist analysis of prehistoric family forms, Engel's (1972) *The Origin of the Family, Private Property, and the State* refutes, once and for all, the notions (convenient for the ruling class to sustain) that patriarchy, class society—and even war—were enduring and unchanging features of society. Leacock (1981) would continue to challenge these ideas emanating from the scientific cloak of evolutionary biology in her *Myths of Male Dominance*. Engels (1972) begins by presenting the development of the family, starting with the Consanguine form. These early families featured what could best be described as group marriage where everyone was essentially someone else's family member and there was no such thing as illegitimate children. Engels makes it clear that this is not that same as promiscuity; something that those adhered to the reveals a key weakness of anarchist ideology—and the pitiful state of the U.S. antiwar movement (Suber 2007)—rather than its evolution. What are correctly perceived to be excesses of the state (i.e., Patriot Act surveillance, globalization, illegal search and seizure, the deadly raids on the Weaver family in 1992 and the 1993 Branch Davidian compound in Waco, Texas) are due to the intensification of government power as a result of global capitalism and the protection of profits, not the government acting independently of the system of capital. Berlet (2009) points out that even Hitler's national socialists once proposed a fusion of left and right in order to "transcend capitalism and communism" (p. 17). By targeting the state itself and building movements around its victims, anarchists run the risk of embracing white supremacists and other violent individuals. Berlet continues:

> Victims of government abuse are not necessarily our role models. They deserve our support against the abuse, but not automatically our support in strategic alliances where their views undermine our existing alliances as a progressive coalition with women, people of color, immigrants, and others. (p. 18)

Sunshine's (2008) analysis of British National Anarchism finds that the movement has become worldwide, mostly assisted by the Internet, and membership is growing. Some of the National Anarchist's recent causes include protesting on behalf of Tibet at the 2008 Olympic torch relay, anti-Zionist marches for solidarity with Palestine, and environmentalism. Indeed, Third Position politics with its rejection of both Marxism and capitalism in favor of racist socialism "have the potential of playing havoc on social movements, drawing activists out from the Left into the Right" (p. 12). In many respects, postmodernism sustains the development of such groups, particularly the concept of autoreferential culture (Antonio 2000). This logic proclaims that "culture operates according to its own autonomous logic, free from modern theory's formative sociological substrates" (p. 50). Antonio also identifies anti-universalism—a rejection of the notion of values across cultures—as a philosophical underpinning behind reactionary tribalism and the reasoning behind right wing anarchist appeals to end affirmative action and workers' rights. Though many postmodernists critique these types of extreme relativistic stances, the ideas are lasting and have infiltrated anarchist and other leftist movements.

Another key motivating ideology behind this form of fascism is the notion of an anarchist society rising out of the ashes of decay and decadence caused by excessive government and minority groups. Sunshine (2008) provides the example of Pat Buchanan, an American paleoconservative presidential candidate and media figure speaking at a Teamsters Union demonstration against the International Monetary Fund and World Bank. Buchanan often invokes anti-immigrant rhetoric juxtaposed with "the decline of civilization" analysis control of both sexes—fell victim to both factionalism and outside attacks by those opposed to their principles of Complex Marriage and women's equality (D'Emilio and Freedman 1997; Hillebrand n.d.). Kaplan's (1995) account of the underground abortion service Jane, relates how eventually the collective's rejection of centralized leadership put the group's underground status in jeopardy while at the same time making it difficult to provide efficient health services to desperate women. With many women in the group rejecting a clear-cut message of connecting reproductive freedom to economics for fear of appearing too "doctrinaire," activist burnout was quite common.

Indeed, Marx (1998) was quite critical about utopian "back to nature" experiments and what he called "monastic economies," declaring them to be yet another facet of bourgeois thought that has existed throughout history (p. 84). In attempting to create free love settlements, anarchists were simply replicating the idealism of the philosophers of the ruling elite. Anarchism can also become a barrier to social change by declaring any social/collective philosophy/action as potential dogma. In this equation, communism is as bad as fascism, the problem being the all-consuming belief systems of both. Taking aim at Max Stirner's "wheel in the head," Marx creates a tongue-in-cheek analysis of this line of thinking:

> Once upon a time a valiant fellow had the idea that men were drowned in water only because they were possessed with the idea of gravity. If they were to get this notion out of their heads, say by avowing it to be superstitious, a religious concept, they would be sublimely proof against any danger from water. (p. 30)

In addition, Marx emphasizes that communism isn't merely a political ideal that has to first be established (i.e., prefigured) in order for the rest of society to follow suit. Instead, communism is the only movement "which abolishes the present state of things" (p. 57).

One major flaw of anarchism is the location of societal problems within the state rather than within the system of capital (although many anarchists also openly oppose capitalism). Anarchists believe that the state is the source of multiple problems and that its removal would mean the end of various forms of oppression. Unfortunately, this has led to recent proposals by anarchists to "unite" both the left and right wings of antistatist movements to combat the excesses of the state, such as with the antiglobalization movement (Antonio 2000; Berlet 2009; Sunshine 2008). Similar to some antiwar activists rallying behind the 2008 presidential campaign of former libertarian and homophobe Ron Paul (oddly enough, one of the two candidates of either major party to openly oppose the invasion and occupation of Iraq; Dennis Kucinich was the other), calls for leftist anarchists to overlook some of the "eccentricities" or ideological differences between themselves and white separatists, devotees of Ayn Rand, Posse Comitatus, and the militia movement

Lynd and Grubiac (2008) address the relationship between anarchism and Marxism regarding social organizing. Grubiac asserts that anarchism is needed to prefigure, or create models of how a new society could be organized even as these microsocieties remain ensconced within a global capitalist system. Lynd finds himself defending Marxist theory as a way for activists to understand how society is structured. Both seek to define a new anarchism that fuses both perspectives, as exemplified by the Zapatistas who have publicly declared that they are not interested in taking power from the state. Both Grubiac and Lynd provide several examples of how prefigured society has worked in different situations, from the Lucasville prison uprising to the Mississippi Freedom schools, to Lynd's own experience living in the Macedonia cooperative community, culminating in Lynd's account of the International Workers of the World (IWW) manifesto:

> The working class must organize itself in such a way...that whenever a strike or lockout is on, affecting one group of workers, all its members in any one industry, or all industries if necessary, can strike together...By organizing industrially, we are forming the structure of the new society within the shell of the old. (p. 16)

In these situations, a reorganization of authority had to be created in order to sustain the resistance, such as racist factions between Lucasville prisoners being set aside for the revolt and workers socializing the provision of meals, child care, and support during long strikes such as teachers' action in Oaxaca. For example, the Zapatistas endorse the concept of providing material support for their leaders so that they are free to take on additional responsibilities. This is one example of accompaniment as praxis, where people work side by side in the trenches, so to speak, rather than working solely on high theory. Instead, anarchist principles would be applied to Marxism, Lynd and Grubiac propose, though they speculate as to whether different social classes could achieve it.

Lynd and Grubiac (2008) also address the low rate of survivability of anarchist experiments. Even trade unions, which were set up to prefigure how social organizing could be transformed post capitalism, eventually became another arm of the state and a "partner" to business. Thousands of prefigured societies were launched, such as Frances Wright and Robert Owen's Nashoba, a community built in the 1820s to prefigure the end of slavery (Best 2003) and John Humphrey Noyes's Oneida which operated from 1849 to 1879 (Hillebrand, n.d.) meant to transform the oppression inherent in marriage. Best (2003) outlines how slaves were to earn their freedom by working to build Nashoba. However, instead of Nashoba serving as the model for social change as its founders intended, the system of slavery itself was microreplicated within the settlement, with exploitation and harsh punishments occurring on a regular basis. Likewise, Oneida—even though it was run via communal presented as unchangeable, when, in fact, it has always reflected the interests of the dominant class by creating a set of scapegoats, even among those of the same social standing:

> In this manner, Piety is made to signify Zeal for the Church or a Sect, Patriotism, Loyalty to a Sovereign, and Purity, Fidelity to the Marriage Bond. In the same manner, Irreligion is identified with Heresy, Treason with the Rights of the People, and Debauchery with the Freedom of the Affections. It suits the Bigot, the Despot, and the Male or Female Prude to foster this confusion of things dissimilar, and to denounce the champions of Freedom as licentious and wicked men—the enemies of mankind. (Intro, par. 9)

Andrews provides the example of adultery, which, during the late 1800s, was pretty broadly and inconsistently defined. If by "adultery" one meant two unattached people deciding to engage in a sexual relationship, then that definition was unfairly given, as was related by labels such as "impure" or "improper": "sexual Purity...is that kind of relation, whatever it be between the sexes, which contributes in the highest degree to their mutual health and happiness, taking into account the remote as well as the immediate results" (Intro, par. 37). For Andrews, the nature of sexual arrangements was also linked to economic relations, as he posited in his proposed Law of Equity. Currently, the economic system of capitalism was not equitably distributing wealth, while marriage arrangements were also unequally distributed between genders along with the social protections that marriage provides. Goldman (1911) likewise argued that marriage "...makes a parasite of woman, an absolute dependent" (par. 23). The ruling class opposes free love because without the bonds of marriage, "Who would fight wars? Who would create wealth? Who would make the policeman, the jailer, if women were to refuse the indiscriminate breeding of children?" (par. 29).

Heywood (1876) carries Andrews' (1889) analysis further by pinpointing the root causes of inequality to the government's role in ensuring that most of the wealth remains in the hands of a few by laws and policies governing property. He does bring a Marxian analysis to his own anarchist views, including seeing labor as the source of wealth in the essay *Yours or Mine*. While he acknowledges that the need for collecting rent will expire when the state no longer sanctions property ownership, he prefigures various solutions in the interim. In his description of renting, if a tenant maintains a property in good condition, then the landlord should expect no money; it is enough that the tenant was a good steward to begin with, since property, over time, decays and requires upkeep to maintain its value. He also advocates a local form of exchange of services, since it is impossible for one person to perform all of the services necessary for society to function. On a smaller scale, child care could be exchanged for house repair and the like.

Echoing Lazarus (1852), Carpenter (1894) separates love—something more reverential and supportive of human freedom—from marriage, which he describes as imprisonment (though not as directly as Kollontai [Holt 1977] who described marriage as resulting in both parties seeking vodka for solace) as this lengthy vignette of two typical young people in love illustrates:

> They marry without misgiving... It is only at a later hour, and with calmer thought, that they realize that it is a life-sentence... not reducible (as in the case of ordinary convicts) even to a term of 20 years. The married life, in so strange and casual a way begun, or drifted into, is hardly, one might think, likely to turn out well. Sometimes, of course, it does; but in many cases, perhaps the majority, there follows a painful awakening. A brief burst of satisfaction, accompanied, probably through sheer ignorance, by gross neglect of the law of transmutation; satiety on the physical plane, followed by vacuity of affection on the higher planes, and that succeeded by boredom, and even nausea; the girl, full perhaps of a tender emotion, and missing the sympathy and consolation she expected in the man's love, only to find its more materialistic side—"This, this then is what I am wanted for ?" The man, who looked for a companion, finding he can rouse no mortal interest in his wife's mind save in the most exasperating trivialities;—whatever the cause may be, a veil has fallen from before their faces, and there they sit, held together now by the least honorable interests, the interests which they themselves can least respect, but to which Law and Religion lend all their weight. The monetary dependence of the woman, the mere sex-needs of the man, the fear of public opinion, all form motives, and motives of the meanest kind, for maintaining the seeming tie; and the relation of the two hardens down into a dull neutrality, in which lives and characters are narrowed and blunted, and deceit becomes the common weapon which guards divided interests. (pp. 6–7).

According to Carpenter, part of what leads to this veil of ignorance surrounding marriage is the unwillingness of society to fully educate young people about sex, in particular females. The focus should be on sexual compatibility and love, rather than on social institutions such as state-sanctioned marriage (Goldman 1911; Fone 2000). Because traditional marriage rests mostly on the outside pressures of church and state, couples do not prioritize the love relationship, making the marriage itself more vulnerable since its foundation is primarily external.

In *Love, Marriage, and Divorce and the Sovereignty of the Individual*, Andrews (1889) outlines a series of dialogues between himself, Henry James, and Horace Greeley. Arguing that the world has been bound by "three cords of superstition,"—the church, government, and marriage—Andrews postulated that just as the church's hold was being shattered in the wake of the Protestant Reformation, so too was the family and government being challenged by new theories of socialism and science. Marriage, like religion, is communities have common features such as equitable workloads, communalization of domestic labor, and shared democratic governance structures. Even within conservatively religious and polygamous Mormon families, women mentioned the sharing of household tasks as the highlight of this kind of marriage arrangement (Coontz 2005; D'Emilio and Freedman 1997; Yalom 2001). The primary difference between the anarchist and social purity movements was that the former was opposed to any type of state entanglement in matters of relationships while the latter tolerated differing degrees of official sanctioning of marriage (D'Emilio and Freedman 1997).

Lazarus (1852) was one of the earlier proponents of the free love movement, which, contrary to conservative portrayals, wasn't about having multiple sexual encounters, but of advocating that people should be free to love whomever they wanted to love, without societal constraints such as marriage. It is no surprise that many conservative familialists, such as Gilder (1986) trace the decline of marriage from its previous function as a social obligation to becoming a relationship where romantic love is the primary incentive. In *Love and Marriage*, Lazarus, much as Thoreau expressed in *Walden* (2005) or Rousseau in *Social Contract* (1762), believed that people and relationships were at their best when they more closely followed nature. As society developed, so did the many things that enslaved humanity. Lazarus (1852) also argued that children of both sexes should be encouraged to form mutual friendships, mostly as a mediating effect against institutionalized schooling. His harshest critique, however, was reserved for legal marriage:

> The unnatural constraint which our marriage institution imposes on the parties, engenders, by reaction, libertinism among those who keep aloof from it, and adultery among those who have accepted it without submitting to it, as it cramps and falsifies the natures of those who submit themselves to it in good faith, exception falling upon the few cases where the parties are spiritually mated to each other. (p. 26)

To Lazarus, marriage created the "sin" of adultery, placing undue hardships onto people as they seek "to escape the compression of the paralyzing marriage form" (p. 27). He viewed marriage as part of a larger, destructive tendency of society to get people to see their own interests as intertwined with the state, where he advises that, "A *man* should never place his powers at the disposal of another man: for the defense of our country virtue suffices to animate...Shall it be the love of his relations, who, perhaps, are persecuting him, or that of a country whose laws oppress him, and whose interests besides are unknown to him?" (p. 34) Likewise, marriage, according to Lazarus, artificially forces people to submit to something that runs counter to their best interests, a view shared by Goldman (1911).

dog-eat-dog sensibility (p. 762). One can easily see the corporate-sponsored Tea Party members and their leader, Glenn Beck, represented in "the crank," a fifth typology that consists of isolated, enraged pockets espousing conspiracy theories (i.e., President Obama not being born in the United States, the secret rule of the government by the Federal Reserve, etc.) (p. 765). Last, we have the most dangerous individual, "the manipulative type," whom Adorno found to be prevalent in the business sector. These people have a detached, mechanical view of life, making them the most merciless of authoritarians. As Adorno maintains, they are the ones to design, not operate, the gas chambers. The only moral compulsion these individuals have is to the concept of loyalty, which they hold above everything else. Here, former Vice President Dick Cheney fits the bill.

In Lakoff's (2002) nation-as-family metaphor, two different perspectives on authority emerge, leading to key policy differences between liberals and conservatives. One is a nurturing morality, centered on parents providing opportunities for children to develop self-control through open exploration of different options. The goal in these families is to develop a person who cares about others and seeks to redress inequality. The other morality is strict-father where obedience and loyalty are the top values learned. The goal is to produce children who will be self-sustaining adults, generally focused on taking care of one's own, no more, no less. Even though conservative beliefs about parenting and child rearing are not supported by research, a pervasive adherence to the necessity of suffering for the greater good ensures their dominance. Lakoff asserts that the contradiction between authoritarian households forming a psychic dependence on authority creates a form of projection where one insists on government staying out of their business. Because conservatism locates the sources of problems in individual people, not social structures, it makes no sense to bring class into the discussion. If bad things happen to people, it is because they must have done something wrong considering that the system of free enterprise is (a) fair and (b) obtainable by anyone. In Lakoff's view, liberals have not been as effective in using logic to counteract these concepts so other strategies need to be used to engage conservatives. Reich (1971) would waste no time with this task because fascism is essentially irrationalism, which can't be confronted in traditional ways. It first has to be properly comprehended in order to be overcome.

Libertarian Anarchist Perspectives

Trahair (1999) provides no fewer than 600 research-supported entries for utopian communities, individuals, and movements worldwide from ancient civilizations to the present. As D'Emilio and Freedman (1997) relate, these settlements ranged from religious-based, such as the Shakers and contemporary Bruderhof/Church Communities International to anarchist/socialist collectives like the Icarian Movement and La Reunion. Most of these internal contradictions, leading to a rise in tension, as Adorno explains in this passage:

> The individual who has been forced to give up basic pleasures and to live under a system of rigid restraints, and who therefore feels put upon is likely not only to seek an object upon which he can "take it out" but also to be particularly annoyed at the idea that another person is getting away with something. (p. 232)

The ultimate aim is an individual who both identifies with those in power and at the same time is unable to criticize those who create great harm. There is also the contradiction of being ever-submissive to authority, but disparaging of the weakness inherent in the act of submitting. Projection also follows, as in the irrational need to punish those seen as violating sexual or gender norms, most likely resulting from those same feared tendencies within themselves. Projection is also commonly aimed at welfare recipients, who are viewed as parasitic and needy, when all along the authoritarian individual is one of the most dependent personalities around. If one can externalize these tendencies away from one's self, then they can be categorized as someone else's problem and neatly filed away. In addition, reactionary individuals not only believe in administering punishment, they also seem to enjoy doing so, whether or not the punishment is viewed as legitimately administered. As children, authoritarians received arbitrary beatings, contributing to an uncertain climate; one can get hit at any time. Thus we have accounts of abused children over-idealizing the very same parents who at any time could do them harm. An attitude of anti-weakness is very intense, especially among males. The need to present one's self as "overcoming the odds" is prominent and relentless, stemming from the need to be on guard as a child, to put on a good face at all times.

Adorno's (1950) typologies of authoritarianism are particularly illuminating when considering today's media. The first, "surface resentment," seeks to blame problems on minorities, women, LGBTQ individuals, immigrants, and so on (p. 753). Here we have the media figure of Lou Dobbs who constantly rails against illegal aliens taking away jobs and government benefits from hard-working Americans. A second typology is "the conventional syndrome," where one prides oneself on the control of impulses and overall holiness (p. 756). The conservative who "has gay friends" but stops short of approving legalized marriage rights applies here. Third, there is "the authoritarian syndrome," one who gains happiness from identifying with authority (p. 759). These individuals manage to transform even apolitical weather events like Hurricane Katrina into deserved punishment for wayward sinners (as in TV evangelists identifying Katrina as a consequence of homosexuality). Fourth, "the rebel and the psychopath," are represented by reactionary attitudes that form in settings such as prisons, embracing a Machiavellian

For Reich (1970), fascism is "the sum total of all the irrational reactions of the natural human character" (p. xiv). There is something in the structure of fascist ideology that transforms people to accept beliefs that are contrary to their economic interests. A revolutionary consciousness, for example, seeks to liberate everyone from immediate suffering, focusing first on practical needs. A reactionary consciousness, on the other hand, will immediately try to halt revolutionary thought by retreating to "natural law" or some form of conservative reasoning as to why it is no use to try to change anything. Freedom from mysticism is an ever-present threat. The problem, however, is that the reactionary "is more keenly aware of this threat than the revolutionary is aware of his goal" (p. 140). By the time that leftists figure out what they want society to resemble, the fascists have already gotten out their message via well-funded propaganda, finding a ready audience in a populace who has been taught to respond to fear.

Sexual sadism is an offshoot of repression, often combined with the trappings of patriotism, sexism, and racism. Davis (1981) describes in great detail the sexual component of lynchings during the Jim Crow era. It is no coincidence that just as rape was a tool used by slave owners to dominate black women, lynchings were bound up with images of white women being violated by black men. In his description of the United States' endorsement of torture, Schell (2009) says that acts of torture "permits one person's body to be translated into a regime's fiction of power. The victim's world is shattered by torture" (par. 10). At the same time, the person doing the torturing, "expands his world—his word—at the expense of the shattered world of the victim … the torturer and the regime have doubled their voice since the prisoner is now speaking their words" (par. 11). However, as with other fascist practices, the power gained from torturing is at its core a form of weakness. As one engages in torture, one distances one's self from a sense of humanity, undoing civilization in the process. This decomposition accelerates in the public at large when they engage in mass denial of acts of torture, labeling such acts instead as "harsh interrogation techniques."

As with Reich (1970, 1971), Adorno's (1950) work was done in the wake of multicontinental fascism in an effort to diagnose and contend with authoritarians and racists, where, he argued, "the marks of social repression are left within the individual soul" (p. 747). The concept of hierarchy pervades the masses, carrying through to behavior and even patterns of thinking. In his section addressing the family and formation of authoritarian viewpoints, Adorno identifies several factors common to these individuals: conventionalism, submission, aggression, anti-interpretation, prone to mysticism, obsession with conventional gender norms, fascination with power, destructiveness and overall cynicism, tendency to project, and sex repression (p. 228). In particular, obedience and respect are the top values that authoritarians believe should be learned in childhood—even if it meant themselves being abused as children in the process. This lays the foundation for a series of people that enables such leaders to hold sway. Nor are people hardwired to "need" leadership and war; to the contrary, Reich (1970) asserts that while this may be the current state of things, this by no means indicates that people can't change.

As the founder of the long-running free school Summerhill, Neill's (1960) writings focused on the role of both the family and education in acclimating children to societal controls:

> Man's slavery is a slavery to hate: he suppresses his family, and in so doing he suppresses his own life. He has to set up courts and prisons to punish the victims of his suppression. And slave women must give her sons to the wars that man called defensive wars, patriotic wars...the fight is not an equal one, for the haters control education, religion, the law, the armies, and prisons. Only a handful of educators strive to allow the good in all children to grow in freedom. (p. 103)

Even so-called progressive education practices utilize a form of consent-getting coercion, under the guise of free choice and self-control. By instilling varying degrees of fear and sexual repression in the child, she or he is taught that one has to have authority in order to remain in control. For Neill, religion was the major culprit in the establishment of repression where one was supposed to have more of a fear of death than an enjoyment of life. Eventually, one becomes a slave, accustomed to suffering in order to make life meaningful, but a slave who doesn't crave freedom: "He is incapable of appreciating freedom. Outside discipline makes men slaves, inferiors, masochists. They hug their chains" (p. 377).

Reich's (1971) *The Invasion of Compulsory Sex-Morality* asserts that early societies were not burdened with sexual repression and the resulting social problems of violence and devaluation of women. These events inserted themselves into family relations at the same time as the advent of private property. Building his analysis off of Malinowski's field work of the Trobriand islanders, he contrasts their open sexual and work practices with our contemporary forms of exploitation and repression. Since "sexual repression is one of the cardinal ideological means by which the ruling class subjugates the working population," the liberation of workers also needs to include sexual freedom, which Reich differentiates from sexual license (p. xxvii). Indeed, Reich makes it very clear that promiscuity and exploitation of women is the flip side of the ascetic coin, not the opposite of sexual morality. In *The Mass Psychology of Fascism*, Reich (1970) takes up the perennial and ever-difficult question of how people identify with the very forces that oppress them. The suppression of sexual urges in children by punishment and religious fear creates a toxic dynamic where people come to accept authoritarianism. Denied a sexual outlet, people turn to religion and fascism as a form of release.

fatherhood advocates endorse. He taught about peace, remained unmarried, hung out with 12 men whose own marital status was up for dispute, and kept company with undesirables. When confronted with these obvious contradictions, Murrow simply states that Jesus was originally a manly man; he was just feminized in Sunday school by ministers who were afraid to offend women, the bulk of whom are church attendees: "Truth is, the Jesus of Scripture is more General Patton than Mister Rogers" (p. 135). The answer to turning around the dangerous tide of feminization and gender role usurpation is for men to regain their central importance in the church, attracting other men with promises of danger and adventure, that is the "real" Jesus' ministry.

Another feature of gender role norms is the insistence on the wife being submissive to varying degrees, from the caustic 1998 Southern Baptist Convention's Resolution on Marriage stating: "wives and husbands should voluntarily yield to their husbands, following Saint Paul's words to husbands and wives" (Hardisty 2008, p. 25) to more subtle appeals from authors such as Arterburn and Stoeker (2001). In one telling passage, the authors describe how the husband is supposed to model his decision-making role not by taking authority by force, but by considering his wife's "essence" while proceeding with plans as usual. In one case, the wife wanted to move back to her home town thinking that it would provide better family connections for herself and her children while the husband did not want to move:

> Yet Brenda remained adamant, so I had to study her closely. Was she expressing her deep emotions, or was something deeper transpiring...her desire to move was based on pure emotion. As leader, I had to find room for her essence, but in doing so I couldn't allow her emotions to determine the outcome. I told her I thought we should stay put...even when things can't go her way, we've at least taken her essence into consideration, and oneness remains strong. (pp. 178–179)

Psychoanalytic Approaches

Drawing upon Freud (who, next to Darwin and Marx is apparently responsible for the collapse of Western civilization if one listens to reactionaries), different concepts of the origins of authoritarianism and fascism are found to stem from the family. Neill (1960), Reich (1970, 1971), Adorno and colleagues (1950), and Lakoff (2002) frame their understanding of the family in light of its role in forming political ideologies that often promote self-subjugation. This section will address key works from these individuals who examine attitudes about authority, punishment, religion, sexuality, and freedom. All of these works share psychosexual assumptions about why people continue to support that which oppresses them. They also ascribe to the notion of collective responsibility; in other words, it isn't a matter of a dynamic leader "brainwashing" the masses, there is something learned in the masses of ordinary

> The men's movement has not evolved an articulated analysis of social power. With no political context, individual growth can as easily lead to abuse of power and more individual aggrandizement as it can to community development and change in power relationships. (p. 6)

Traditional gender norms are a prominent feature of the fatherhood movement, both in terms of the appropriate roles for men and women within a family and a constant appeal to these roles being inscribed as eternal. The implication is that the troubles families are facing today are due to these norms being violated. Part of the upheaval relates to what Yalom (2001) identifies as the role of the wife no longer consisting of a stable set of characteristics, such as needing to be married in order to have children. Smaller practices, like women choosing to keep their maiden names, are read as outright subversions to established practice. Even with the understanding that early band societies had more informal sexual and child-rearing conventions, conservative proponents view this history as a violation of civilization, calling for colonial correction. For them, modern alternative family forms and gender roles are overturning the hard-fought civilizing qualities of the nuclear family (Coontz 2005).

Widespread misconceptions about gender roles and marriage are presented as anthropological fact, when "it is simply a projection of 1950s marital norms onto the past," including that marriage was developed to protect women, with the man as the hunter and woman as the gatherer (Coontz 2005, p. 40). Leacock's (1981) own work challenges the protector theory of marriage held by authors such as Gilder (1986) and Gutmann (1999). Gutmann (1999) argues that the tribal male had to break away from the mother bond by engaging in warfare and hunting. For modern males, the over-feminization of society drives them to create comparable activities to replace what they are no longer able to do (a variation of "boys will be boys"). Or, worse yet, they will choose the homosexual lifestyle, an outcome Gutmann views as inevitable if masculinity doesn't intervene.

Murrow (2005), in his lighthearted book *Why Men Hate Going to Church*, employs fatherhood rhetoric to explain the reasons behind a lack of male involvement in most mainline congregations, despite the leadership of these churches being overwhelmingly male. Employing Gutmann's (1999) separation/individuation theory, he emphasizes:

> A boy must reject his mother and her feminine ways, and take his place in society as a man. A male who fails to separate psychologically from his mother faces a lifetime of gender confusion, abuse, or dysfunction. Think Norman Bates in the film *Psycho*. (p. 90)

Murrow, like many conservatives, also struggles with the figure of Jesus, who, if taken at face value, definitely does not meet the masculine norms that the hands of a few families. Indeed, Gilder (1986) laments, "in any disintegrating society, the family is reduced to the lowest terms of mother and child" (p. 85), revealing his belief that only the father confirms legitimacy and status onto the child. Lest the guys feel left out of Gilder's portrayal of doom, consider his description of unmarried males:

> The single man is poor and neurotic. He is disposed to criminality, drugs, and violence. He is irresponsible about his debts, alcoholic, accident-prone, and susceptible to disease. Unless he can marry, he is often destined to a troubled and abbreviated life. (p. 62)

Gilder does not explain how such a man would be an attractive marriage partner to further the dominance of the nuclear family ideal. If anything, a woman would be demonstrating her intelligence to steer clear of marriage if this was her only option for a potential father!

Canfield (1999) views the Promise Keepers, a hybrid of men's movement philosophy and Christian evangelism, as a potential ally in the reestablishment of the nuclear family. The average Promise Keeper participant is married, in his late thirties, and has a college degree, yet 60 percent of Promise Keepers' wives work outside of the home, though mostly part time (p. 51). As Kivel (1990) points out, this is a movement where the majority of participants are white and upper middle class, making it more of a backlash phenomenon against perceived unfair gains by women and minorities:

> When one is working with groups of people who already have a measure of social power, such as the power white middle class men have at work and in their families, the potential for abuse is great. (p. 2)

The Promise Keepers, like other men's movements, assert that men have been made powerless by the system of gender. They also claim that men suffer from a lack of closeness to their own fathers, or that fathers are completely absent from their lives (Canfield 1999). By contrast, Kivel (1990) responds to this claim, based on his own experience counseling men that their fathers were frighteningly *present* in their lives, overshadowing all other aspects of their existence, through religious, emotional, and physical abuse that carried over into adulthood. Indeed, Canfield's (1999) summary of a National Center for Fathering survey indicated that 33 percent of Promise Keepers had substance abuse and emotional turmoil in their households growing up. In addition, 10 percent indicated being physically abused with 6 percent reporting sexual abuse (p. 51). As Kivel (1990) states, men are not "disempowered" by women, instead, they are "empowered" to be abusive and aggressive toward those they perceive to be weaker or even gay. The answer to this legacy of abuse that the Promise Keepers propose is individual self-help. In his strongest passage, Kivel concludes:

nuclear family form intact (removing the burden on elites) while (b) allowing these private sector solutions to impose religious restrictions on those seeking much-needed aid (Eberly 1999).

Even the way most K–12 curricula are structured favor the maintenance of the nuclear family above other forms. Coontz (1997) describes an experiment that a teacher conducted regarding the timing of homework assignments. She found that the traditional one-earner nuclear families preferred getting schoolwork done during the week so that the weekends would be free for recreational activities. Most homework given follows the weeknight deadline structure. However, this format tended to penalize single-parent, dual-earner, and divorced families since weeknights were high stress times. The teacher designed homework for weekend completion, along with creating activities that were more natural for the alternative families (such as having a child read out loud to a parent while he or she was making dinner or in the car on the way to run errands). Another math assignment involved finding the best prices across different brands. Children of single-parent families found this assignment easier to complete since they were used to accompanying a parent to the store, versus children of two-parent families where kids were typically left at home while the parent ran errands. Within just a couple of weeks, the traditional nuclear families lodged several complaints demanding that the homework return to the regular format. Authors such as Coates (1999) and Popenoe (1999) would consider this teacher's efforts to be subverting the inherent superiority of the nuclear family and that she was unfairly penalizing these families by making things too easy for alternative families.

Other challenges to the nuclear family include step-parenting and even technologies such as in vitro fertilization, where a child can have as many as five parents (not counting any future remarriages)! These changes have been building for the past 100 years, not emerging suddenly as conservatives claim. Currently, one-fourth of U.S. households consist of just one person, reflecting a decline of marriage in general as a form of organizing one's life (Yalom 2001, p. 397). According to Yalom, women are waiting an average of five years later than their counterparts in the 1950s to marry while at the same time experiencing more premarital sex, independence on the job front, divorce, and remarriage (p. 354). Oddly enough, rates of divorce and unmarried births are the highest in the South, where survey results indicate the highest levels of disapproval of these practices (Coontz 2005). The extended life span to age 85 means that "the average married couple will live for more than three decades after their kids have left home" (p. 268). This only increases the tendency toward divorce.

These developments strike at the heart of the propertied interests in conservatives' obsession with occurrence and legitimacy of children. Bebel (1910), Coontz (1997), Engels (1972), Kollontai (Holt, 1977), Leacock (1981), Trotsky (1970), and Yalom (2001) trace the interest in establishing bloodlines in children as part and parcel of the concentration of wealth into can marry her boss, can bear his children. She can go out with the young executive. Her social and sexual value, though influenced by class, is relatively little affected by her earnings and occupational status. (p. 40)

It is interesting to compare Gilder's passage above with an excerpt from the blog of George Sodoni, who killed four women at a Pittsburgh health club (sic):

> Good luck to Obama! He will be successful. The liberal media LOVES him. Amerika has chosen The Black Man. Good! In light of this I got ideas outside of Obama's plans for the economy and such. Here it is: Every black man should get a young white girl hoe to hone up on. Kinda a reverse indentured servitude thing. Long ago, many a older white male landowner had a young Negro wench girl for his desires. Bout' time tables are turned on that shit. Besides, dem young white hoez dig da bruthrs! LOL. More so than they dig the white dudes! (Adkins 2009, par. 2).

Though Sodoni brings race into the discussion where Gilder (1986) holds back for the time being, the implications are the same: women (along with African Americans) are powerful and selfish, taking away opportunities from beleaguered (white) males. For Gilder, when women don't place limits on men, everything in society disintegrates. However, the answer can't be equality, because this also causes cultural disintegration by taking away the social Darwinism that makes men thrive. Finally, young, attractive women, who hold all the power, are money seekers who collectively reject victimized men, who are then forced to "act out" in the only way they know how via crime, violence, rape, sexual harassment, and so on. It is the woman's job to tame the man, and when they divert from this goal (such as seeking a career) they run the risk of contributing to a growing degree of male delinquency, unemployment, and corruption. Only marriage and the conventional nuclear family offer salvation.

Coates (1999) elevates the role of the male-headed nuclear family "as the source not only of our personal values and security, but of our political, economic, and social order as well. Any nation whose families are weak will eventually find itself without strong institutions of any kind" (p. 119). For Coates, social context is entirely ignored, with the weakening of the family being not a part of the collapse of the social safety net in the 1970s, but of a trending away from marriage and bad personal choices. This is much like the fall of Rome being blamed by conservatives on a lack of family values, not extreme imperial overreach. For conservatives, the government is the culprit behind the collapse of the nuclear family, because its policies encourage wayward behavior along with giving social support to deviant family forms. Instead, if we just relied on the churches and charities to provide for needs, all would be well because this would have the double effect of (a) keeping the privatized victimization of males by females/femininity/feminism with women portrayed as having the real power; the decay of the nuclear family as the source of societal problems; and a reassertion of essentialist gender norms, often using science to prop up these concepts. The fatherhood movement shares Moynihanian beliefs about poverty as a cultural/personal defect, though some in the movement are more religiously oriented than others who adopt a more secular stance.

Feminism is a primary target of fatherhood movement writers. The overall worsening of economic conditions for all families is reflected in backlash rhetoric aimed at women who are viewed as "taking" opportunities away from men who are the real victims (D'Emilio and Freedman 1997). In response to the increased visibility of women in the public sphere, fatherhood movement writers craft arguments that children are harmed by women working outside of the home, along with dangerous exposure to neglect in day care facilities (Warner 2006). Coontz (1997) debunks another common line of thought that maternal neglect and a lack of a strong father in the home leads to gang activity; in fact, the rate of gang memberships is roughly the same for intact families, suggesting other important factors than marital status alone. There is no evidence that women working outside of the home have a harmful impact on children. What is harmful is a lack of affordable and reliable day care for all families.

Horn (1999) states three core beliefs of the fatherhood movement, none of which address structural, material problems: masculinity should be by default about fatherhood and responsibility, mothers and fathers are different, and children should have a bond with their fathers. Not satisfied with consciousness raising, this particular movement seeks to alter behavior through the application of social policy. For example, Gallagher (1999) portrays the single mother as selfishly choosing "rights" and her lifestyle while consciously depriving her children of a male role model. For Gallagher and Whitehead (1999), men withdraw from family responsibilities because they are "tired" of the hostility from women combined with women taking over their role. Coontz (1997) identifies this as the zero-sum "package deal" construct where either a woman totally submits to the fatherhood ideal, gaining a responsible male as part of the bargain who protects and legitimates her children, or she has to face the consequences of abandonment and take all the blame; a view pronounced by Henry (1999). Likewise, the message is that a man can't be a good father unless he is married to the mother of his child. Many families are assumed to be "fatherless" when in fact the fathers are highly involved.

Using conservative logic, Gilder (1986) outlines how men are victimized by all-powerful women. Because of the sexual revolution, women no longer place behavioral limits on men, who are wired to be more sexual by nature. This situates women in a position of power, because men are and have always been vulnerable to sexual attraction:

> Where single men and women work at the same jobs, the women are socially superior. They know it and the men know it. The young secretary

According to Gilder's (1981) logic, the working class in California is voluntarily reciprocating the generosity of the capitalists by giving up social services. It also makes no sense to attempt to understand this reality. In fact, one of the main flaws of socialism is that it is too rational and doesn't rely on faith. Instead, we need to acknowledge that the world is too complex and people are too ignorant to grasp what is happening to them. It is the mysticism of capital that is needed to propel families out of poverty and into the endless upward trajectory of supply-side economics as in this particularly fanciful passage:

> Matter melts, but mind and will can flash for a while ahead of the uncertain crowd, beam visions across the sky and induce their incarnation in silicon and cement before the competition gathers. (p. 76)

Marx (1998) launched a pointed critique of individualism, in particular, the "self-made" man developed by Kant. As Marx maintained, people make each other, but not themselves. What Kant did was to invent "free will" as a force apart from the interests of the ruling elite, making both above reproach. Part of the mysticism of the lone individual fighting against "the competition" involves a total divorce of the populace from historical understanding, giving, as Marx wrote "full reign...to the speculative steed" (p. 70). What the conservative movement does is "present its interest as the common interest of all the members of society...it has to give its ideas the form of universality, and present them as the only rational, universally valid ones" (p. 68). In addition, Marx notes, conservatives frame the abolition of private property as a personal attack on their livelihood:

> When the theoreticians of the bourgeoisie come forward and give a general expression to this assertion, when they equate the bourgeoisie's property with individuality in theory as well and want to give a logical justification for this equation, then this nonsense begins to become solemn and holy. (p. 246)

If the working class experiences an inevitable disconnect between what they are told about free will and what they experience, they are supposed to bear this quietly. Thankfully, Marx maintains that the large numbers of workers in the world think differently, "and they will prove this in time, when they bring their 'being' into harmony with their 'essence' in a practical way, by means of revolution" (p. 66).

The Fatherhood Movement

For the fatherhood movement, the nuclear family becomes an ideological weapon that is simultaneously used to exclude single parents and LGBTQ individuals as well as to outline regulations applied to sexuality within marriage. This particular backlash subset of conservative familialism contains several common elements, as Kivel (1990), a critic of the men's movement, describes: perceived (pp.128–129), something that doesn't seem to trouble Coates (1999), since, as he maintains, "our greatest challenge is not a lack of resources, but a lack of conscience and character and integrity (p. 119). In an essay, Horn (1999) interviews several conservatives such as Stephen Goldsmith, who take the character directive to the extreme by pressuring schools to release statistics on the number of unwed teen births along with test score reports and sponsoring a program entitled "Jobs or Jail." In this program, Goldsmith describes how unemployed, unmarried fathers would be referred to a local resource center, where they would be compelled to take parenting classes and work in community service. If they refused, they would be jailed.

Coontz (1997) believes that such policies based on stigmatizing people are ineffective and result in punishment for only the poorest and most vulnerable, similar to what happened during Prohibition in the United States. It also re-introduces the concept of a "deserving" poor and an "undeserving" poor, creating even more bureaucracy than ever before; imagine the oversight needed to review each welfare case to determine if a child was "intended" or an "accident," going all the way back to the motives of the mother in selecting a sexual partner (was she a victim or did she contribute to the situation?). Because many of the conservative politicians themselves have broken the very rules that they would easily impose on the public (like Newt Gingrich being married multiple times, Dick Cheney's daughter having a child with her lesbian partner, Mark Sanford having a high-profile affair and lying about it, etc.), they would be likely to issue exceptions/protections for themselves, their families, and their elite buddies while launching punishments at the poor.

Gilder (1981) is considered one of the key philosophers of conservative familialism. His work centers on policy implications of right wing ideology, in particular a reliance on capitalism and supply-side economics as the solution to poverty. He presents capitalism as eternal, even religious, calling upon workers to be willing to give (i.e., work) without reward. He declares, "the only dependable route from poverty is always work, family, and faith...the poor must not only work, they must work harder than the classes above them" (p. 87). At the same time, "entrepreneurs must be allowed to retain wealth for the practical reason that only they, collectively, can possibly know where it should go, to whom it should be given" (p. 31). The worst that can happen is "forced redistribution" via government, which compels helpless business owners to seek advantage by using overly aggressive practices such as price gouging or cheating. Instead, it is better if redistributionist policies are voluntary, especially if it is the working class that antes up, as with California's current budget crisis:

> Without raising taxes on the rich, states are going to attempt to solve their expanding budget crisis...by expanding taxes on consumption, which fall chiefly on the working class. These taxes will be coupled with enormous cuts in social benefits that the working class depends on—education, school lunches, Medicaid, and other social programs. (Geier 2009, p. 32)

(p. 7). In many respects, they want the 1950s family without the government funding that created it. This type of funding was not simply for immediate sustenance but for long-term development, such as the GI Bill and postwar housing grants, meaning that even if a family hit hard times, they weren't likely to retreat back into poverty. The overall message of the conservative movement is that income doesn't matter. They regularly point to Sweden, one of the most successful social welfare systems in existence today, as an example of the failure of liberalism despite the prosperity of that nation, all because there is a high rate of single parent families (Popenoe 1999). Conservatives also ignore the accumulated material effects of poverty, such as Coontz's (1997) discussion of lead exposure and its impact on school-aged children, including classroom behavior and cognition.

Conservative familialists have made several suggestions for changing government policy, none of which offer any concrete solutions. Popenoe (1999) presents the following: encouraging widespread abstinence and marriage/parenting classes in K–12 schools, counteracting the excesses of the sexual revolution and steering laws toward privileging the married, nuclear family. While he does acknowledge that capitalism has excesses that can conflict with building family values, he portrays this as more a matter of requiring increased self control on the part of individual consumers than an inherent feature of the free market. Mincy and Pouncy (1999) propose a more open concept that could start with diverse family forms, albeit steering them toward the ideal nuclear model. For example, they suggest that the unmarried, custodial father should be included as part of the definition of "family" and alternatives to cash child support payments could be made in the form of services, lessening the financial burden on unemployed, noncustodial fathers. Once a father obtained regular employment, cash payments could resume. The fact that many of the jobs these fathers would hold are likely to be minimum wage work is overlooked.

Coates (1999) is more strident:

> Government policy should communicate a clear, public preference for marriage and family on matters such as public housing, the tax code, family planning, and divorce law. Rewarding intact families is not, as some argue, a form of discrimination. It is a form of self-preservation. (p. 124)

Coates also advocates work requirements for welfare, arguing that this will lead to the creation of more entry-level jobs and will reduce the societal burden of welfare recipients. Coontz (1997) would counter that such programs make up only 5 percent of federal spending (even though nearly half of Americans place the percentage between 20 and 50 percent) and that after five years, cash payments cease, throwing many women into dead-end, low-paying jobs. In the meantime, the top 20 percent of asset holders received nearly 100 percent of the total gain in wealth between 1983 and 1989 alone as a lynchpin against groups perceived to be dangerous, such as enshrining the 1950s nuclear family amidst discrimination against non-Christians, communists, minorities, and LGBTQ individuals (Coontz 1997).

As D'Emilio and Freedman (1997) explain, the gains of the social liberation movements of the 1960s created a panic within conservative circles, resulting in a revival of late 1800s purity movements. At the same time, sex became a commercial product and was marketed as a form of individual identity, as easily purchased as a new outfit or car. Even with the ready availability of symbolic sex-as-commerce, conservatives have taken aim recently not at "indecency" (after all, indecency contributes mightily to the bottom line), but government spending, as it promotes immorality (as in the welfare system being seen as "rewarding" single motherhood): "The republican leadership has used its new clout to implement the economic goals of free-market capitalism: cutting taxes, reducing social welfare spending, and eliminating government regulation of business" (p. 366). This culminates in authors such as Charles Murray (1994, with Herstein 1996, 2006) using neo-eugenic language to claim that welfare is promoting the reproduction of intellectually inferior minorities who are burdensome to society.

Coontz (1997) maintains that the welfare system actually benefits the ruling elite because it is much cheaper than the creation of good-paying jobs, serving as a safe measure for a capitalist economy that is unable to handle full employment. While the welfare system should rightly be critiqued as a form of social control that merely maintains surviving and not thriving, as Piven and Cloward (1993) do, the source of conservative ire appears to be the public nature of social support, not the concept of social control per se. Note that Piven and Cloward view welfare negatively because it is used in lieu of genuine socialist alternatives, while conservatives launch attacks on the perceived immorality of a "culture of poverty," as Cloud (2001) explains:

> The ideology of the family thus "solves" the actual economic contradictions of capitalism for the capitalists, as familialism can explain and justify exploitation, crisis, and suffering in private terms that absolve the industrial capitalism of blame... In reducing economic and political problems to matters of family responsibility, familialism serves as an ideological linchpin of the New Right. Familialism simultaneously justifies the erosion of welfare spending, and backlash against women's and gay and lesbian liberation movements" (pp. 76–77).

As Coontz (1997) reminds us, there has never been a time in class-based society when any one family form was able to prevent poverty and other social ills. By taking a historical perspective, personal relationships can be placed into a larger social context, which alleviates the blame that is often hurled at the family by conservative politicians. Instead of concrete social changes, ideologues offer "feel good homilies about cleanliness, chastity, and charity"

1997; Kelly 2002). This would eventually form the backdrop of the "welfare queen" construct that was a powerful manifestation of false consciousness during the Reagan administration. As Coontz (1997) points out, the mythology of the immoral single black female who bore children just to receive a welfare check is not borne out in reality. The intensified police patrolling of minority neighborhoods and soaring incarceration rate of African American males alone contributes to the "fatherlessness" that right wingers often decry. The states that receive the highest welfare benefits have lower rates of children born out of wedlock while larger welfare payments are correlated with faster returns to the workplace, yet the myths persist. In addition, poverty is growing among all families, not those with single parents; three-fourths of poor households in New York City either have two parents or have a male as head of the household—and this with welfare spending having decreased by a third since 1970 (p. 90)!

In most cases, dire poverty and a lack of employment opportunities precede a rise in teen pregnancy, which is what happened with the youth pregnancy rate in the United States during the 1990s. Though 50 percent of all babies in Sweden are born to unmarried women, Swedish children do not experience the dire poverty that U.S. children do (Coontz 1997, p. 89). Coontz posits that even if all African American children lived with both parents (as conservative pundits pose as being the only economic solution), the equality gap would still remain between white and black Americans, with nearly 67 percent of black families likely classified as poor (p. 140). Though there are some family values writers who admit that more structural policies like paid leave and child care need to be implemented (Mincy and Pouncy 1999; Wallerstein 1999), most policy makers are reluctant to move in this direction (Coontz 1997). As Kelly (2002) argues, it is convenient for politicians to be able to blame social ills on those who don't conform to the nuclear family ideal.

The nuclear family is the incubation chamber of capitalism and the prospect of LGBTQ individuals getting married provides an important challenge to the privatized nuclear family preferred by the ruling elite. For example, GOP chairman Michael Steele and Republican Governor Mark Sanford of South Carolina both argue that gay marriage would be an economic liability because the government and businesses would have to widen the social safety net to accommodate spouses (Bynum 2009; Dickinson 2009). Both Steele and Sanford calculate that presenting gay marriage as "taking away from" other married couples could be an important campaign strategy to reach voters and provide a convenient way out of appearing to oppose gay marriage on religious grounds alone, a position growing more tenuous as tolerance for diverse family forms grows. Despite record-breaking bank bailouts in the trillions that happened in a matter of weeks, Sanford claimed that government "can't afford the current level of promises" so gay couples need not apply (Dickinson 2009, p. 45). This follows the traditional line of using the family while conducting interviews, four- to five-year-old children would constantly run up to their mothers, apparently unable to sustain independent play for longer than five minutes because of all the floor time their moms spent with them one on one. This is evidence of what Neill (1960) identified as too much stimulation or pressure placed on children to be at the top of their game, "the coercive quality of our civilization" (p. 28). The addition of high-stakes testing contributes to a competitive culture of surveillance, where parents no longer trust other families, schools, or teachers. Instead, other adults are potential barriers to push past as teachers report increased difficulties recruiting parents to be part of a team to work on their child's academic or behavioral problems (Warner 2006). The extracurricular activities that are part of the overscheduled child's life can't just be for creative or recreational enrichment, but for the college application and resume. The constant exposure to structured, "productive" activities means that young children have a harder time transitioning from preschool to first grade where 24/7 entertainment and one-one-one adult-child time isn't practical.

The most frightening development that Warner (2006) describes is a management consulting firm specializing in helping families better oversee their human resources. The firm collects survey data and generates a "growth summary report" so that families can more realistically examine executive strengths and weaknesses in developing a list of targeted areas of improvement for each family member. This vision, combined with the hypervigilance surrounding issues of child safety, only heightens parental anxiety. The media, using the rubric of Bowlby's (1983) attachment theory, raises the specter of the neglected child (witness the constant coverage of Romanian orphans in the mid-1980s) to the extent that parents rarely allow their child to do anything unsupervised, even within the home, for fear that they will be damaged for life. Tragic news stories like that of Susan Smith drowning her two young children are turned into cautionary tales against maternal selfishness. Warner (2006) is correct when she concludes, "motherhood has been made into a production of the highest consequence" (p. 56).

Conservative Familialism

Moynihan's Wake

The heart of Johnson's War on Poverty legislation, Moynihan's (1965) government report, *The Negro Family, A Case for National Action*, is often cited by right wing ideologues as a rationale for conceiving of poverty as a personal defect rather than the result of years of capitalist policy infused with systemic racism (Davis 1981). Even though Moynihan intended the report to buttress efforts toward redressing poverty and inequality, it had the unintended effect of indicting family structures other than the nuclear model as inherently pathological and dysfunctional (Coontz 1997; D'Emilio and Freedman of production, one of Warner's interviewees came to the rational conclusion that eventually alienation had to be confronted as one couldn't outsource companionship and sex and still have a functioning marriage.

A distinct feature of late-stage capitalism is the production of the super-child. As in all aspects of alienation, the rearing of children is also further removed from the experience of parents in that it is no longer enough to successfully see a child through high school graduation and to independence as it was in the past. Today's parents are responsible for overseeing a college degree and even writing resumes or calling bosses to negotiate a first post–college graduation job for their child (Erwin 2008). This seismic shift to a staggering form of affluent child-centeredness has taken place against a background of entrenched divestment and privatized retreat from the very concept of society, which leaves most children out of the loop (Coontz 1997). As a result, many families within the middle class feel compelled to make sure their kid gets the best since few outside options exist. So the alienation effect is doubled; beginning with the impact of capitalism creating a climate of fear and anxiety within the household, and culminating in dramatic social class differences in terms of which children receive educational and health benefits and which do not. In fact, Warner (2006) notes that this newest cult of motherhood hit in the 1990s at the same time as the Clinton administration's welfare reform legislation.

Historically, children were viewed as contributors to the economic unit of the household and the overall expense of seeing them to adulthood was minimal (Coontz 1997; Minge 1986; Warner 2006). With today's prolonging of childhood well into their twenties for educational reasons, children have become a financial liability and a source of marital disruption as they reach adolescence (Kelly 1986; Minge 1986; Shashahani 1986; Yalom 2001). This is due in part to an overall lack of living wage work and affordable housing for young adults, forcing them to remain at home longer than previous generations. As Willett (2002) explains, parenting has been changed into "intensive labor...involving the cultivation of requisite social, cognitive, and cultural skills in children so that they may succeed in an increasingly competitive labor market" (p. 120). Instead of viewing the family as a unit, capitalism structures the equation where one has to choose between competing rights—the mother's or the child's. If not enough time and investment is placed into the child, a mother runs the risk of condemning her child to possible destitution. To hedge one's bets, one reaches for the enrichment flashcards and *Baby Einstein* DVDs even if the child is not yet two years old.

Warner (2006) addresses the extremes of child-centeredness that is part of the production of the super-child and "winner-take-all" parenting. Unlike her experience in France, American homes that Warner visited were dominated by children's toys and equipment with no apparent boundaries between child and adult spaces. The suburban backyards looked like parks as the parents transformed their home to replace public spaces. Warner found that

Warner (2006) documents in an honest and painful manner the psychological effects of alienation within marriage and how it radiates outward to form societal attitudes. Beginning by contrasting her experience with the social safety net in France after her child was born with her current situation in the United States, Warner paints a grim picture of the context underlying marital tensions. As she interviewed stressed women about their lives, she found that as soon as she referred to her time in France, they would become defensive and refused to look to socialist resolutions for their problems. Warner's account of parents who refused to provide a neighbor's address for an emergency readiness plan at a local school is quite illustrative; they would rather have family members drive across town than organize a plan in their local community. In essence, these women used an irrational cult of motherhood to privatize their problems. With a lack of dialectical materialist analysis, they would instead turn to self-help books, media personalities, and counseling. In one case, Warner explains how as these women were losing their status and decision-making powers by choosing to stay home to raise kids (mostly due to escalating child care costs), they would rationalize it as follows:

> No, this wasn't tantamount to sucking it up, they insisted...they weren't giving in. They were just expecting the strengths or weaknesses of your spouse or the good and bad in your relationship, as one woman who prided herself on her co-parenting arrangements put it. It was just a matter, said others, of accepting that men and women were different. (p. 124)

The "men are from Mars" line of self-helpisms furthers the privatization process by making alienation seem only a matter of changing one's perspective on the situation or focusing on the wide array of "choices" that one has; choice being a fetish straight out of the logic of Ayn Rand. If, in the end, you couldn't make it all work, then it had to be because you didn't try hard enough or that you chose poorly: "the term choice was attractive...because it offered American women rights lite, a package many perceived as less threatening than unadulterated rights, thus easier to sell" (p. 180).

Victorian women would justify their lack of status by pointing to the "special" status of motherhood. Likewise, Warner's (2006) women are taking refuge in the "Zen" of selfless parenting and the production of the super-child. Warner found that women would engage in psychological deflection where it was okay to talk about anything but anger. Instead, conversations would be about how men were helpless and at the same time annoying, or that the deprivations they were experiencing—particularly sex—were signs that they as mothers were doing something worthwhile. Gender differences were constantly evoked to justify the lack of meaningful sexual connections within their marriages with children serving as the rationale, as in one mother who stated that her children were a sufficient substitute for an adult relationship. Ultimately, while privatization relies on the outsourcing of nearly all means replaced personal connections with large, distant bureaucracies that function in a top-down manner. The worker's political activism is thus reduced to being a consumer, with product choices providing the illusion of free will.

Contemporary marriage, existing within the rubric of the idealized nuclear family, produces effects of alienation in the form of shattered expectations. It is now no longer enough to have the goal of a home ownership and a stable job. Now couples are supposed to combine income earning prowess with sexual compatibility and keeping up one's erotic appeal through diet and exercise (D'Emilio and Freedman 1997; Warner 2006). Since marriage is often the only social support network available, the combination of media images of the perfect family and constant psychological challenges of turning tensions resulting from employment to relationship confrontations can produce its own form of alienation, as Coontz (1997, 2005) and Warner (2006) document in great detail. The solution lies not in trying to pinpoint gender differences between men and women as television personalities maintain, but in addressing a society that refuses to provide essentials such as health care and education. Queen (2008) notes that it isn't just gay couples who face barriers; even heterosexuals face an uphill climb if they seek to create a decent partnership in the wake of capitalism. Marriage itself can become an alien institution since it is the state-sponsored arrangement that formally legitimates the production of the next generation of laborers.

Within the household, alienation results from the constant accumulation and repetition of unpaid domestic labor, even if a woman works outside of the household. The more financially strapped a particular family is, the less likely it will be able to outsource domestic tasks, leaving many women working a second shift (Croll 1986; Wang 2001; Smith 2005). Women who are employed outside of the home spend roughly 35 hours a week on household chores while also resenting traditional female roles and their decided lack of status (Yalom 2001, p. 359). For example, the more money a husband earns, the more "bargaining power" he has in the home including not being as likely to do chores (Coontz 1997). Even when men wish to take on child rearing, their workplaces tend to view their requests for family leave as reflecting a lack of ambition.

Yalom (2001) also describes how the acquisition of "time saving" appliances like dishwashers only end up further placing families in debt as they attempt to relieve the ever growing domestic workload; much as the arrival of newer machines on the assembly line results in speed-ups: "The time saved by the mechanical devices [is] wiped out by increased standards of performance" (Warner 2006, p.35). Divorce can also be considered a symptom of alienation as the nuclear family can no longer withstand the pressure of capitalist expectations. Finally, if these remedies don't provide relief, there is a ready market of pharmaceutical products for kids and adults such as Xanax and Dexedrine to make things more bearable and counteract increases in teen suicide, depression, eating disorders, and stress, as Warner reminds us.

Tilly's (1986) historical analysis of female worker resistance during the Victorian era reveals how the workplace has always been structured to minimize dissent and that the household has to be considered as a critical component of revolution. According to Tilly, working class women will be more inclined to take collective action (a) when they are able to meet with others who have similar grievances and turn these commonalities into a more formal organizing; (b) when they have access to resources; (c) when the labor they produce is essential to employers and the subsequent threat of withdrawing their labor can lead to workers' gain; (d) when they are not vulnerable or when they have some autonomy in the household; (e) and when the household itself can be mobilized and generalized into the larger social sphere. Tilly describes an organized, militant food riot in 1911 where women destroyed merchant property and obtained necessary goods as a result of solidarity and an orientation to human rights rather than a struggle over consumer rights (another form of female activism during this era). What made the food riot successful was that the women combined familial concerns that extended to other members of the household (i.e., viewing their husband's labor as linked to their own role) with a sense of solidarity with other households. This shatters the myth that women aren't capable of mobilizing resistance because their labor within the household is unproductive or because of the barrier of patriarchy

Alienation and the Super-Child

According to Marx (1973, 1977), a central feature of the worker's life under capitalism is alienation, or the removal of one's labor from one's self. This is a totalizing process, as Marx (1977) outlines in the first volume of *Capital*:

> They alienate from him the intellectual potentialities of the labor process in the same proportion as science is incorporated in it as an independent power; they deform the conditions under which he works, subject him during the labor process to a despotism the more hateful for its meanness; they transform his life-time into working time, and drag his wife and child beneath the wheels of the juggernaut of capital. (p. 799)

While Marx and followers have discussed the impact of capitalism on the worker, not as much has been written about the impact of alienation on the family unit. When labor no longer belongs to the worker, it contributes to the feelings of insecurity, especially as the threat of layoffs and unemployment loom. Just as labor under capital is no longer for the purpose of meeting human needs, neither is the privatized labor of the family (Davis 1981; Wang 2001). For women especially, industrialization transformed labor from productive to unproductive, as in Shashahani's (1986) description of agri-business's recent co-optation of locally grown food and crafts, once under the direction of women in many countries. Neill (1960) explains that industrialization kept in the realm of low wages/no benefits (Hammam 1986, Kelly 1986; Mullings 1986; Wang 2001). More indirectly, sexism and sexual harassment serve capital by playing on the key role of the gendered nuclear family in labor reproduction while keeping wages low (Hickey 2006). The gender divisions within the family coupled with enforced gendered norms via harassment help to police the system overall.

Expanding outward, the work that women do, especially in third world countries, is also afforded lower status as a way of managing wages (Croll 1986; Hammam 1986). Safa's (1986) analysis of the hiring practices of multinationals reveals a shift from utilizing mostly male factory workers to now recruiting females, "the availability of cheap labor" as "the prime determining factor for investment" (p. 58). NAFTA and other free trade incentives have opened up the global market to a largely female workforce, this being the prime strategy of late-stage capitalism. Arizipe and Aranda (1986) found that younger women are considered ideal employees due to their dependent status (living at home with parents) and not tending to resist oppressive workplace conditions. Younger women also contribute to high staff turnover as they leave work to get married, and high turnover saves factory owners from having to pay extended benefits. Croll, (1986), Kelly (2002), and Hammam (1986) conclude that the exploitation of women is directly linked to the overall exploitation of the working class, with the role of the nuclear family assisting in the establishment of this relation through reproduction while maintaining a reserve army of labor.

Leacock (1986) views with skepticism calls for modernizing majority world countries, such as initiatives for educating select groups of disadvantaged female workers at the expense of community solutions. These proposals are often wrapped in the language of liberation, but are still in thrall to a system of profit. The markers for development include Western standards, not the specific needs of the people. Any form of resistance to this cultural and economic co-optation is blamed, for example, on the "backwardness" of gender relations (as in Middle Eastern countries). Leacock continues:

> A real contradiction exists between women's liberation as a goal and the enormous expense entailed by the socializing of domestic labor...industrialized countries already have forms for socializing domestic labor and child care. All manner of food preparation and dispensing services, cleaning services, and formal and informal child care arrangements are at hand. All that is required is to make them more healthy, accessible, and cheap by removing them from a profit-making structure and making them responsive to the needs of the people using them. (p.256)

While it is understandable that activists might want to begin by confronting sexism, a revolutionary program that focuses only on addressing patriarchy will not succeed unless it is coupled with an attack on capitalism itself.

heads home at the end of the day to clean his or her own home, then he or she is doing unproductive labor. Today's dual earner family is thus composed of a mixture of productive and unproductive laborers, creating a contradictory situation. Beneria and Sen (1986) view the wage as not simply payment for being productive, but part of the overall calculus of the cost of reproducing and monitoring workers, a concept that fits in nicely with patriarchy's prioritizing of the male wage.

Production both requires consumption and creates consumers (Marx, 1977, 1998). The means of production have to be replaced, and in the case of workers the family is the site of reproducing the future workforce:

> Hence, both the capitalist and his ideologist, the political economist, consider only that part of the worker's individual consumption to be productive which is required for the perpetuation of the working class, and which therefore must take place in order that the capitalist may have labor power to consume. (Marx 1977, p. 718)

The family oversees and bears the expense of reproducing future workers. A major part of this process involves securing education and training to place one's children at an advantage, and individual households are responsible for handling these costs. Yet capitalists regard this transmission process of education and training as belonging to them, one of many variables they employ to turn things to their advantage (like having the pick of overqualified job applicants). In the end, production and reproduction are "a total, connected process," producing "not only commodities, not only surplus-value, but it produces and reproduces the capital-relation itself" (p. 724). Croll (1986) further differentiates between physical reproduction (tasks related to children and household) and social reproduction (ideologies to maintain the system). Forrest and Ellis (2006) view capitalism as coercing people into the nuclear family form just long enough to not only obtain a new generation of workers, but to instill ideologies of homophobia and heterosexism that ensures the maintenance of the system.

Davis (1981) and Beneria and Sen (1986) outline how capitalism has always been resistant to the concept of socializing domestic unproductive labor since its expense would be incalculable at the outset. Many feminists have also opposed the idea of socializing housework, considering it degrading, calling instead for the use of wages-for-housework as a strategy for raising the status of women's work. Pointing out that African American women have been on the losing end of this wages-for-housework strategy for decades as hired domestics, Davis (1981), along with Beneria and Sen (1986), considers this solution inadequate to meeting the needs of families in the form of universal child care, among other services that could be provided socially. The free labor of housewives is also used as a leveraging device against raising the wages of domestics. It is because this labor is essential that it is deliberately despite an overall conservative message of containment coming from authorities. As Coontz (1997) illustrates, the close spacing of children by young families also meant that while gender roles were more established, women could work outside of the home sooner or even return to school once the children were grown. Government support and policies buttressed the survival of the middle class via the nuclear family in the form of higher wages, home ownership, and college education, which would last until the economic downturn of the 1960s and 1970s.

D'Emilio and Freedman (1997) outline the reversal of 1950s trends. Within just 20 years, the rate of marriage declined by 25 percent while the median age of first marriage rose to 25 years for males and 23 for females, along with a dramatic drop in the number of couples choosing to have children (p. 330). Since the 1970s, uncertainty of employment and falling wages have placed the nuclear family form into question (Warner 2006). The final and most fearsome challenge to the dominance of the nuclear family, at least according to conservatives, was the fight for recognition of gay rights. Once able to use the family as a way to exclude LGBTQ individuals, activists are now challenging the very definition of marriage itself (Yalom 2001). Coontz (2005) summarizes the seismic shift as the nuclear family fades into the past:

> It took more than 150 years to establish the love-based, male breadwinner marriage as the dominant model in North American and Western Europe. It took less than 25 years to dismantle it.... In barely two decades, marriage lost its role as the master event that governed young people's sexuality, their assumption of adult roles, their job choices, and their transition to parenthood. People began marrying later. Divorce rates soared. Premarital sex became the norm. And the division of labor between husband as breadwinner and wife as homemaker, which sociologists in the 1950s had believed was vital for industrial society, fell apart. (p. 247)

Production/Reproduction

In *The German Ideology*, Marx (1998) outlines essential concepts behind production and reproduction that are also relevant to the family. Production "is always a certain body politic, a social personality that is engaged on a larger or smaller aggregate of branches of production" (p. 3). In the case of the family, we are looking at its role in creating the future workforce. Since unpaid domestic work is part of any household and its results are immediately consumed (i.e., the repetition of doing dishes and laundry), Marx (1977) considers it unproductive labor. Contrary to the feminist interpretation of Marx as dismissive of "women's work," his is not a commentary on the importance of the labor, but rather its unpaid status (consumed for use-value, not creating exchange value). A housecleaner, on the other hand, doing the same sort of labor but for a wage, would be designated productive. But when he or she seeking waged work (Coontz 2005). Within the household, gender divisions became more pronounced as males worked outside of the home, creating the role of housewife for women and an array of flowery justifications for the cult of motherhood and the home as a retreat from the hostile world (D'Emilio and Freedman 1997; Yalom 2001). The double standard was reflected in state policy toward prostitution and the "fallen woman" (D'Emilio and Freedman 1997). Kollontai's (Holt 1977) socialist parable, which tells of a wife who initially resents the prostitute her husband hires and ends with the wife herself being abandoned and forced to walk the streets, exemplifies the way that the nuclear family was used to create repressive categories of womanhood, often resulting in higher risks associated with desertion (Coontz 1997). Wives, being removed from wage labor, became more financially dependent on their husbands, a dilemma that Gilman (1994) and feminists protested: "husband and wife are one person, and that person is the husband" (Yalom 2001, p. 185). In addition to being isolated economically, women were isolated communally, since the nuclear family was supposed to provide all of the emotional support one would need. Social scientists at the time argued that the nuclear family form would be globalized with the spread of industrialization; and this was considered the height of progress (Coontz, 2005).

The Jazz Age and the invention of adolescence as a unique social category transformed the significance of marriage and family with marriage being delayed until the completion of high school, which was now compulsory. Women were able to access contraception and family planning, further removing sexuality from procreation (D'Emilio and Freedman 1997). In addition, psychoanalysis revolutionized concepts of the self and sexuality, which were popularized in widely available films, radio programs, and magazines. As D'Emilio and Freedman point out, at the same time that the family was a privatized entity, sex hit the public marketplace with full force, alarming many. The purity movements of the Victorian era experienced a minirevival at this time, though they couldn't keep up with women's suffrage and a rise in female employment outside of the home. World War II would provide for an even greater entry of women into wage labor and even brought about temporary socialization of child care at major war plants (Yalom 2001). Illustrating the links between capital and the family, any ideological reluctance to allow women to enter the workforce disappeared as soon as the glaring need for workers was felt, though it was spun as a matter of being patriotic and supporting one's family (Eskridge 2008).

In the postwar era, women were expected to once again return to the home and take on the role of housewife, but the number of working mothers continued to increase at a dramatic rate, predating the effects of the women's movement (Warner 2006). During the apex of the nuclear family, couples married younger and the adolescent pregnancy rate soared (D'Emilio and Freedman 1997; Eskridge 2008). Teen culture and the automobile, along with rampant consumerism and temporary prosperity, redefined dating would join convents or, if they lived in northern Europe, could become part of a commune of other single women who made their living by producing goods and services, abolishing the myth that marriage was a universal norm throughout human history (Coontz 2005). For the nobility, marriage was a way to ensure the transfer of property, and the Church became involved in sanctifying these unions (Yalom 2001). At the same time, romantic love became a pastime for the wealthy, with courtly love being expressed by minstrels and through elaborate poetry, but not associated with marriage per se.

An important transition in the family began to appear during the 1600s along with the adoption of capitalism and wage labor: marriage for love. The notion that people could base a marriage on romantic love was at first glance a liberating trend. But state regulations soon took the place of parental permission and the legitimacy of children became more important (Coontz 1997, 2005; D'Emilio and Freedman 1997; Yalom 2001). A case example of these changes is found in the Puritans who, contrary to the stereotype of frigid and intolerant people, were merely intolerant, as D'Emilio and Freedman (1997) maintain. Puritans advocated regular sexual intercourse as an essential part of a healthy marriage. At the same time, they sanctioned harsh punishments for adultery and unwed motherhood, believing that the sex drive should be channeled into state-sanctioned marriage and that all community members had a responsibility to monitor each other's actions. The establishment of romantic love as a foundation of marriage also hastened the privatization of the emerging nuclear family by placing primary importance on a couple's relationship rather than on their social role within a larger network of households (Coontz 2005). Yet within families, fathers, not mothers, were the ones who assumed the education of children (Coontz 1997).

British colonists in the Americas justified imperialist acts by pointing to Native American sexual practices as heathen and depraved, linking communal land ownership, gender equality, and the like to sinfulness (D'Emilio and Freedman 1997). The slave economy brought about white fears of miscegenation and a hyper-idealization of white females, particularly in the South. Like Native Americans, African Americans were portrayed as sexually immoral, intensifying the double standard where black females were denied the protections given to white women while being forced to be sexually available to white males (Davis 1981). Black couples attempted to maintain stability despite the underlying fact that neither party owned her or his own body and their marriages could be dissolved at any time (Davis 1981; D'Emilio and Freedman 1997). The nuclear family form (including subordinate roles for women) was imposed on these groups, "to strengthen the mechanism of societal control over slaves, and to justify the appropriation of Indian and Mexican lands through the destruction of native peoples and their cultures" (D'Emilio and Freedman 1997, p. 86).

By the Victorian era, the nuclear family was gaining strength, especially as people moved from agricultural settlements to crowded urban centers, by a discussion of Marx's views on the family's role in production/reproduction and alienation, which is supported by forces impacting the family today, on both a global and a local level. Various theories of the family that originate from conservative, psychoanalytical, anarchist, and materialist perspectives are described. Especially relevant to educators is an examination of the child as a "super-product" of today's nuclear family and the effect this has on schooling.

The Emergence of the Nuclear Family

Contrary to conservative laments that the nuclear family has been both unchanging as well as an ideal to strive for, a simple historical overview demonstrates that the family has always been highly diverse, with its appearance and sexual dynamics shaped by economic forces (Cloud 2001; Coontz 1997,2005; D'Emilio and Freedman 1997). As class-based society entrenched itself in ancient Greece and Rome, the family became a way to organize the work necessary to run an agricultural economy. A typical household included husband, wife, children, older relatives, slaves, and possibly concubines. At the same time, for the majority of families without major property holdings, marriage was relatively unregulated by the state and Rome allowed no-fault divorce (Coontz 2005; Yalom 2001). Yalom (2001) describes how the account of Adam and Eve in the Book of Genesis reflected the societal concern with establishing both lineage of children and the subordinate role of wives. These traditions would carry over into the New Testament with divorce and remarriage considered adultery according to Mark 10:11. In some cases, Yalom continues, marriages could be dissolved if a wife could not produce a child, as in Athenian society, indicating the growing concern with the legitimacy of children and inheritance.

The Medieval era continued the intensification of consolidation of property with married peasant households forming an essential production unit (Coontz 2005). At the same time as the Church took on an ascetic stance toward sexual relations, even within marriage, peasant communities had a more tolerant and varying range of attitudes toward sexuality such as unwed pregnancy, which could be accommodated into their network of interdependent households and midwives (Yalom 2001). Until the Victorian era, women were viewed as having a greater sex drive than men, a total inversion of today's beliefs. For peasants, marriage was largely an economic and practical arrangement, and until the Church took on a regulatory role, communities monitored their own marriages; if they even occurred at all. Pedersen (2009) recounts the Celtic tradition of Lammas Day, a festival that included allowing couples to start trial marriages of 366 days, or until the following festival, when either the man or woman could walk away if things didn't work out. Due to economic conditions and the shorter human life span requiring large families, women married and had children at a young age (Coontz 2005; Yalom 2001). Unmarried women only child care available for this woman is church-based, and that only covers two days a month plus having to attend speakers or craft activities which might not interest all mothers. There is no universal health care, and as any parent of a child with special needs understands, the advocacy required to obtain the most minimal of quality services is Herculean. This mother has no access to family members or friends; just other moms who are most likely facing the same child care situation. She has to put in a lot of time just to find support groups and secure a spot for herself and her child. Public transportation is unavailable, requiring the purchase of a second car, causing further financial burden. The husband appears detached from the situation, existing as he does in the sphere of wage labor while his wife works in the unpaid sector, crucial for propping up capitalism.

This decaying nuclear family structure can no longer bear the stresses of capitalism and has far outlived its usefulness with just under 30 percent of families resembling two-parent household of the 1950s (Kelly 2002, p. 229). Yet until we figure out a way to effectively address what the Victorians termed "the woman question," that is, the socialization of the domestic sphere, Marx's vision will fail (Cloud 2001; Kelly 2002; Wang 2001). In all of the previous experiments with communism, one of the first hurdles was the enormous expense of making housework paid work. Whether in Russia or Cuba, the provision of universal child care, laundries, communal kitchens, education, and house cleaning proved to be too expensive to implement on a large scale across the board. This left many women in the same situation and pointed to the futility of putting women into the paid workforce in the hopes that it would "solve the problem" of gender segregation (Croll 1986; Davis 1981; Holt 1977). Without the underlying social support for handling necessary domestic labor, the nuclear family, in effect, continues to exist on a shoestring, placing psychological and physical barriers to the full realization of socialism.

Often Marxists are focused on the big picture, which is indeed necessary. We need to develop and communicate a vision for a saner society, which will involve the dismantling of capitalism. At the same time, in developing this vision, we have to, as Reich (1970) reminds us, pay attention to the "everyday things," while asking ourselves what socialism will look like enacted *within* the family, not just outside of it. This, of course, touches upon the dilemma of permanent revolution. Do we start to rearrange the family now, hoping it will lead to change? Even if the necessity of revolution is acknowledged, there is still the problem of the family. So how does one handle both issues simultaneously? Contrary to the advice columns, this situation obviously requires a better solution than having dad do the dishes once in a while!

This chapter addresses the centrality of the family as an issue involving not just sexuality, but economics. Indeed, the family plays a key role in perpetuating hegemony, because it normalizes oppression through a variety of intimate, day-to-day practices (Moeller 2002). A presentation of the history of the family from the ancient era to the present starts the chapter followed watch your kids while you get 3 hours to socialize, listen to a speaker, or do arts/crafts. Frankly I think some of the speakers and arts/crafts are stupid, but being able to have the babysitting and socializing with other moms is still worth it.

Getting a network of friends, or failing that, acquaintances (I don't really have friends currently, but I have a lot of other moms that I hang out with in various clubs/groups/church I belong to) is important to keep your sanity.

Do you have family nearby? You don't necessarily have to have the best relationship to ask them for help. Having them watch your kids for even a few hours a week, once per week, might help you get some sanity back.

I know it's hard having fibromyalgia. I have something similar. Your body aches all the time. Your joints hurt or your muscles hurt. And you have chronic fatigue that is quite debilitating. And no one else seems to understand or thinks you're just being melodramatic. And for some unexplained reason, you seem to get sick a lot, and don't have good immunity towards anything. People will say things to you like "just get over it" or "toughen up" or older people will say things like "you're more than half my age, what's wrong with you?!" Or you try to do activities with your kids, and by the time your husband gets home, he wants dinner on the table and the kid(s) screaming and all you want to do is go to bed, too tired to take your clothes or shoes off your body.

Just try to ask for help. It's hard. Especially if you don't know who to ask or have no one to ask. Finding your network is a chore in and of itself. I am about to move, and my network that took me awhile to set up I will have to reestablish. I won't have family where we're going. But, I've already gone on to the web and found a couple groups. My husband wanted to go to a church that didn't really have kids' activities at it and didn't have many members. I stood my ground and said that our church had to have kids' Sunday school by age, and have family groups to interact with, and be not too far of a drive from home. I am not sure if he's still mad at me for putting down all those rules. But that's what I needed. We had several weeks of debate. We haven't moved up there yet, but it was based on a couple visits when he interviewed and then we house hunted. And visiting web sites. I also found the town's "Newcomers" group which has a subgroup for moms.

I know there are lupus support groups. You might get ideas from other moms who have fibro, and what they are doing to cope or what they did if their kids are older. I don't know if you have two cars or not. My husband and I did with one car for so many years, but when my daughter was 1, I needed a car for myself since public transit around here is unreliable...and where we are moving, there just are not public buses.

While this mother has a chronic medical condition that arguably brings depression and exhaustion into the picture, the frustration expressed and futile attempts to cope only point to the need for socialism—and fast! The

Chapter Three

The Family: Conservative, Psychoanalytical, Anarchist, and Materialist Readings

If anyone doubts the relevancy of Marx for today's world, or that his analysis is "inadequate" when applied to the domestic sphere, one should take a look at the numerous posting boards that are visited daily by countless moms—both stay-at-homes and working moms. The following post from www.medhelp.org illustrates the direct impact of capitalist society on the family and mothers in particular. It also makes glaringly clear the absolute lack of any kind of social safety net, leaving families to fend for themselves:

> I am a stay-at-home mom with an autoimmune disease.... I know that if I stayed at home without doing activities with my daughter, I would go nuts. She is a late talker (diagnosed with pervasive development disorder) and most of the time screeches, whines, or drags me around by the hand. Unless of course we are outside playing, and then she is thoroughly happy and stops annoying me (the high pitch screeching has been really getting on my nerves and she does it because it bothers me).
>
> I like being outside as well so going outside makes her happy, which makes me happy. I also get bored "playing" in the house. We go to the parks a lot. When it rains or is too cold, we go to the toy store. And I of course get followed around by the store clerk getting the evil eye while my 2 year old plays with every single electronic toy that you can press buttons in the packaging. I go anyways. Maybe I'm rude. I don't know. I'm just trying to keep my sanity.
>
> I joined a Mothers of Preschoolers group 2 years ago. It's run through various churches, though you do not need to be part of the church or even Christian to join. They meet during the school year, but usually groups will have activities over the summer at homes, like play dates and stuff. Twice a month during the school year, they have meetings where other people

definition of who is or isn't a homosexual. The research itself might be based on random sampling and careful definitions of who constitutes a particular category of sexuality. For many conservative groups, "homosexual" means anyone engaging in same-sex acts, including bisexuals or heterosexuals who experimented once or twice. By including these groups, the numbers/percentages can go from ho-hum to alarming with the click of a mouse. Sometimes bisexuals will be excluded from the numbers, depending on which better portrays LGBTQ individuals in the worst light. This was the case with Ford and Jasinski's (2006) study of drug use and sexual orientation in college students. Burroway (2006) explained that conservative groups added together the percentages of categories of sexual orientation, making it appear that homosexuals were decidedly more prone to drug abuse than heterosexuals.

Another famous research article cited in conservative circles, known as "the Dutch study" concluded that same-sex partnerships were less likely to last due to promiscuity and other factors (Xiridou et al. 2003). Burroway (2008) notes that those who cite the article make it appear that gay marriages do not last and that this study provides scientific proof, therefore why make it legal? However, the participants who were studied were overwhelmingly young (those over 30 were excluded), urban, and unmarried. The age of the participants decreased their chances of being in a long-term relationship, because they didn't have as much likelihood of building one. Monogamous couples were also not included, skewing the sample from the start. The study's target population wasn't representative and this is made clear in the original study. However, conservative groups have quoted liberally from this article to further their case against gay marriage.

Conclusion

The origins of homophobia are complex and bound to economic and social factors. Rather than simply being an individualized, psychological matter of "intolerance," homophobia is inextricably linked to the capitalist system. As we have seen, with the rise of property ownership came the appearance of homophobic ideology, benefitting the ruling elite. Whether it was used to justify the seizure of land and wealth during the Middle Ages or currently providing a means to prevent the extension of government benefits to workers, homophobia is another way of keeping people from forming solidarity across identity lines.

The growing nonsustainability of the nuclear family in the face of mass economic decline is presenting the working class with a crisis of ideology and identity. The following chapter examines theories and philosophies about the family and ideas about sexuality, gender identity, economics, and religion. Just as homophobia has been used to promote capitalism, so have certain ideas about the nuclear family been upheld to support privatization of social services and the ongoing oppression of women and LGBTQ individuals.

injections, and drug therapies. LGBTQ individuals were confined to mental institutions to be warehoused or eventually made straight, sometimes against their will or for fear of social retributions (Fone 2000). These harsh and abusive policies were defended by scientists and anxious parents as desirable alternatives to making life easier for deviants who were a threat to society or the best thing for curing "sick" individuals (Eisenbach 2006). In the 1940s California hospitalized and sterilized men considered perverts and 29 states enacted expansions of punishments of existing sex crime laws with homosexuals in mind; only three states limited these severe punishments to sex involving minors (Eskridge 2008, 81). All of this took place against the backdrop of a profession that barred gays and lesbians, leading to many disguising their identities and refusing to take a stance against the American Psychiatric Association's classification of homosexuality as a pathology (Eisenbach 2006).

The language of pseudo-science and fields such as evolutionary biology and sociobiology rely on essentialist readings of gender roles as fixed and innate to human societies (Fone 2000; Gaspar 2004; Leacock 1981). Women, in many cases, were and are left out of such research (Eskridge 2008). Charles Socarides (1995), author of *A Freedom Too Far*, presents the argument that homosexuality results from neuroses brought on by domineering and overprotective mothers and absent or weak fathers. Similar theories are brought about by Jeffrey Satinover (1996), a follower of Jung. Using secular-sounding rhetoric, both assert against homosexuality being inborn and view it as a problem of subverting "natural" gender roles. Even A.S. Neill (1960), founder of Summerhill, viewed homosexuality as a form of neuroses originating in early childhood, seeing it as "unproductive" and a result of guilt over forbidden heterosexual sex play by punitive parents (p. 208).

The AIDS epidemic of the early 1980s unleashed many gay-sex studies, which were used by the media to stoke conservative backlash (Cameron 1989; Cameron, Cameron and Proctor 1989; Diggs 2002). The promotion of gay rights was seen as a threat to public health and rationales were provided since no one knew how the virus would mutate next (Eisenbach 2006). William Buckley Jr. (2005) reiterated his original 1986 eugenics suggestion of tattooing individuals (though he mentions only gay people) with HIV/AIDS as a matter of public health: "Murderers need to be stopped, and if this means opening their mail, well, such things happen and you can take comfort that you may be saving a life. The objective is to identify the carrier and warn his victim" (par. 3–4). In 1986, 14-year-old Ryan White was expelled from school for having HIV, which he acquired from a blood transfusion for his hemophilia. White's social ostracizing and expulsion launched a debate in the media about the privacy rights of individuals diagnosed with HIV/AIDS. Though he eventually won the right to attend school, prejudices lingered about how HIV/AIDS was transmitted, leading to panic in many educational settings (Johnson 1990).

Burroway (2006, 2008) dissects how pseudo-science operates: Antigay writers will cite a study, and then adjust the numbers to suit their own particular language. Turning the tables, Burroway draws from sex research the same way that homophobic publications do, to build the case against a looming heterosexual menace. In one passage, he quotes, "Militant heterosexual researcher Paul Cameron conducted a survey that found 36 percent of all men who behaved heterosexually engaged in anal intercourse with women, and 20 percent of women who behaved heterosexually reported anal sex with men" (p. 2). An epilogue of *The Heterosexual Agenda* includes "How to Write an Anti-Gay Tract in Fifteen Easy Steps," including such strategies as citing less reliable surveys from gay magazines or adding behavioral statistics using convenience samples, along with manipulating the data. After mentioning threats to health, children, the family, and society, it's important to "close on a compassionate note" (p. 27).

In the 1800s, the use of scientific language regarding sexuality demonstrates that authors had progressive intentions. Homosexuality emerged in late nineteenth-century scientific discourse as part of an overall push to move away from social policy based on religion alone. As part of the growing field of psychology and psychotherapy, studies began to emerge in the early 1860s referring to same-sex attraction as a psychological condition apart from longstanding notions of sinful behaviors (Fone 2000; Forrest and Ellis, 2006). Nineteenth-century figures such as Karl Heinrich Ulrichs were opposed to the legal and religious persecution of homosexuals, and thought that scientific study of sexual behavior could help to change these practices.

However, though medical writings began to take homosexuality out of the realm of "choice" and place it into a category of involuntary human behavior, capitalist society at large transformed this language into one of homosexuality as pathology: "The focus on sexual acts rather than any idea of sexual identities did nothing to challenge critical perceptions of homosexuality as deviancy" (Forrest and Ellis, 2006, 91). In many ways, the addition of scientific language furthered the cause of those who sought to connect homosexuality to pedophilia and other forms of deviance in need of a "cure." For example, Wolf (2009a, 2009c) outlines how early gay activism in the 1950s was held back by a form of internalized homophobia where members of homophile groups sought out medical attention and therapies to change their condition rather than challenging discriminatory policies. Former Mattachine Society member Edward Sagarin (a.k.a. Donald Webster Cory) underwent years of psychiatric treatment and conversion therapy and came to the conclusion that such treatments didn't work; however, years later, he reversed his position and warned that parents should intervene when young people exhibited homosexual tendencies or they would face a world of prostitution and sexual excess (Eisenbach 2006). Sagarin's (1969) *Odd Man In: Societies of Deviance in America*, promoted the still-lingering image of gay men as highly promiscuous and seeking to recruit younger males, providing ammunition for those promoting conversion therapy as an answer.

Such "therapies" included electric shock where wires were placed on the groin while patients were shown gay pornography, lobotomies, hormone spouse via Social Security, the chances of three or four Social Security payments going to polygamous families is practically nil, so there is no danger.

There are religious voices that argue against homophobia. Kahn (1989) presents a counter-argument to Prager's (1990) interpretation of Judaism. Since heterosexual couples aren't subjected to proof of fertility prior to marriage, then why are LGBTQ people denied marriage rights based on their inability to have children? Instead, the strength of marriage, as Kahn (1989) and Spong (1990) envision it, includes more than procreation. It has a value that can apply to multiple family forms, not just limited to the nuclear model. John (1993) also rebuts the fundamentalist view of marriage by suggesting that fidelity and companionship should be important features of all relationships, not just limited to heterosexual ones. Jesus and even Paul did not talk about procreation or gender differences when they spoke of marriage, yet these are common arguments against allowing gays and lesbians to marry. Early Muslim cultures encouraged homoerotic literature among a wide range of intellectual ideas that were tolerated, making them understandable targets for Christian imperialism during the Middle Ages (Fone 2000).

Conservative gay columnist Andrew Sullivan (2004) turns the tables on an imaginary dialogue with Patrick Buchanan, culture warrior and gay rights opponent, by presenting the following question: If procreation is a central feature of marriage, then why don't the Buchanans themselves have children? The common religious response to this contradiction is to reply that sterile couples could be hit with a miracle and eventually conceive. Sullivan continues the solitary conversation by responding, "But, if it's a miracle you're counting on, why couldn't it happen to two gay people?" (p. 84). He concludes that there appears to be a double standard when it comes to marriages that don't produce children. Infertile couples are viewed sympathetically because they are "trying," while gay and lesbian couples are portrayed as sinful and lonely by society at large.

Pseudo-Science and Secular Homophobia

In the wake of the 1973 elimination of homosexuality as a category of mental illness, there are a number of antigay arguments circulating that are noticeably nonreligious in nature. However, these pseudo-scientific ideologies rely heavily on the religion and family values defenders as much as these groups depend upon scientific-sounding language in order to "medicalize" their arguments, making them appear more trustworthy and neutral (Page 2006). Sometimes pseudo-science ideology can also work in the form of censorship, as Feldt (2004) and Goode (2003) outline how Bush administration officials pressured researchers studying the spread of HIV/AIDS to remove wording such as "gay," "homosexual," or "transgender" from federal grant applications. Government Web sites were also "edited" to remove sexually explicit content from HIV/AIDS prevention sections.

Burroway's (2006) satirical take-off on antigay pamphlets, *The Heterosexual Agenda: Exposing the Myths*, skewers the contemporary use of pseudoscientific that homosexuals should be respected for being different, but they should understand that it is "a difference with real consequences" (p. 56). "Marriage necessarily presupposes sexual differentiation, for human procreation itself presupposes sexual differentiation" (p. 56). Apparently marriage is the mechanism through which sexuality is appropriately restricted, both in terms of intimacy and property. During the Middle Ages, heretics were commonly accused of being homosexual, providing a convenient justification for enriching the Church with confiscated lands and wealth (Fone 2000). Today, gender roles and differences are only the visible manifestation of things working the way they should be.

One example of religion's use as a convenient way out of tough philosophical and policy debates is Prager's (1990) stance: "Jews or Christians who take the Bible's views on homosexuality seriously are not obligated to prove that they are not fundamentalists or literalists, let alone bigots.... The onus is on those who view homosexuality as compatible with Judaism or Christianity to reconcile this view with the Bible" (p. 62). Those who take the Bible literally can retreat to its authority, challenging the rest of society to attempt to change their minds. Common Biblical scriptures referenced in support of homophobia include the Old Testament (Leviticus), especially the Sodom story, and Pauline doctrine in the New Testament (Romans). As Kotulski (2004) points out, if we were to follow Biblical prescriptions on marriage, this would mean endorsing a union between one man and one or more women (Gen. 29:17), validating a marriage only if the wife were a virgin, under threat of her execution (Deut. 22:13–31), not allowing divorce (Deut. 22:19; Mark 10:9); or assuming that the brother of a man who dies will marry the widow if there were no children produced (Gen. 38:6–10; Deut. 5:10) (pp. 6–7). Fone's (2000) detailed analysis of the Sodom story demonstrates that the "big sin" wasn't homosexuality per se, but refusing hospitality to guests (fundamentalists seem to overlook Lot's offering of his two daughters to be raped by the angels as a minor side plot) while Paul's view of homosexuality is part of a larger asceticism that shuns sexual pleasure in general (Eskridge 2008; Fone 2000). Homosexuality simply provides a convenient symbolic scapegoat for sex outside the bounds of procreation.

Prager (1990) argues that homosexuality "denies life" because it denies God's plan for humanity as laid out in the Torah, which "juxtaposes child sacrifice with homosexuality...both represent death" (p. 64). Homosexuals, by not procreating, prevent children from being born. In this fundamentalist vision, males and females are not complete persons until they marry and have children and marital sex is the only correct form of sexual expression. Prager presents an unusual final argument: to accept homosexuality would mean subverting "the extremely hard-won, millennia-old battle for a family-based, sexually monogamous society" (p. 66). Instead of appealing to the age-old tradition of the nuclear family, Prager takes the long view by presenting the struggle for monogamy as a victory that took a thousand years to establish, so don't let the other side reassert itself. Coontz (2005) takes a practical stance and considers how stingy the government is about granting benefits to a single surviving

Opposition to antibullying efforts in schools is linked to the rationale for blocking hate crimes legislation; if one were to have legal means for preventing harassment, then that would be endorsing something that goes against God (Record 2005). Kupfer (2005) describes resistance to a Pride Prom event at his school as something that would promote the spread of HIV/AIDS. Beach (2005) recounts a parent who wanted to present "the other side" of homosexuality in his classroom. When Beach pointed out that this was a public school and that all kids and their families were to be respected, the man organized a picket outside the school, with signs stating "Homosexuality— an abomination against God" (p. 214). Beach's students eventually confronted the picketers by posing thoughtful and difficult questions that took to task the protesters' views of religion and sexuality.

Religion often provides a cloaking device for gays and lesbians who wish to remain closeted for fear of being discovered or fired. Lyons (2005) relates how he developed "The Strategy" for working as a teacher in a religious school. Convinced that he could do more to promote a broader view of Christianity from within the closet, Lyons was more concerned that he didn't rock the boat by being pushy, thereby turning kids away from God. "The Strategy" necessitated Lyons being silent about his identity, so he wouldn't make things uncomfortable for the parents and students who came from privileged, heterosexual, Christian backgrounds. These individuals would have benefited from Bishop Spong's (1990) sharp critique of discrimination against LGBTQ people: if we can bless MX missiles by calling them peacemakers, or name a submarine *Corpus Christi* (meaning body of Christ), then why not bless gay marriages? Lyons (2005) would eventually gain the courage to come out, after hearing a gay minister endorsing a wider interpretation of religious beliefs.

Gay marriage as it relates to the role of the church has been the premier lightning rod issue for the religious freedom crowd. Reverends Charron and Skylstad (1996) present the rationale that marriage is a special form of relationship that deserves promotion and protection, both religiously and secularly. Citing the decline of the family, the reverends appeal to the defense of marriage in order to protect women, children, and ultimately society. However, gays need not apply: "No same-sex union can realize the unique and full potential which the marital relationship expresses. For this reason, our opposition to same-sex marriage is not an instance of unjust discrimination or animosity toward homosexual persons" (p. 53). At minimum, this view advances the notion that while LGBTQ individuals shouldn't be attacked or abused, it's perfectly fine to not allow them to have important civil rights because all of civilization is at stake.

Elshtain (1991) believes that if same-sex marriage is allowed, it will threaten marriage's place on the societal hierarchy. Extending the reach of discriminatory language, Elshtain defends the exclusivity of heterosexual marriage by linking it to the reasonableness of society showing a preference for two-parent families. Echoing Charron and Skylstad (1996), Elshtain (1991) reminds readers media, Kotulski (2004) humorously deconstructs the bestiality argument:

> I think most people will recognize the difference between two loving adults wanting to share a lifetime commitment and Old MacDonald wanting to tie the knot with his chickens, his cow, or all five of his sisters.... I hear these arguments so much, it makes me wonder why these guys are always thinking about marrying their dogs and if there is another meaning to "a dog is a man's best friend." (p. 95)

The Religious Freedom Argument

As Head on Radio Network Internet talk show host Bob Kinkaid maintains, the First Amendment has become nothing more than a life support system for the religion industry. "Religious freedom" is used as a shield to permit discrimination to occur unchallenged within the classroom and in society at large. Religious beliefs about sexuality are freely allowed to interfere with the rights and protections of LGBTQ and straight communities. While the state can't impose many sanctions on churches, the church is able to have the collective ear of politicians who will forsake the Constitution in favor of moral values or not offending "their base" (Kotulski 2004). At the same time, the religion industry makes enormous profits from homophobic books, films, pamphlets, "educational" materials, and speaking tours, in fact, one suspects that if it were not for homosexuality, the bank rolls of many evangelists and their attendant spin-off businesses would be endangered. In a sense, Fone (2000) is absolutely correct when he opines that not only constitutional law, but moral law holds sway over the lives of LGBTQ individuals; moral law is even more insidious because there is no clear ability to appeal.

The selection of Reverend Rick Warren to speak at Obama's inauguration rightfully enraged many LGBTQ people and allies. Schulte (2009) describes how Warren's Saddleback Church denies membership to gays and lesbians unless they apologize for their sins; required repentance being a warmed-over version of "love the sinner, hate the sin" (Fone 2000). Warren himself worked to pass Proposition 8 and seeks a constitutional amendment banning same-sex marriage. In an effort to kick off an administration that "welcomes all kinds of political thought," Obama essentially approved of Warren and those he inspires, including Martin Ssempa, a Ugandan pastor who burned condoms in support of promoting faith healing and abstinence to combat HIV/AIDS (Schulte 2009, par. 10–16). When these individuals are challenged in the media, their homophobia is often downplayed, as Rampton (2008) sums up in a common quote heard from the religious freedom crowd: "these people and these religious institutions are not promoting hate, they are just not agreeing that marriage can be between a man and a man or a woman and woman. This is a cultural difference of opinion" (par. 7). Church doctrine can be a convenient way to preserve the status quo and avoid the unpleasantness of confronting discrimination, as with the Mormon Church's endorsement of racism until the 1970s.

Hardisty 2008; Kotulski 2004). The problem is never the falling wages that lead to the necessity for both parents to work outside of the home; it is feminism and selfish self-fulfillment. "Threats to the family" aren't the lack of universal health care or access to affordable child care, it is LGBTQ couples who seek to marry and adopt children. Cloud (2001) elaborates:

> Conservative familialism scapegoats and vilifies unconventional families...whose arrangements challenge the modern nuclear family ideal necessary to the reproduction of labor under capitalism. In addition, conservative familialism encourages people to see their intimate, domestic relations as both the source of their most difficult problems and the site of those problems' resolution—even if there are external, social, structural, causes of personal crisis. (p. 71)

One can see this play out in conservative lesbian talk show host Tammy Bruce's (2004) exhortation against gay marriage, an institution cast as the government acting as mom and dad and smothering one's life choices: "Let's get real—the only thing that will make gay people whole is personal acceptance of ourselves by ourselves...society has been the benevolent parent for a very long time" (p. 202). In Bruce's view, LGBTQ individuals who insist on equal marriage rights are looking for a government handout. Instead, they need to leave marriage to the heterosexuals, respect the wishes of the majority, and be content with civil unions as they are currently defined.

The breakdown of the family ultimately becomes the scapegoat for capitalism's failures, revealing the true contest as between work and the obligations of government (Cloud 2001; Forrest and Ellis, 2006. What Cloud (2001) calls the "fantasy" of the nuclear family is a means to keep people from dialectical materialist analysis as they spend time on false nostalgia for a lifestyle they cannot attain. As Coontz (1997, 2005) points out, politicians want the 1950s family but not the attendant government policies that sustained it. The supposed appeal to the protection of the traditional family and patriarchy becomes a way for the ruling elite to attack not just LGBTQ relationships, but all domestic partnerships in their efforts to contract the welfare state (Coontz 1997; Kotulski 2004).

The opposition to anything that deviates from the nuclear family is so extreme, that divergent family forms are instantly equated with polygamy, incest, or even bestiality as part of a slippery slope line of thinking: make gay marriage legal and you'll have men marrying their own children or joining the North American Man-Boy Love Association (Arkes 1993; Bennett 1996; Krauthammer 1996; Wilson 1996). Reacting negatively to the 2003 decision in *Lawrence v. Texas*, Republican senator Rick Santorum stated that repealing sodomy laws would lead to "man-on-dog" sex (Blumenthal 2006). Most likely in response to this and similar sentiments expressed in the

Neither this Constitution or the constitution of any state, nor state or federal law, shall be construed to require that marital status or the legal incidents thereof be conferred upon unmarried couples or groups. (par. 7)

If passed, this Amendment would target not only LGBTQ individuals, but straights and their children, bringing back old notions of "illegitimacy" and "bastards." Building off of the momentum of the proposed amendment, October 12–18, 2003, was set aside for Marriage Protection Week, endorsed by George W. Bush (Marriage Protection Week 2003). Supporters were encouraged to stand up for heterosexual marriage and demand that their politicians publicly oppose homosexual marriage, all in the name of preserving the sanctity of the nuclear family. Since gay marriage is prohibited in nearly every state, or can be revoked via the whim of a referendum, these policies are inherently discriminatory since they only allow married people to be sexual. In the United States, LGBTQ individuals are de facto condemned to eternal abstinence by their own government.

This delegitimizing of sexual rights is justified on the belief that people "choose" to be gay or "choose" to have children. The choice argument is used to divide oppressed groups from each other as conservatives enact the recent strategy of posturing as supporters of "real victims" of oppression, namely African Americans, since they were obviously "born black," unlike gays who "choose" their lifestyle. The implication is that it is offensive for LGBTQ people to equate their oppression with those of other minority groups. Of course, the fact that religion is technically "chosen" yet is also considered a protected class is conveniently overlooked as is the history of the same politicians and religious leaders who pretend to be in solidarity with minorities. Conservatives actively opposed integration and interracial marriage in the 1950s and 1960s as antigay crusader Jerry Falwell did during the civil rights era

Torosyan's (2007) reading of an interview on John Stewart's *Daily Show* with William Bennett on the subject of gay marriage touches on the conflict between choice and human condition in terms of sexuality. Bennett attempted to turn the conversation about gay marriage into a "slippery slope" discourse and presented the common example of polygamy as the next step after gay marriage. Stewart pointed out that a polygamist chooses to marry two to three women, but being gay is something different. Stewart "speaks to the larger question of what it means to be human…[while] Bennett treats differences of human condition as subject to choice and, hence, regulation" (p. 112). Torosyan compares Bennett to Edmund Burke in that both share a great distaste for anything that subverts habit and tradition whereas Stewart presents the argument that society has to adapt to changing human needs and family forms.

Patriarchy and the naturalization of the nuclear family is part of homophobic discourse that ratchets up fears regarding the collapse of the traditional two-parent household (Cloud 2001; Eskridge 2008; Forrest and Ellis 2006;

a common view promoted in the ex-gay or "restorative psychology" movement (Chambers 2006; Dallas 2003; Exodus International n.d.; Santinover 1996). How can aggressive gays who continue to advocate for marriage do so in the face of such reasonable alternatives? Blog commentary following MacDonald's (2009) post on the Free Republic Web site confirms that abstinent gays are to be "admired." One poster describes an "old friend" who is now "a retired abstainer with an antique shop" who finds the whole gay scene just too overwhelming and dangerous.

The sheer arrogance of these comments is reminiscent of women under the thumb of Comstock laws in the early 1900s being told by their doctors that the only option available if they didn't want to have children was to abstain from sex with their husbands. Again, this "sex isn't necessary" attitude assumes that sexuality outside of being "productive" (producing both children and the future workforce) is something of a luxury and therefore it is perfectly acceptable to limit these rights since they are not legitimate in the first place. Of course, this carries over into common rationales behind not providing universal health care, child care, or paid leave: "They should have thought of THAT before having children." For those operating outside the norms of the nuclear family, monumental reproductive decisions are thus reduced to a trivial level akin to picking out a wallpaper pattern. As Ettelbrick (1989) asserts, "Marriage creates a two-tier system," becoming "...a facile mechanism for employers to dole out benefits, for businesses to provide special deals and incentives, and for the law to make distinctions in distributing meager public funds" (p. 127).

If it isn't enough to limit rights through religious beliefs, federal guidelines for schools impose abstinence on the LGBTQ community by advancing this definition for all sex education programs that seek funding:

> **Abstinence means voluntarily choosing not to engage in sexual activity until marriage.** Sexual activity refers to any type of genital contact or sexual stimulation between two persons including, but not limited to, sexual intercourse.
>
> Throughout the entire curriculum, **the term "marriage" must be defined as "only a legal union between one man and one woman as a husband and wife**," and the word "spouse" refers only to a person of the opposite sex who is a husband or a wife (U.S. Department of Health & Human Services, 2006).

Similar prohibitions underlie the proposed *Federal Marriage Amendment*, H.J. Res 56 (2003), which could have been lifted right out of Margaret Atwood's (1998) Republic of Gilead in *The Handmaid's Tale*:

> Marriage in the United States shall consist only of the union of a man and a woman.

gay couples are able to, for example, participate in gentrifying neighborhoods by getting rid of lower income LGBTQ people, many of whom are transgendered youth, the elderly, and minority groups.

Antigay messaging is specifically used to raise funds for organizing, appealing to popular and patriotic notions of discipline and authority through the family, even when one's own family is outside of the nuclear family norm (Fone 2000; Warnick 2009). The idea is that granting "special rights" to LGBTQ and unmarried straight families by removing sources of discrimination will somehow contribute to the instability of the heterosexual nuclear family and threaten the rights of heterosexual parents, the theme of Anita Bryant's Save Our Children campaign in the 1970s (Eisenbach 2006). Advancing civil rights is thus opposed on a Puritan reading of the law where the government is responsible for maintaining a moral society, individual liberties must bend to the common good, and the natural order of things must be followed, even private matters such as sexuality (Eskridge 2008; Kotulski 2004; Lakoff 2002).

At the heart of these arguments, sexuality outside of the confines of the capitalist nuclear family is presented as a "luxury" or "excess" that one can choose to succumb to or not. Even conservative gay marriage supporters such as Sullivan (1996, 2004), Rauch (1996), Bawer (1994), Brooks (2003), and the editors of *The Economist* (1996) utilize these perspectives in their writings, urging marriage activists to not push social changes too quickly or think of gay marriage as a civil right (i.e., don't use the courts, use the legislative process, etc.). Marriage becomes a way to contain deviant sexualities while adhering to the privatization that conservatives love dearly. Those who agitate for civil rights through gay marriage, rather than joint property ownership alone, are presented as wonton hedonists who are pushing their agendas onto a hapless public whose heterosexuality is apparently so fragile that the very thought of legitimating LGBTQ relationships can lead to visions of a collapsing Western civilization (Bruce 2004). People, especially the LGBTQ community, are supposed to be willing to sacrifice sexuality for the good of the larger cause of decent, God-fearing heterosexuals who are the foundation of society via the nuclear family. Witness the comments of MacDonald (2009), an ex-gay Christian singer who suggests that contrary to popular opinion, God loves homosexuals and will show mercy. All they have to do is remain abstinent and sin no more:

> The culture of pushing sex down people's throats is not working. There is nothing wrong with abstinence from sex. This goes for everyone who is not married to someone of the opposite sex. I've been single and chaste for many years after having left the gay community. You don't die from not having sex. It's not like air or water. (par. 9)

Note that this is considered a TOLERANT position within the religious right. Abstinence is kindly presented as a way around hellfire and brimstone, through thoughtless political correctness. Unspoken, but nonetheless clear as a bell, is the implication that LGBTQ individuals cannot be moral (Lakoff 2002). Scalia's dissent also includes the usual slippery slope argument that bigamy and polygamy were prevented by laws and the removal of the ability to discriminate based on sexual orientation would loose a wave of multiple marriages, including to relatives, children, or even animals.

According to Eskridge (2008) "the politics of criminal sodomy, which had become a politics of lies, racism, fag bashing, and constant dose of denial, came to an end in 2003" (p. 301). Overriding the decision in *Bowers v. Hardwick* (1986), which upheld Georgia sodomy statutes, *Lawrence v. Texas* (2003) overturned state laws prohibiting sodomy. Predictably, the reaction of the right wing was one of moral panic. The day of the Court's decision, a group of conservatives held a prayer vigil in the Court, reading Scalia's *Lawrence* dissent declaring the decision "judicial activism" (not the laws against sodomy, of course) (Eskridge 2008). Scalia and other conservatives claimed that the precedent set in Lawrence would "constitutionalize" homosexuality, leading to same-sex marriage rights:

> Indeed, state justifications for denying marriage rights to same-sex couples—signaling traditional family values and protecting old-fashioned marriage—are not hugely different from those...advanced in Romer or...suggested in Lawrence. And in both cases, the Court ruled the justifications did not even constitute a rational basis. (p. 356)

The Lawrence decision sets the stage for an analysis of contemporary opposition to LGBTQ rights.

Ideologies of Homophobia

The Threat to Family and Society

Kahn (2009) maintains that the focus on sexuality within the family rather than opposition to sodomy represents an important shift in how the right wing frames "family values." The long view of the right includes rolling back legal protections against discrimination for all individuals, but LGBTQ people represent an easier target due to the power of homophobic ideology. However, straight families are also affected, as was seen in 2008 when Arkansas removed the right of gay and most straight unmarried couples to adopt or become foster parents (Brown 2008). These events prompted Sycamore (2008) to urge activists to widen the fight for same-sex marriage to include demands for health care, housing, citizenship, tax policies, and so forth for all people, not just married ones. By narrowly tying rights to marriage alone, or viewing marriage as the ultimate solution, LGBTQ individuals are still supporting a discriminatory capitalist system where wealthier married

Eisenhower signed an executive order in 1953 making sodomy and other acts of "sexual perversion" grounds for being fired from public sector jobs, including teaching (Wolf 2009a). These employment records could be shared with the private sector, along with reported "concerns" from neighbors or co-workers, creating a powerful blacklist that would follow a worker throughout their life. Making matters worse, vice squads were rewarded for their "productivity," or the number of arrests and prosecutions made (Eskridge 2008). By January 1955, over 8,000 people were declared security risks and were removed from their jobs, 600 of whom were involved in sexual crimes (Fone 2000). Ironically, Roy Cohn, McCarthy's aid and participant in the purges, was himself gay (Eisenbach 2006; Fone 2000). In many ways, the Clinton compromise of "don't ask/don't tell" is the McCarthy legacy; between 1998 and 1999, over 1,600 military careers ended due to either discharge or resignations because of social pressures (Fone 2000, 415).

In the wake of the Stonewall rebellion, homosexuality was removed from the American Psychiatric Association's list of mental disorders in 1973 (Kotulski 2004). Sodomy decriminalization soon followed in several states (Eskridge 2008). However, the repeal of sodomy laws was not the same as acceptance of homosexuality. Many communities "compromised" by repealing sodomy laws but making gay marriage off limits, or they fought to maintain the criminalization of sodomy to draw a line in the sand as part of a campaign of backlash. In 1991 Denver, like Dade County, Florida, in the 1970s, adopted laws prohibiting discrimination based on sexual orientation. Much like Florida, it didn't take long for televangelists to lead a crusade against these laws. Echoing Anita Bryant's Save Our Children campaign, hysterical stereotypes were promoted, leading to Colorado voters supporting an amendment by a margin of 53 percent to bar relief for victims of same-sex discrimination (p. 280). If LGBTQ individuals were barred from jury duty or the right to vote due to having a criminal record related to sodomy laws, according to the language of the amendment, they could not seek relief from Colorado courts. This amendment would be challenged in 1995, forming the basis of *Romer v. Evans*. The amendment was overturned in 1996 by the Supreme Court.

In his dissenting opinion, Justice Scalia famously declared:

> The court has mistaken a Kulturkampf for a fit of spite. The constitutional amendment before us here is not the manifestation of a bare...desire to harm homosexuals, but rather a modest attempt by seemingly tolerant Coloradans to preserve traditional sexual mores against the efforts of a politically powerful minority to revise ideals and mores through the use of laws (Scalia's Dissent, 1995, par.1).

Scalia's statement buttresses the homophobic rhetoric of the modern era which presents LGBTQ individuals as a "powerful" minority unfairly pressuring a decent and victimized majority who want to preserve family values (no matter how sexual) as not raising eyebrows due to the hidden nature of a homosexual culture as a whole. Many LGBTQ individuals remained hidden by choice due to continuing societal consequences.

An emerging sense of homosexuality as an identity began to form during the latter half of the nineteenth century, bringing with it a spirit of growing solidarity and resistance. Much of this included scientific studies that sought to examine homosexuality as a different form of sexual expression. However, this was not without cultural backlash. After Oscar Wilde's conviction, many works challenging homosexual persecution were suppressed (Fone 2000). Purity movements became a craze that targeted prostitution and alcohol (Eskridge 2008; Coontz 2005). Between 1881 and 1921, parallel to the nadir of race relations in the United States (Loewen 2005), the media at the time promoted what Eskridge (2008) calls "the lynching narrative" where black males preyed upon white women (p. 49). Along with this narrative were cautionary tales of gay men looking to recruit innocent lads. The combination of presenting African Americans as sexually rapacious with the "activist gay male" as recruiter made several prominent African Americans such as W.E.B. DuBois leery of joining with homosexual causes. The ruling elite was succeeding at keeping minorities and whites and gays and straights apart just at the time the labor movement was accelerating.

In the twentieth century, after the definition of homosexuality as a sexual identity was established, scientific publications appeared alongside novels and popular writing that featured stereotypical gay characters (Fone 2000). Dr. George Henry's (1948) *Sex Variants: A Study of Homosexual Patterns* took on the legal and social practices of entrapment, surveillance, and arrests of LGBTQ individuals as the source of blame, not the homosexuals themselves. Kinsey's (1998) groundbreaking report on American sexuality placed the number of men engaging in exclusively homosexual acts as between 4 and 10 percent of the population (D'Emilio 1992; Fone 2000). Kinsey argued that a significant minority of the male population in the United States had experienced homosexual sex from the time they were teenagers. As a result of his research, Kinsey opposed laws against consensual sodomy and he was the first prominent person in the United States to take a no-compromise position on the matter (Eskridge 2008). A more libertarian reading of consensual sex acts began to influence policy makers who distinguished between pedophilia and rape as deserving of criminal status while recognizing the inefficiency of prosecuting what happened between two adults regarding private sex acts.

However, as Eskridge (2008) points out, the social limits of privacy rights were displayed as judges hesitated to grant them to LGBTQ individuals, holding off until 2003's *Lawrence v. Texas* when state sodomy laws were ruled unconstitutional. The reluctance to apply libertarian principles to homosexuals was felt during the McCarthy era, where communism was linked to homosexuality and sodomy, subjecting LGBTQ individuals to witch hunts (Eskridge 2008; Fone 2000). After the purge of gays from the military, existence of the law served more as a means of enforcing social control and a convenient way to seize property from those accused: "the state claimed the right to judge the sexual activities of its subjects and also the right to define what was natural in terms of secular law, thus giving it unprecedented power over the bodies and behavior of its subjects" (p. 217).

During the Enlightenment, the transformation from religious offenses to secular ones was complete. However, social class played an important role: those with more wealth were less likely to be persecuted by the law (Fone 2000). The development of an early gay male counterculture in urban areas of England, including cross-dressing and Molly houses where men could meet each other, only heightened social anxieties. The image of a predatory homosexual recruiting one's children expanded to include more effeminate males as potential threats. At the same time, however, major upheavals, such as the French Revolution, resulted in decriminalization of homosexual acts in several countries in Europe, including the removal of the death penalty for sodomy in England in 1861. It is interesting to note that once decriminalized, sodomy could always be reinscribed as a crime at any time, as was the case during Napoleon's reign.

In the American colonies, sodomy was considered a capital crime. As in Europe, there were few executions performed, but the law's import as a means of social control remained intact (Eskridge 2008; Fone 2000). After independence, while the death penalty was removed in the original 13 states, laws against sodomy were maintained for the following reasons: the protection of minors (pedophilia), protection of women from sexual assault, protection from atheism, maintenance of community, and "to protect the institution of procreative marriage generally" (Eskridge 2008, 18). Even Thomas Jefferson, hero to liberals and libertarians alike, recommended castration for males who committed sodomy (Fone 2000). Sodomy was also bound up with notions of emerging nationalism and racial and religious superiority, as colonists suspected Native tribes of cross-dressing and engaging in homosexual acts, along with committing heretical practices such as land sharing (Fone 2000; Leacock 1981).

In the nineteenth century, a wave of writers began to challenge convention regarding sexuality in general and homosexual acts in particular. John Addington Symonds and Oscar Wilde presented homosexuality in a positive light, claiming it had moral virtue. In his writings, Symonds deconstructed the common religious arguments against homosexuality, exposing them as a sham and pointing to growing medical and psychiatric evidence that those who performed homosexual acts were not criminals (Fone 2000). Symonds also stated that gay males could not be detected and that many looked just like ordinary citizens, a radical idea at the time. Edward Carpenter, a socialist, viewed homosexuals as part of the overall struggle for working class rights, which placed him apart from other pro-homosexual perspectives that sought to resist more subtly by using effeminate mannerisms as hidden signals or retreating into Arcadia. Coontz (2005) presents same-sex friendships (prostitution, sodomy, or sex not resulting in children). From *The Laws*, Book VIII:

> Was I not just now saying that I had a way to make men use natural love and abstain from unnatural, not intentionally destroying the seeds of human increase, or sowing them in stony places, in which they will take no root; and that I would command them to abstain too from any female field of increase in which that which is sown is not likely to grow? (par. 35)

Plato begins to craft the traditional arguments against homosexuality: sodomy is unnatural, animals are not gay, gay sex is for gratification only and not for having children, and homosexuality and other unnatural acts lead to the destruction of society. Note that the problem is not sodomy between males; it is that this kind of sex act does not produce children and can lead to intemperance and a lack of virtue (Fone 2000).

In the Middle Ages, the interpretation of Plato's *Laws* was more narrow and the legal argument began to be constructed against homosexual relations. In 342 C.E., an edict mandated punishment for male prostitution. This was followed by a banning of all homosexual acts by the Theodosian code in 390 C.E. and a follow-up condemnation by the Church, stating that sodomy was a sin because it was unnatural (Fone 2000). In 533 C.E., emperor Justinian proscribed the death penalty for sodomy and linked natural disasters to homosexual acts, much as televangelist John Hagee did in the aftermath of Hurricane Katrina when he claimed that Katrina was God's way of disrupting a gay pride event (B.J.L. 2008). In the twelfth century, Alain of Lille claimed that sodomy and murder were the two most serious crimes (Fone 2000). The scope of sodomy laws spread beyond same-sex relations; since sodomy included any act not leading to procreation, it also affected heterosexual relationships. More importantly, with these laws on the books, rulers could use the threat of punishment as a means of controlling populations and dissident actions, since ties to heresy were implicit in these laws. Sexuality outside of procreation was now linked to civil disruption and treason and the marriage ceremony, reified, was the means of protecting the population from such treachery and sin (Forrest and Ellis 2006; Fone 2000).

Another important shift in legal and societal policies during the late Middle Ages was from one of detection and punishment to outright prevention and persecution. By the fifteenth century, some cities experienced mass panic about the family and society. Fone (2000) describes how Florence regularly dispatched police to locate those suspected of committing homosexual acts. In the span of 70 years, 15,000 males were tried, resulting in over 2,000 convictions for sodomy (pp. 193–4). In the other major urban areas during the Renaissance, roughly 16,000 people were accused of committing homosexual acts and of those tried 400 were executed (p. 214). Even though the number of executions was far fewer than the number of people tried, the

Brooks (2005) explains that homophobia became internalized to the point where he didn't consider his same-sex partnership to be as important as those of his heterosexual co-workers who were able to talk openly about their families. He was willing to forgo talking about his partnership as if it were equal to "legitimate" heterosexual marriages.. Heterosexuality is normalized to such a degree that everyday expressions are taken for granted. In more family-related work settings such as the classroom, this becomes pervasive, as Petr (2005) carefully explains:

> Every year the social committee collects $20 from us. This is to pay for new baby gifts, wedding gifts, anniversary gifts, basically any celebration of meeting societal norms and expectations. Every year I fail to pay by the deadline. Every year some poor fool in charge of collecting the funds hunts me down. Every year I tell myself I'm going to finally step up and say, "Hey, I'm never going to benefit from this fund. I'm not going to get married if the current government has its way; I'm not planning on having a baby, and there's no way I'm going to accidentally become pregnant. Therefore I'm not giving you my money." Every year I hand over a check for $20. Introverts don't like confrontation. (p. 210)

Homosexuality in Ancient and Modern Times

According to Eskridge (2008), laws against sodomy provide a de facto history of sexuality for both straights and gays, revealing an overall discomfort with "the morality of pleasure," especially in the United States (p. 2). History also reflects the development of homosexuality as a modern construct, made possible by the growth of capitalism (D'Emilio 2009). As capitalism intensified through industrialization, the notion of a sexual identity called "homosexual" took hold at the same time the nuclear family developed (Coontz 2005; D'Emilio 1983, 1992, 2002, 2009). This led to anxiety and tension surrounding the family and perceived threats to the family that continue today.

In Ancient Greece and Rome, masculinity and femininity along with attendant roles were standards not to be violated; sexuality was not an identity per se (Coontz 2005; Forrest and Ellis 2006). There were no laws in ancient Greece or Rome condemning sex between males until the third century c.e. outlawing gay prostitution. If one engaged in homosexual acts, what mattered more was being masculine, in control of one's actions, and not disruptive to society (Fone 2000). Laws regulating sexuality also focused on relations between slaves and freeborn citizens, tying sexual rights and protections to property ownership. Fone discusses the emergence of homophobic discourse in *The Laws* of Plato, centered on an emerging sense of pedophilia as a crime and distinguishing natural sex (procreation) from unnatural sex relationships and making them more vulnerable. The advent of AIDS and its association with "promiscuous" gay males permitted mainstream Americans to support such extreme acts as police searching the bedrooms of consenting adults. Ironically, Washington, D.C.'s law against sodomy didn't stop the city from having one of the highest HIV infection rates in the United States (p. 218). Laws and policies that imposed the closet on many LGBTQ individuals and straight allies harmed public health initiatives to stop the spread of HIV/AIDS, creating misconceptions that only gay white men could get the disease. The Reagan administration continued their crusade against teaching young people about safe sex and refused to fund research on HIV/AIDS until the death of Rock Hudson in 1985, forcing them to face the matter.

Homophobia: The Personal Impact

Homophobia is divisive in two ways: it serves to separate people from each other, further benefiting the ruling class and it separates a person from his- or herself, creating intense alienation and a the belief that assimilation and the closet is the lesser of two evils (Fone 2000). This form of alienation also benefits the elite because closeted individuals are less likely to speak out or draw attention to themselves, making challenging workplace inequalities more rare. Ironically, extreme stereotypes of LGBTQ individuals (i.e., effeminate gay males or masculine lesbians) can serve to make it easier for the closeted person to remain hidden, since they often do not exhibit these characteristics (Eisenbach 2006).

In some cases, prejudice within the gay community itself against "extreme displays" of gender identity by the LGBTQ community play on the notion that antigay discrimination from straights is justified, a viewpoint that the "gay right" endorses through the Independent Gay Forum. Bawer's (1994) classic work, *A Place at the Table*, seeks assimilation as the solution for homophobia, rejecting collectivism. Debates about including transgendered individuals emerged in 2002 with New York's Sexual Orientation Non-Discrimination Act and in the Employment Non-Discrimination Act of 2007, revealing that some gays and lesbians (along with liberal straight "allies") did not want to extend these protections to gender identity for fear that the sexual orientation component would be compromised if it "went too far" (Rotello 2007; Spade 2008).

In the classroom, sex is already a contentious issue, so many LGBTQ teachers are reluctant to disclose their sexuality, even when asked. The book *One Teacher in Ten* (Jennings 2005) includes several accounts of LGBTQ teachers who struggle with homophobia. Lyons (2005) describes the constant hiding of his homosexual identity to be exhausting, causing him to use different tactics such as changing the subject when sexual issues came up in the classroom or simply not speaking about it. For self-protection, Gold (2005) would scan the newspapers before handing them out for a class project, making sure there were no gay-related articles, "This way I could have some kind of control over the conversations" (p. 261).

a deviant and immoral behavior; gay males in particular are predatory and spread disease such as HIV/AIDS; and gays are atheistic, seeking to destroy the nuclear family (Eskridge 2008; Forrest 2006; Lakoff 2002). Eskridge (2008) also connects homophobic laws to racism and an overall promotion of second-class status for women. Allowing homophobia to flourish sends a message that weakness will not be tolerated—meaning displays of femininity by males— promoting hatred against women on another front. In the United States, religion feeds into legal consequences, as 69 percent of those who identify as Christian believe that homosexual relationships should be illegal, compared to 44 percent of non-Christians (Fone 2000, 420). In 1998, 64 percent of Christians were unwilling to extend legal protections in the workplace to homosexuals (p. 420). Fone reveals that only 11 states make discrimination against LGBTQ individuals illegal and only half have enacted hate crime laws in the wake of Matthew Shepard's death. This contributes to growing violence against not only homosexual whites, but gay minorities; vulnerable younger LGBTQ populations who reside on the streets, often cast out of their homes due to their sexual orientation; and people with HIV/AIDS (O'Brien 2008). In many ways the refusal to enact hate crimes legislation promotes state-sponsored homophobia, and feeds into assumptions that the government has the right to monitor the activities of LGBTQ people.

The timing of landmark legislation and court cases concerning same-sex rights also heightened the legal impact of homophobia. Far from being a strange footnote in history, the persecution of the LGBTQ community was a logical extension of Cold War containment policy: "Just as hidden enemies imperiled the security of the nation, dangerous criminals lurking in the shadows menaced the postwar family" (D'Emilio 1992, 68). During McCarthy's communist witch hunts, California began investigating the public school system for the presence of homosexual teachers. Laws against sodomy were used to justify the firing of teachers, since educators were not allowed to be "law breakers." From 1958 to 1964, suspected homosexual public school teachers, college students, and professors resigned or were dismissed (Eskridge 2008). Colleges and universities willingly took part in suppressing gay rights and exposing homosexuals. California law was especially harsh as sodomy was included on the list of crimes for which a second offense could mean automatic life in prison. In the 1970s district judges upheld statutes that forbade "disruptive pro-gay advocacy" by teachers, even if they were straight. What was considered "disruptive" was pretty broad and left up to school boards. The purpose of these statutes was to create a climate of fear in conservative states like Oklahoma, where teachers were already under intense pressures to conform to community and religious norms.

In 1986, in the early days of the HIV/AIDS epidemic, *Bowers v. Hardwick* upheld the right of a state to maintain its sodomy laws, refusing to extend privacy rights to LGBTQ individuals. Eskridge (2008) explains that legal arguments were made supporting the criminalization of only homosexual sodomy since marital sexual acts were considered procreative, thus isolating same-sex and Hawaii and were on the way to modest success when the Defense of Marriage Act (DOMA) passed during the Clinton administration in 1996. DOMA assured that states would not have to recognize out-of-state same-sex marriages or the attendant rights that come with marriage. DOMA marked a significant shift in antigay policies. Up until its passage, the focus was on maintaining antisodomy laws. After DOMA, sodomy was no longer the main threat to traditional family values; gay marriage was (Eskridge 2008). This presented gay rights groups with a dilemma since they had put the majority of their support behind fighting sodomy laws in court, not confronting the right-wing political machine. When conservative states began to overturn sodomy laws, these groups were not prepared to switch to the fight for marriage and other gay civil rights. Indeed, many activists kept voting for the same Democratic politicians who tended to waffle on important issues, fearful that support for gay rights would bring on accusations of not being tough on child molesters, one of the issues that emerged after McGovern's public endorsement of gay rights in the 1972 election (Eisenbach 2006).

Homophobia: the Legal Impact

Homophobia presents a contradictory problem for capitalism. It negatively impacts the workplace, reducing productivity and morale but at the same time can be used to successfully divide and control workers over issues such as benefits and same-sex marriage or even civil unions (Morton 2001; Nowlan 2001). In this sense Nowlan (2001) is correct when he describes homophobia as "a structural relation of inequality produced in the social relations of production under the capitalist mode of production" (p. 137). Morton (2001) asserts, "to end homophobia, one must first end capitalism" (p. 2). This is far different than framing homophobia as an issue of tolerance that workplace diversity training can address. In many cases, better paid LGBTQ workers are content to depend on the granting of same-sex union rights by the companies they work for, which is tenuous at best (companies can renege on benefits at any time, as Kotulski [2004] points out), only serving to confuse key differences between those who own the means of production and those who do not. Morton (2001) sees this approach as workers seeking "not the end of wage labor, but capitalism with a smile" (p. 2). The tenuousness of relying on minimal reforms for human rights is starkly evident in larger struggles, such as Proposition 8, where a May 26, 2009, court decision upheld the ballot initiative results banning gay marriage while leaving the already 18,000 such granted marriage licenses valid. This was presented as a compromise. Robin Tyler, one of the plaintiffs in the lawsuit, expressed that she took no comfort in her marriage being preserved while other marriages were denied to LGBTQ individuals (Schulte 2009b).

Moving to the legal impact of homophobia, laws and policies are influenced by stereotypical beliefs about LGBTQ individuals: homosexuality is that research has become ideological in its search for truth. This viewpoint has persisted as part of the backlash against sexuality rights, with the assumption that, of course, truth claims aren't "ideological," particularly those emanating from the ruling class. As Moeller asserts, it isn't political correctness (PC), but "conservative correctness" (CC) that still dominates. This CC movement is far more established, more organized, and better funded than the liberal PC side. As an example, funding sources for the passage of Proposition 8 included Focus on the Family, Concerned Women for America, Family Research Council, Church of Jesus Christ of Latter Day Saints, Knights of Columbus, along with individual donations from wealthy elites like Elsa Broekhuizen, mother of Blackwater's Erik Prince, and Howard Ahmanson Jr., banker and follower of Christian Reconstructionism guru Rushdoony (who called for the implementation of Biblical law including stoning for homosexuals) (Khan 2009). Rampton (2008) notes that having a gay son who resigned from the church due to its stance on gay rights didn't stop Mormon pollster Gary Lawrence from becoming a campaigner in support of Proposition 8.

Illustrating the power of the religion industrial complex (Digby 2008), Berkowitz (2009) outlines a strategy used by David Kuo, former special assistant to President George W. Bush. Kuo helped to create a series of faith-based conferences in congressional districts where Republican incumbents were under threat in the 2002 elections. The conferences used populist language to promote faith-based charities as friends of the disadvantaged, providing food and other basic provisions so that government wouldn't have to. Nineteen of the 21 targeted GOP candidates won after the implementation of Kuo's strategy. Even though Congress has not officially enacted faith-based initiatives, 35 governors and more than 70 mayors from both political parties have created organizations philosophically oriented to the federal Faith-Based and Community Initiatives Program (p.11).

These efforts would have no salience if not for backlash, which reacts against the successes and gains in civil rights. Wolf (2009a) elaborates:

> In a strange sense, the right wing's fears that gay visibility would encourage others to either experiment with homosexuality or at least be tolerant of it turned out to be accurate. While the Right may shudder at that fact, the widening visibility and confidence of a gay movement did pave the way for others to come out and has transformed public consciousness ever since (p. 44).

Indeed, over the course of 20 years (1987–2007), support for recognizing LGBTQ relationships grew from 43 to 59 percent. When it comes to supporting equal rights on the job, the same poll shows that 89 percent of Americans are in favor of them (p. 44).

However, powerful backlash began building in the late 1980s when Denmark took the plunge and legalized same-sex relationships. Immediately LGBTQ couples began to challenge their exclusion in the District of Columbia analysis: "Dominant group members are vehement in their defending their own privilege, even as they deny its existence. Dominant group members claim that attention to the specifics of social location keeps them from being perceived and treated as individuals" (Moeller 2002, 173). Individualism is the default mode whenever followers of dominant ideology (ultimately the ideas of the ruling class) are confronted with mounting social evidence that throws a wrench in the works. The media plays its part in holding this back by insisting on "balance," whereby credible spokespeople for civil rights causes have to share airtime with extremists, such as abortion doctor murderer Paul Hill, thus legitimizing the fringe element as having a plausible "side" to present (Feldt 2004).

Eskridge (2008) explains that after the civil rights movement and subsequent legal enforcement in the workplace, expression of outright racism, sexism, and bigotry became less acceptable. However, it was still appropriate and one could be met with little resistance for holding homophobic beliefs. Homophobia and participation in backlash movements with its emphasis on hyperindividualism, fundamentalist religion, and narrow concepts of morality provide a convenient way for people to explain falling wages, a disappearing social safety net, crumbling schools, and so forth. "Family values" then becomes a rallying cry for backlash politics, taking the heat from where it belongs, on capitalism and the unsustainable nuclear family structure, and placing it on homosexuality or divorced/co-habiting straights as the source of social decline. D'Emilio (1992) considers the targeting of lesbians and gays during the McCarthy era as backlash against perceived social upheavals impacting the family (women in the workforce during WWII, women and men forming same-sex relationships while in sex-segregated war situations). According to McCarthyites, homosexuality could "spread," much as a disease spreads in a weakened body. The stresses faced by the family in the postwar era were tied to this "weakness" narrative, which affected national policy

Democratic Party victories in 2006 and 2008 have contributed to the belief of many progressives that conservatism is under siege and its legitimacy possibly permanently in question. But as Berkowitz (2009) points out, the Heritage Foundation, one of the premier right-wing think tanks, has exceeded its two-year Leadership for America fundraising campaign by almost 25 percent, raising a total of nearly $105,000,000 between 2005 and 2007 (p. 11). This will enable the Heritage Foundation to remain in the media even if it has to rely literally on "buying its way in," which means spokespersons appearing regularly as guests on cable TV news shows and policy papers ending up as direct feeds to newspapers. Kotulski (2004) echoes Berkowitz's concerns in explaining that these groups wield financial prowess to such an extent that Focus on the Family, one of the 29 organizations taking part in Marriage Protection Week, has a yearly working budget of $126 million compared to a $54 million dollar budget of one of the more prominent LGBTQ rights groups.

Moeller (2002) discusses the National Association of Scholars (NAS), a conservative group that argues that diversity has led to a lack of standards and in her opposition to hate crimes legislation, claimed that Matthew Shepard was merely the victim of a robbery, not killed because he was gay (Linkins 2009). To Foxx, Shepard's murder was nothing more than "a hoax that continues to be used as an excuse for passing these bills" (par. 1). Though Foxx later apologized, her comments reveal much about how homophobia dismisses the legitimate concerns of the LGBTQ community. All in all, homophobia stands "as the last acceptable prejudice" (Fone 2000, 411).

A Problem of Backlash

It is impossible to consider modern forms of homophobia without a working understanding of how backlash operates to shape political discourse. Backlash refers to a period of intensified racism, sexism, classism, homophobia, or nationalism following the actual and/or perceived gains of an oppressed group. Usually these gains are hyperinflated to make it appear that the majority now becomes the persecuted minority. Civil rights are transformed into "special rights," implying that a minority group is receiving "preferential treatment," always at the expense of the majority (as if rights are subtractive and a limited commodity). Even though the social context of a particular form of discrimination might be unique based on time, place, and who is involved, the tactics of backlash are relatively predictable (Kotulkski 2004). Typically, the dominant social group presents itself as neutral and its values as normative, implying that the group seeking civil rights is unreasonable, petty, pushy, or even dangerous; they have an agenda (Cudd 2002; Kotulski 2004; Moeller 2002; Nowlan 2001). Cudd (2002) asserts that backlash is essentially an effort to deny a group of people civil rights while Nowlan (2001) posits that backlash benefits ruling elites who use divisive rhetoric to keep workers apart by normalizing capitalism as a natural organizing feature of human society. Roberts (2005) explains how this operates in the context of heterosexism and being an educator in capitalism's schools:

> Until recently intense pressure was put on everyone in society to be normal. "Normal" basically meant everyone was expected to participate in our school system, become a productive member of society, marry, have children, and not deviate from societal expectations of heterosexuality and appropriate gender role behavior. Teachers for their part were expected to be society's role models and, as such, could not deviate far from the norm, especially in such crucial aspects of their lives as sexuality and gender identity. (p. 170)

The decision of many educators to come out and refuse to hide their sexual identity in the classroom violates the norms of society; norms that also serve to hold back the working class as a whole. According to backlash logic, it isn't the dominant society that is ideological; it is the minority groups who have an agenda. When oppressed groups begin to name different forms of oppression, the dominant group cries foul and refuses social context as a point of mentions, it seems to be more prevalent in Western cultures or former colonies of Western cultures. In addition, it is linked to rabid sexism and the use of sexual violence against perceived offenders, in particular trans-individuals (O'Brien 2008). Homophobia is also a result of growing anxiety about oral and anal sex among heterosexuals, making gay sodomy the sole sexual target of legal action in order to safely segregate such practices away from "wholesome" nuclear families (Eskridge 2008). Most tragically, homosexuals themselves can be homophobic, as is the case with Idaho Senator Larry Craig and televangelist Ted Haggard. D'Emilio (2009), Ettelbrick (1989), and Browning (1996) locate the oppression of sexual minorities in capitalism and the economy in general and the institution of the nuclear family in particular. Homophobia excludes a broader range of family forms and sexual practices, considering them "unacceptable," permitting discrimination to flourish in the form of exclusive social and legal protections under heterosexual marriage, negatively impacting working class gays and straights as a whole.

The concept of homophobia was groundbreaking because it made the "problem" of homosexuality rest with the holder of the prejudice, not the LGBTQ individual (Eisenbach 2006; Forrest 2006). This placed things in a dramatically different light. Up until the late 1960s, it was the responsibility of the LGBTQ person to "prove" that she/he was not a child molester, a recruiter to the gay lifestyle, or even a communist. With the problem now squarely on the shoulders of the homophobe, it was up to them to demonstrate their tolerance (or lack thereof). With the tables turned, reactionary rhetoric in the wake of this situation was predictably fierce. Fone (2000) continues:

> Many homophobes allege the truth of fundamentalist biblical arguments that are said to prove and condemn homosexual sinfulness and substantiate its status as a perversion. Some Christian fundamentalists even insist that homosexuals afflicted with AIDS are sinners appropriately punished by divine retribution. Many homophobes, whose political views are colored by their religious convictions, crusade to deprive gay people certain civil rights that at present are guaranteed to all, gay and straight...for homophobes, the terms homosexual and heterosexual register absolute concepts of sexual difference, and pit abnormality against normality. (p .11)

Despite growing social pressures toward tolerance, homophobes continue to assert that they are not the ones with the problem. In many cases, they argue that it is part of the "gay agenda" to approve of homosexuality through the invention of the term "homophobia" and the removal of homosexuality from designation as a mental illness in the 1970s. Reversing the definition of victimhood, homophobes commonly place themselves in the role of the victim, fighting against an overwhelming gay menace or being a meek defender of traditional values against a hostile coalition of homosexual activists who will stop at nothing to get their way. Even North Carolina representative Virginia Foxx, of pedophilia and reminders that even if you have gay friends, don't let them near your children. At this point Wurzelbacher is most likely experiencing cognitive dissonance with his past appeals to lack of government interference and libertarianism, so he ends with "...they're people, and they're going to do their thing."

What Is Homophobia?

Even as late as 1999, 40 states allowed LGBTQ individuals to be fired from their jobs without cause. In the late 1990s, 54 percent of Americans identified homosexuality as a sin, with 59 percent seeing it as morally wrong. Forty-four percent believed that homosexual relationships should be illegal (Fone 2000, 12–13). Though the numbers are changing for the better, homophobic beliefs are quite salient.

Eisenbach (2006) traces the first use of the term "homophobia" to George Weinberg, a straight psychologist who worked with gay rights activists in the late 1960swhen. Weinberg and later K.T. Smith hypothesized that the irrational hate toward mostly male manifestations of homosexuality was a result of repressed homosexual tendencies in those perpetrating the hate (Fone 2000). Until recently, the invisibility of lesbians has in some way shielded them from the homophobia experienced by gay males. Indeed, homophobic individuals tend to level their wrath on gender defiance in general; that is, effeminate men or masculine women. Part of this derives from myths that differentiate between gay men who initiate sexual contact and those who are "innocent victims" of seduction; in other words, recruiter and recruited:

> ...one basis for this fear, many argue, is the perception that homosexuality and homosexuals disrupt the sexual and gender order supposedly established by what is often called natural law. Adverse reactions to homosexuals...are founded upon fear and dislike of the sexual difference that homosexual individuals allegedly embody....Another source of homophobia is the fear that the social conduct of homosexuals...disrupts the social, legal, political, ethical, and moral order of society. (p. 5)

Fone proceeds to differentiate between three different forms of prejudice, arguing that homophobia is unique in that it fits into all three categories. Obsessional prejudice sees the homosexual as a dangerous enemy who seeks the destruction of society and therefore must be destroyed. Hysterical prejudice views LGBTQ people as inferior and a threat to "normal" sexuality. Those who hold narcissistic prejudice can't bear the existence of people who are not like them.

Homophobia tends to emerge with ferocity when there are perceived threats: economic, foreign, religious, health (i.e., AIDS/HIV), or security related (Eskridge 2008). It is also tied to nationalism. As Fone (2000) not reproduce! They recruit! Many of them are after my children and your children" (Eskridge 2008, 215).

The effects of the Dade County referendum were immediate. The Florida legislature banned homosexuals from adopting children and old "crime against nature" laws were reinstated, making sodomy a felony. Other cities, including Wichita, St. Paul, and Eugene repealed their gay rights ordinances by using the referendum tactic (Eskridge 2008, 212). And the right wing consolidated their power by using an antigay platform to support a wider antiworker series of policies and politicians.

The Nature of Homophobia

These right-wing "Dogpatch demagogues" (Freeman, 2009, par. 38) and "Mayberry Machiavellians" (Suskind, 2003, par. 12) have expanded their attack to include not just perceived threats to U.S. safety and security but LGBTQ individuals and allies as well. In an interview with *Christianity Today*, faux populist media creation Joe the Plumber (Samuel Wurzelbacher) was asked how he felt about same-sex marriage being legalized in Iowa:

> At a state level, it's up to them. I don't want it to be a federal thing. I personally still think it's wrong. People don't understand the dictionary—it's called queer. Queer means strange and unusual. It's not like a slur, like you would call a white person a honky or something like that. You know, God is pretty explicit in what we're supposed to do—what man and woman are for. Now, at the same time, we're supposed to love everybody and accept people, and preach against the sins. I've had some friends that are actually homosexual. And, I mean, they know where I stand, and they know that I wouldn't have them anywhere near my children. But at the same time, they're people, and they're going to do their thing. (Slack, 2009, par. 4)

Wurzelbacher's answer is remarkable for including all of the essential elements of contemporary homophobic rhetoric in one paragraph. First, there's the appeal to state's rights, a device used throughout history to deny civil rights to African Americans (Morris 1984; Schaller 2008). Then notions of normalcy are brought up through the use of the word "queer" to imply biological inferiority under the surface while still downplaying any reasons that LGBTQ individuals might be upset at that kind of derogatory labeling ("it's not like a slur..."). Bringing in reverse racism always helps, as in the suggestion that a white person being called "honky" is a slur on par with "nigger" or other derivations. Wurzelbacher then moves to essentialist/Biblical gender roles, or "what man and woman are for." However, to avoid turning off too many people at this point, a slight detour is introduced, reminding everyone that after all, we have to still love gays and it is even better if one can point to having gay friends. Finally, it never hurts to follow up with hints an affirmation of traditional marriage"), making the No side appear paranoid when they tried to reply to these assertions.

The No on 8 campaign did not present advertisements with positive images of LGBTQ couples. Dickinson (2008) describes the approach as one of "if we hide, they'll give us our rights" (p. 47). The ads that were used tended to project confusing messages, such as a spot portraying a woman trying to convince her friend that while she "understood" her reluctance to support gay marriage, she should consider it a good idea anyway. The No campaign also failed to target minority communities, most likely relying on the assumption that they were Obama voters and, by default, gay marriage supporters. The result of this negligence was turning a strong 17-point lead for the No on 8 campaign into a 15-point lead for the Yes on 8 supporters (p. 47). When the No campaign finally decided to respond, they called on a former director of the Log Cabin Republicans, one of the most conservative/assimilationist gay rights groups in the country, to lead the opposition.

When one delves into recent history, similar patterns are revealed. Eisenbach (2006) asserts that Reagan's victory in 1980 demonstrated to Republicans that they could win more conservative Democratic votes simply by launching an attack on gay rights. Throughout the latter half of the twentieth century and into the twenty-first, bipartisan language (in differing degrees of severity) has linked homosexuality with communism, secret organizing, and threats to the family (D'Emilio 1992; Eskridge 2008). Eskridge (2008) explains that from 1977 to 1993 revocations of protections for LGBTQ individuals through the use of local referenda were 79 percent successful. After 1993, however, local politics became more tolerant of gay rights, so sights were set on state referenda, where city vote counts could be easily overwhelmed by more conservative suburban and rural votes (p. 279).

In 1976, an effort to introduce a bill that would prohibit sexual orientation discrimination in the workplace in Dade County, Florida, was met with immediate resistance led by Anita Bryant, former Oklahoma beauty queen and orange juice saleswoman. It didn't matter that Dade county was "a liberal bastion" (Eisenbach 2006, 285) just as it didn't matter that California supported Obama in 2008. As with the fight against Proposition 8, Florida's attempt to overturn antigay policies was met with unexpected defeat. This was, after all, on the heels of the famous Stonewall rebellion and the mass exodus out of closets across the nation. Aware that the proposed bill would impact LGBTQ school teachers, Bryant's core argument resonated with a backlash-ready public: that the antidiscrimination ordinance would violate the rights of heterosexual Bible believers and harm the children. Familiar "no Promo Homo" rhetoric of gay men "recruiting" unsuspecting children permeated advertisements, suggesting a legal slippery slope starting with acceptance of gays via the antidiscrimination bill. "Kill a Queer for Christ" bumper stickers appeared soon after, accompanying Jerry Falwell's ominous mass-mailer warning that "Homosexuals do

2009, par. 22). Schulte argues that the fear of offending conservatives has been detrimental to labor unions, civil rights, and activism, placing these issues perpetually on hold. These factors played a pivotal role in the outcome of Proposition 8.

Two months prior to the election, Proposition 8 was behind by 17 points in the polls (Dickinson 2008, 45). The "No on 8" forces decided to portray marriage equality as a single issue not connected to health care, civil rights, immigration concerns, minority rights, or impact on straights (Kahn 2009). It didn't target its organization efforts on low-income communities of color; the very communities harmed most by antifamily policies. The *No* side also rested on their $20 million war chest that was meant to stop the gay marriage ban, rather than including LGBTQ grassroots organizations in the fight. The summer before the election, the Yes on 8 side had 100,000 volunteers working door to door, including on election day, comprising a density of 5 people per precinct to check voter rolls and contact people who hadn't yet voted (Khan 2009, 5). The No side could only organize 11,000 people to hold up "No on 8" signs near polling places (Dickinson 2008, 46).

When it comes to effectiveness, the political advertisements about Proposition 8 are also a study in contrast. The Yes on 8 campaign translated TV ads, informational brochures, and materials into 14 languages (including the often neglected Asian audiences); utilized free viral advertising through a prominent presence on social networking sites (Facebook, Twitter, MySpace); and took a more pragmatic approach, conceding on certain issues such as agreeing to allow hospital visitation for LGBTQ partners so as not to appear overtly homophobic and attract moderate voters (Khan 2009). High-profile media events such as a pastor's 40-day fast and "100 days of prayer" mobilized multicultural congregations. These events were fused with an immediate succession of effective advertisements, though misleading in content:

> Multiple advertisements told voters that without Proposition 8, their churches would be forced to perform same sex unions and be stripped of their tax-exempt status; that schools would teach children to practice homosexuality; and that even Obama had stated during his campaign that he did not favor gay marriage...Obama's statement against gay marriage was circulated in a flier...targeting African-American households. The campaign also used Obama's voice in a state-wide robo-call. (p. 5)

Kahn also mentions that the overall tone of the advertisements, especially those enhanced by Obama's own, was one of compassion toward LGBTQ individuals. People were told that this was all about protecting marriage and children, not removing existing rights from gays and lesbians. This made it difficult for counter-arguments via the media. Yes on 8 supporters could simply play the compassion card ("We aren't taking rights away—all we want is arguments about traditional family and gender roles being the foundation of society. The family and society proponents often work hand-in-glove with religious institutions. Yet each of these three ideologies has unique facets that are part of their function, as readily seen in Proposition 8.

Implications of Proposition 8

After a tumultuous, if not entertaining, two-year campaign, the presidential election of Barack Obama seemed, at least for the moment, to represent an interruption of the most rabid reactionary elements of conservative ideology, which had had its grip on the United States since the Reagan era. As the returns came in and Democrats braced for a monumental televised victory celebration centered on Obama's acceptance speech and hip hop star Will.I.am's hologram appearance, every major media outlet had cameras poised and commentary at the ready.

California's Proposition 8, a ballot proposition that sought to end gay marriage rights by changing the state constitution to bar same-sex marriage, wasn't a source of liberal concern at the time. This may have been because formerly Republican states such as Indiana and Florida were leaning toward Obama in the election and the projected electoral vote continued to tally in his favor, liberals assumed that any of the positions of the old guard would be swept away, bringing in a new progressive future. Chamberlain's (2008) warning should have been heeded: "keeping a conservative campaign on the defensive is not the same as a decisive victory over it" (p. 27). By the same token, winning a constitutional right does not mean that another group fighting for the same thing will prevail, as Eskridge (2008) describes in his analysis of *Bowers v. Hardwick* (1986). This case upheld Georgia's sodomy law criminalizing homosexuality, despite earlier rulings such as *Griswold v. Connecticut* (1965) and *Roe v. Wade* (1973). In these cases, even "sympathetic" commentary by Supreme Court justices didn't do much to improve LGBTQ rights.

The morning after the 2008 election, a different picture for LGBTQ and straight rights emerged. Proposition 8 passed by a narrow, but not miniscule, margin—52 percent voted for it (Khan 2009, 3). More than 2 million Obama supporters also voted in favor of Proposition 8, effectively endorsing the restriction of marriage to one man and one woman (Dickinson 2008, 45). Forgetting that the Christian Right makes up about 15 percent of the voting public and that their presence can throw an election (Chamberlain 2008), or that a majority of voters age 30 and under (including minority youth) opposed Proposition 8, the media immediately blamed African Americans for the success of the ballot initiative (Dickinson 2008; Kahn 2009). Also ignored was the long-standing practice of Democrats who have continually "...held out an olive branch to the religious right, knowing they have the support of those on the left of them safely secured" (Schulte

Chapter Two

Sources of Opposition to Sexuality and LGBTQ Rights in the Schools

This chapter examines the ideologies behind the resistance to sexuality education and LGBTQ rights and perspectives in K–12 education. As Hypolitio (2008) maintains, "one cannot separate the economic and political aspects of the neoliberal project in education from the cultural component of the conservative restoration" (p. 149). By necessity, this chapter focuses on larger social struggles, beginning with the implications of and lessons learned from the passage of Proposition 8 in California. A discussion of the nature of homophobia includes definitions of homophobia, how it operates primarily through backlash, and the legal and personal impact on teachers, students, and families. Because dialectic materialism is the theoretical frame of this analysis, a history of homophobia is explored, from ancient Greece through modern times. Finally, the following three powerful ideological sources of opposition are presented: the threat to the family and society, the religious freedom argument, and pseudo-science and secular homophobia. These categories are loosely based on Eisenbach's (2006) historical overview of homophobia:

> American homosexuals, however, still faced tremendous resistance from two powerful sectors that have long competed over the American mind: religion and science. Many religious leaders and institutions continue to regard homosexuality as a sinful state that individuals choose, while the American psychiatric community diagnosed homosexuality as a psychological sickness that needed to be cured. (p. 219)

It is important to note that these three categories are not entirely discrete. These ideologies intertwine when it is convenient for their proponents. For example, the religious freedom crowd will often invoke secular/scientific language in efforts to broaden their base and legitimize their cause for the mass media. The pseudo-scientists regularly appeal to well-worn and unproven

The Shepards made the right choice, and we can learn something about how best to handle intolerance in all of its forms. This family was faced with the choice to request the death penalty for McKinney or to accept an agreement of life without parole. As part of the agreement, McKinney could not ever talk about the murder through media interviews or profit from the case in any way. In choosing the latter, the Shepards opted for the ultimate nightmare for radio preachers and Rush Limbaugh followers who are usually given carte blanche media access in the name of "free speech" to spread homophobic rhetoric. McKinney's imposed silence meant there would be no "gay panic" arguments endlessly discussed via a video feed from prison by his attorneys on daytime talk shows, no Dr. Dobson publicity stunts visiting McKinney on death row to convert him to Christianity while making a case for how he was "recruited" by being molested by a perverted homosexual, and no "other side of the story" lovingly crafted by Glenn Beck and transmitted to his perpetually frightened audience. In addition to life without parole, McKinney's secondary—though no less powerful—sentence was enforced irrelevance.

These two high-profile tragedies—Matthew Shepard and Bobby Griffith—encapsulate different aspects of the culminating effects of a society. This is a society that refuses to overtly protect sexuality rights, from a lack of K–12 educational policies that openly prohibit gender-based bullying to opposition to hate crimes legislation. It seems there is more hand-wringing over the possibility that those promoting homophobic rhetoric would be unfairly persecuted than a need to rally around the enabling of young people—gay or straight—to exist in an environment that takes a stand against physical and emotional harm. Even though hate crimes laws and anti-harassment policies won't prevent all actions targeting LGBTQ individuals, there were many important opportunities in the lives of both Bobby Griffith and Matthew Shepard where, at the very least, life could have been made more bearable. The same is true for Shepard's killers. If the schools of any of these individuals had perhaps included a gay-straight alliance or integrated, positive depictions of LGBTQ individuals in the curriculum, the pathology of homophobia would have been placed on the shoulders of society at large rather than internalized within.

Matthew Shepard didn't have to contend with intolerance within his own family, as Bobby Griffith had to. His parents were fortunate to be raised in relatively open-minded households where religion wasn't overt and oppressive. Though never repulsed by or scared of gay people, Judy Shepard (2009) recalls not noticing them while she was a college student, indicating a general climate of silence:

> When I was growing up in Glenrock, I didn't know anyone who was gay (or at least I didn't know that I did). The very concept of "gay" was something I never really thought about—and I don't think the people around me thought about it either. It simply wasn't on anyone's radar. (p. 33)

Much like Mary Griffith, Judy Shepard did not have much access to information about the LGBTQ community. As she began to wonder about Matthew's sexual orientation, she didn't have much to go on other than recent coverage of HIV/AIDS. "I really couldn't expect to find anything in the Caspar library, and there weren't (and still aren't) any gay bookstores in Wyoming" (p. 35). She and her husband, Dennis, Matthew's father, pretty much let things be.

Suffering from posttraumatic stress after a rape and robbery while traveling, Matthew found it difficult to tell his parents about his sexual orientation, though they did not react negatively. Instead, most of the struggle related to his bouts with extreme depression and chemical dependency. At the time, those were attributed to growing pains and his need to work things out with minimal parental meddling. After a few close calls—including an attack by a bartender who thought Matthew was making a pass at him—Judy eventually was convinced that Matthew was back on track after getting established and making friends at the local college in Laramie. Only later, when the autopsy results came in would Judy and Dennis discover that Matthew had recently become HIV positive.

The details of Matthew's murder are well known and Judy describes in several moving accounts the unfolding of events in her memoir, *The Meaning of Matthew*. She reveals how, during the trials of the murder suspects, Russell Henderson and Aaron McKinney, attorneys attempted to use the "gay panic" tactic in crafting an argument as to why both men became overcome by rage when they thought Matthew was coming on to them. In McKinney's case, his attorneys tried to build on the homophobic fears of "gay predators" by recounting how McKinney had been molested at an early age, resulting in his not being able to tolerate the presence of gay men. The presiding judge would not allow this type of defense to stand, which facilitated a guilty verdict for both. Henderson was given life in prison for murder, robbery, and kidnapping. During the trials and shortly after, Judy established the Matthew Shepard Foundation and became active in supporting hate crimes legislation.

Nothing seemed to work and after a futile attempt at self improvement, Bobby withdrew even further, eventually dropping out of high school. The inability to "cure" himself created even more depression and shame, as Aaron (1995) explains: "Bobby was loved, but was simply not okay. To be acceptable and accepted, he would have to change. It occurred to no one at the time that it might have been the family's responsibility to change, not Bobby's" (p. 98). Bobby filled his journal with romantic sentiments of longing, to find a man who would love him. Even when he met other gay people at work who were more secure with their sexual orientations, he didn't feel like he could talk about what was happening or learn from their advice. Eventually, he became part of the gay scene and focused intensely on his appearance. Only able to find menial work as a high school dropout, he took a short-lived job as a prostitute, experiences he recounted in his journal. Even after moving to Portland with his cousin Jeanette, who was a lesbian, Bobby was unable to see himself as capable of living a normal life.

Right after the suicide, Mary attempted to reconcile the Bible she knew with her son's death. Initially, she was concerned about the afterlife. After reading as far as she could on her own, she consulted a more liberal minister who explained the origins of the Bible and how Old Testament accounts were misinterpreted, such as the Sodom account. Mary's response was shock—why hadn't anyone revealed this information to her before? Couldn't homosexuality be one of many behaviors no longer recognized as sinful? Perhaps God had failed to heal Bobby because He didn't need to. After the shock wore off, Mary began to despair that her beliefs had led to his suicide and that all of her efforts to change him meant that she had missed out on getting to know her son as a person. At that point, Mary began to transform her internal grief into anger directed toward social context as a point of analysis, leading to her, as Aaron (1995) describes, crossing the Rubicon.

After joining PFLAG and a local group of gay and lesbian educators, Mary began to call for accountability of the church for its narrow vision of heterosexuality and for imposing that vision onto vulnerable people, gay or straight. Mary's role as a speaker was important because she was a representation of what could happen if parents were not accepting of their child's sexual orientation. But Mary also understood that it wasn't a simple matter of "intolerance," but an overall lack of institutional support for LGBTQ people and their families that was the real problem. Project 10 in Los Angeles was the perfect vehicle for Mary to become involved with, as it sought to create a safe space in schools for LGBTQ and straight students. Mary and Bob also created the Bobby Griffith Memorial Scholarship to bring awareness to the positive side of being gay, and participated in PFLAG marches. In an afterward of the book, Aarons (1995) interviewed Mary about her altered views concerning religion. She stated, "Once my beliefs were my reality. Now my reality forms my belief" (p. 230).

The Griffiths lived in Los Lomos, California, during the time of growing gay activism in San Francisco. But in the sheltered existence of their household, these events seemed miles away. For Mary, homosexuality represented decadence and sin, and was condemned outright by her church, which espoused a more conservative wing of Presbyterianism. She recalled seeing news coverage of a rally where a mother was carrying a sign reading "we love our gay children" and being confused at the sight. This only heightened her fears as once-dormant questions about Bobby's sexuality emerged in her mind:

> The idea of Bobby being somebody society didn't approve of scared her, and not only for Bobby's sake but also for her own...Mary strove to discourage anything feminine...a vague sense of threat to her carefully ordered existence caused her to wish this problem away...in the early days she never equated sissy with homosexual. She simply feared difference. (pp. 45–46)

As puberty approached, Mary had to confront more and more aspects of Bobby's personality that she initially chalked up to his being a withdrawn, bored, and sometimes hostile adolescent. Bob had his own difficulties connecting to his son, but he was not opposed to homosexuality. For him, it was one more aspect of something unfamiliar and not part of his own experience.

At this time, Bobby began to keep a journal, which Aarons (1995) excerpts throughout the book. While Mary was beginning to consult Bible verses about homosexuality in attempts to assail her own fears, Bobby wrote:

> I can't ever let anyone find out that I'm not straight. It would be so humiliating. My friends would hate me. They might even want to beat me up. And my family? I've overheard them. They said they hate gays, and even God hates gays, too. Gays are bad, and God sends bad people to hell. It really scares me when they talk that way because now they are talking about me. (pp. 55–57)

After Bobby made an initial attempt at suicide, he eventually revealed that he was struggling with homosexuality. Mary's immediate response was to roll up her sleeves and attempt to cure him of his problem, though she later wrote "I did it all wrong with Bobby, Lord, you know it and I know it" (p. 70). With a conservative minister who didn't want to hear about homosexuality in his flock, and a lack of counseling services available in town or at Bobby's high school, Mary turned to books by Tim LaHaye and other fundamentalist authors. Bobby went through his own self-induced "improvement" plan of Bible readings and prayer. Except for Bob, the entire family got into the act of posting Bible verses and inspirational messages around the house. Mary even turned up the Christian radio station so its message could be heard in Bobby's room.

rejected by their families because of their sexual orientation, makes them targets for the police in the name of neighborhood gentrification (Rosado 2008). These LGBTQ youth are routinely rounded up by police or chased out of high-rent neighborhoods, often with the approval of more well-to-do gay and lesbian households and businesses concerned about the homeless "menace" in their communities. Goldstein's (2003) equation is correct: "the onus grows greater the poorer or queerer you get" (p. 18). The passage of this federal hate crimes expansion would bring important representation to those lower down the social food chain.

Bobby and Matthew

High-profile cases of ordinary people caught in tragic and extraordinary circumstances can provide important insights into larger social dynamics, as with Bobby Griffith and Matthew Shepard, both victims of homophobia. Countless individuals—teachers like Gold (2005)—attribute their decisions to come out as a result of these tragedies. However, as Perrotti and Westheimer (2001) remind us, there are many other victims, like Steen Fenrich, who never obtain this status, which unintentionally contributes to the misconception that being gay is a problem affecting only young, white, affluent males. In 1999 Fenrich, an African American, was found dead with racist and homophobic words written on his skull (Steen Fenrich, n.d.). Fenrich had been killed by his stepfather, John, who was enraged about his stepson's sexual orientation. (John later committed suicide at his "failure" as a parent.)

Aarons (1995) writes, "Unable to reconcile his gay sexual orientation with his family's religious and moral beliefs, Bobby had leaped to his death from a freeway bridge in 1983" (p. 2). So begins an agonizing and honest account told through the eyes of Bobby's mother, Mary, in *Letters to Bobby*. Aarons documents how Mary, a woman with little education and her own psychologically abusive upbringing to contend with, underwent a profound transformation at an unimaginable cost. As Aarons explains, Bobby was never thrown out of his house (as with many LGBTQ teens who reveal their sexual orientation), which makes his story all the more tragic and moving. Mary's motives arose from her childhood, where she was surrounded by images of a harsh and punishing God. She learned from her mother that "Damnation was a living reality for those who do not reconcile themselves with God...it was a struggle to stay within the narrow safety zone" (p. 29). After getting married and having children, Mary consulted only Christian books on parenting, in which discipline became a matter of saving future souls. Her husband, Bob, wanted nothing to do with Mary's Christianity, and tended to remain in the background, thinking the kids would make up their own minds about religion. As a young child, Bobby eagerly joined in the activities of the church, even wearing a ring with Jesus on it as outward evidence of his religious commitment.

end during Reconstruction in the late 1800s, the elimination of remaining sodomy statutes in 2003 removed important legal barriers between groups of workers that capitalism has always used to keep wages low. Goldstein (2003) describes the impact of removing the stigma of illegality from sodomy:

> I'm talking here about the effect on heterosexual men who could no longer count on faggots to affirm the sexual code. Every institution that depended on male hierarchy felt the shock of queer refusal, and the reaction was typical of the way men respond to panic: hysterical rage of a sort I'd encountered only in my nightmares of what it meant to be gay. (pp. 38–39)

Despite these instances of violence against LGBTQ individuals, Goldstein discusses a growing conservative gay subculture, which he calls the "homocon" movement. While the homocons may support same-sex marriage, they often want to keep it a state-to-state matter, as with the gay rights component of hate crimes legislation, in order to gently "bring along" more reluctant straight conservatives. In order to gain acceptance from the heterosexual population, the homocons believe that the LGBTQ community needs to not focus on activism, instead highlighting their economic prowess, which includes rejecting any semblance of victimhood through hate crimes legislation. This position severely narrows the range of causes to those that only impact more well-to-do LGBTQ folks, as well as feeding into right-wing arguments that gay rights are "special rights."

In October 2009, an expansion of the existing 1969 federal hate crimes law, the Matthew Shepard and James Byrd Jr. Act, was finally passed by Congress and is awaiting Obama's signature (H.R. 1592, 2007). Once signed, the act will be the largest expansion of hate crimes protections in over 30 years. It includes federal grants for local and state law enforcement and increased sentencing for bias crimes. What makes the act important is that it includes sexual orientation and gender identity status under federal protections. The act will also support kindergarten-to-adult settings, facilitating the development of safe schools programs:

> In implementing the grant program under this subsection, the Office of Justice Programs shall work closely with grantees to ensure that the concerns and needs of all affected parties, including community groups and schools, colleges, and universities, are addressed through the local infrastructure developed under the grants. (par. 5)

As with antiharassment policies, hate crimes laws will not prevent all assaults. But their presence makes it easier to enforce consequences while protecting those who might not be able to protect themselves, including straights who are perceived to violate gender norms. Even in densely populated urban areas where it is assumed that an LGBTQ presence is more readily accepted, the vulnerability of young people, many of whom are homeless due to being unlikely sources. For example, the Federal Equal Access Act (1984) was instigated by Utah Republican Senator Orrin Hatch in order to insert Bible clubs in public schools. This particular law has been referenced by the courts in support of groups establishing gay and lesbian student clubs. Schools have been found liable for punitive damages for failure to stop gender orientation and anti-gay harassment in *Nabozny v. Podlesny* (1996), *Henkle v. Gregory* (2001), and *Montgomery v. Independent School District No. 709* (2000). One particular case, *Doe v. Bellefonte Area School District* (2004), points to how a school can take important steps to have and enforce clear-cut policies against harassment. Because the school had responded to and documented each reported incident, the court decided in their favor. The opposite happened in *Flores v. Morgan Hill* (2003), in which the school perpetually ignored student-reported incidents of peer-to-peer harassment based on sexual orientation. The most recent issues to be addressed in the courts include the right of transgendered students to wear clothing expressive of their identity (*Doe v. Brockton School Community* 2000) and carving out a legal precedent for cyber-bullying (Meyer 2009).

Hate Crimes Legislation

When Matthew Shepard was murdered in 1998, a federal-level hate crimes law had been on the books for 30 years, though it doesn't specifically mention sexual orientation as a protected category even though anti-gay crimes stand at 16 percent (Shepard 2009, 258). In many respects, rising violence against LGBTQ individuals has a lot to do with an encouraging climate brought on by conservative backlash encapsulated in measures such as the repeal of state anti-discrimination laws in the 1970s, policies like Don't Ask, Don't Tell, and a variety of state referenda, such as the Briggs Initiative and Proposition 8. All of these legislative efforts were followed by a spike in anti-gay assaults and even LGBTQ suicide (D'Emilio 1992; Eisenbach 2006; Eskridge 2008; Forrest and Ellis 2006; Johnson 2009). In Britain, the passage of Section 28 in 1988 added to the already tenuous position of gay men who were perceived as responsible for HIV/AIDS, and resulted in an increase in anti-gay attitudes as documented by numerous political surveys (Forrest and Ellis 2006). As Eskridge (2008) explains in his overview of the social purpose of sodomy laws in the modern era, "...they situated homosexuals outside the normal protections of the law...many individuals felt they could victimize homosexuals with legal impunity" (p. 67). It wasn't until 1990, as part of the Hate Crimes Statistics Act, that violence against LGBTQ individuals was tallied and categorized as such in the United States (D'Emilio 2003).

As D'Emilio (1992) maintains, "...while capitalism has knocked the material foundation away from family life, lesbians, gay men, and heterosexual feminists have become the scapegoats for the social instability of the system" (p. 12). As with the intensified violence that took place after slavery's change their policies. For K–12 settings, this would include the Boy Scouts and religious organizations that discriminate against LGBTQ individuals while meeting on public school grounds.

When it comes to professional educational organizations, including the National School Board Association, safe school policies are not controversial and are readily endorsed. Yet Massachusetts and Wisconsin are the only states that require the inclusion of sexual orientation as part of anti-bullying school policies, and Minnesota is the only state that protects transgendered individuals against gender identity harassment, but not in schools (Perrotti and Westheimer 2001). Meyer (2009) explains that while many schools may implement policies against bullying and sexual harassment, these policies tend to be very broad and general, and do not tend to address many of the problems that lie behind negative and violent behaviors. In schools that do have clearly listed policies, the process of enforcement might not be clear, or teacher workloads are so great that often offending behaviors go unnoticed or unenforced. Or, as GLSEN and Harris Interactive (2008) found, a minority of secondary and elementary schools have specifically targeted LGBTQ students for protection in their anti-bullying policies, workshops, and curriculum. Part of the resistance to the inclusion of sexual orientation as a protected class has to do with the right-wing backlash against gay and lesbian issues in general. Part of this backlash includes beliefs that there aren't many (or any) LGBTQ people at a particular school, and if there are, there isn't evidence of harassment being a big problem since no one comes forward to complain (Perrotti and Westheimer 2001).

It isn't enough to assume that people will be treated fairly because it's the right thing to do. Clearly stated policies that stem from more powerful laws are important for several reasons, as Aarons (1995) and Perrotti and Westheimer (2001) list. Federal and state laws can provide important financial incentives for public schools to enforce consequences. Laws can also raise awareness and inspire people to become involved in activism. More vulnerable individuals know that laws and policies are speaking for them, and they might eventually feel inclined to come forward in the long run. Teachers can point to a policy and simply state that they have to enforce it as part of their job with no further explanation necessary, which would make enforcement more likely. "At the most fundamental level, if sexual orientation is part of school policies, then communities must not ignore that gay and lesbian students exist" (Perrotti and Westheimer 2001, 37). Conversely, a lack of clear-cut policies that protect sexual orientation sends the message that LGBTQ people are not worth protection or are invisible. Of course, having a safe schools policy can protect all students, contradicting the common backlash notion that civil rights are "special" rights.

The courts have continually referred to Title IX as a source of protection against gender identity and sexual orientation harassment, as Meyer (2009) addresses in her review of several key cases. Sometimes precedent comes from

Research on administrators' response to gender-orientation bullying is rare. The most definitive source is the 2008 GLSEN and Harris Interactive study, which addresses issues of teacher in-services regarding gender-based harassment. The role of teachers figures prominently in safe schools research, and patterns are similar between both K-12 administrators and teachers. Yet it is quite telling that a majority of LGBTQ students are not out to teachers and school staff, though most are out to their peers (Kosciw, Diaz, and Greytak 2008). In most surveys, the majority of teachers and administrators demonstrate awareness of verbal and physical harassment related to gender identity, though many don't perceive this to be a frequent problem (Forrest 2006; GLSEN and Harris Interactive 2008; Harris Interactive and GLSEN 2005; Kosciw, Diaz, and Greytak 2008; Meyer 2009). Participants also state that they consistently respond to such events, and that a majority of LGBTQ students would feel safe at school (GLSEN and Harris Interactive, 2008; Harris Interactive and GLSEN, 2005). However, LGBTQ students overwhelmingly state that teachers and administrators are either not present when homophobic instances happen or, if they are, they do not respond to such instances effectively (Forrest 2006; Harris Interactive and GLSEN 2005; Kosciw, Diaz, and Greytak 2008; Meyer 2009).

Dolor (2008) found that in her sample of well-educated and more experienced teachers—a group assumed to be more highly tolerant overall—nearly half were resistant to confronting homophobia in the classroom and one-fifth did not feel that it was a teacher's obligation to include LGBTQ students as a protected class (p. 171). Her conclusions raised the question: If this group of teachers was not willing to address homophobia, would LGBTQ and straight students who had less educated or inexperienced classroom teachers be even worse off? Indeed, in terms of teacher intervention with regard to regions of the United States, students in the West reported more teacher involvement in stopping gender-orientation and other forms of harassment. Students in the South were less likely than any other region to report teachers intervening for any type of harassment (Kosciw, Diaz, and Greytak 2008).

Aarons (1995) and D'Emilio (1992) recommend that to combat homophobia within school settings, policy has to set the standard, followed by administrative enforcement, whether one is talking about K–12 settings or colleges and universities. Related to policy are federal-level protections for all people, gay or straight. But policy and enforcement isn't enough. Students need gay-straight organizations, diverse activities, life skills classes, and an intellectual and integrated curriculum. Teachers, dorm directors, counselors, and other staff should be made aware of the unique circumstances facing LGBTQ youth through in-services, study groups, and training. Universities need to continue to develop and support gay and lesbian studies along with promoting the selection of such topics for dissertations. D'Emilio (1992) argues that any university-based group whose membership requirements promote discrimination (like the ROTC) should be banned from all campuses until they

The Safe Schools Movement

> I work for a school system that prohibits discrimination on the basis of race, color, national origin marital status, religion, sex, age, disability, or sexual orientation in employment or any of its educational programs or activities. They offer domestic partner benefits to employees. It's all very promising on a highfalutin' legal level, but "faggot" is still the least corrected slang thrown at other students in the hallways, and "gay" is used at least 1,000 times per day to refer to something that is weak. (Petr 2005, 199)

This salient observation by a gay teacher is an important reminder why policies including the protection of sexual orientation and gender identity are needed in all educational settings. While large-scale statistics on school experiences of LGBTQ students has only been recorded since the 1990s, research is steadily mounting that supports the need for explicit policies. For example, Ellis and High (2004) found that since 2001, there has been a steady increase in reported problems at school for LGBTQ students, especially in the key areas of conformity pressures, being outcast from the group, assault, and verbal abuse. This culminates in higher suicide and attempted suicide rates for LGBTQ teens—three times the general rate (Aarons 1995; Perrotti and Westheimer 2001). LGBTQ students are more likely to skip classes, drop out of school, or forgo college attendance altogether (Aarons 1995; Kosciw, Diaz, and Greytak 2008; Perrotti and Westheimer 2001).

In terms of frequency of homophobic bullying, a majority of LGBTQ students have experienced some form of harassment, from name calling to physical assault, at a higher rate than the rest of the student population (Aarons 1995; Harris Interactive & GLSEN 2005; Meyer 2009). A newer form of harassment, cyber bullying, was experienced by 41 percent of LGBTQ students, which is four times the national average (Meyer 2009, 21). Most of the perpetrators of gender-orientation and sexual harassment are groups of males, followed by mixed girl/boy groups (Forrest 2006). Forty-eight percent of students under age 18 have been violently attacked, and 40 percent of those attacks took place in school and by other students (Forrest 2006, 117; Perrotti and Westheimer 2001). Appearance-based terms like "gay" (used in a derogatory manner) or "faggot" were the most common forms of verbal harassment, followed by sexist language. Most surveys of LGBTQ and straight students report the frequency of hearing such terms in the 90 percent range (GLSEN 2008; Harris Interactive and GLSEN 2005; Kosciw, Diaz, and Greytak 2008; Meyer 2009). What is most disturbing is that when it comes to more insidious forms of constant harassment or assault, Kosciw, Diaz, and Greytak (2008) found that nearly 61 percent of students who experienced it did not tell school staff for fear that no action would be taken or that their situation would worsen (p. xiii).

My own homophobia was a self-imposed barrier that I didn't know how to break. (pp. 17–18)

Before coming out, Nicolari couldn't even network with other known gay and lesbian teachers because they were in the same situation. Davis (2005) lists fears related to employment status, with an uncertain economy leading administrators to look for any reason to let a teacher go. Other teachers throw themselves into their academic roles, avoiding dealing with internal matters (Roberts 2005). Dolor (2008) argues that because of the negative school climate, many LGBTQ students will remain closeted, often hiding their personal problems related to homophobia in order to avoid detection.

Related to the problem of the closet, Sykes' (2004) research on sexual orientation harassment in K–12 sports and physical education revealed that teachers and coaches are often caught in a dilemma about how to respond. Some utilized a "pedagogy of censorship" in which they attempted to cease student use of anti-gay slurs while others maintained a "pedagogy of inquiry" in showing students the harmful effects of name calling (p. 76). In the case of the inquiry pedagogy, LGBTQ teachers often used their own life histories as examples, putting themselves at risk of rejection or worse. Sykes was keenly interested in the third pedagogy employed, a "pedagogy of masochism" where LGBTQ teachers took incredible and sometimes irrational risks in order to confront homophobia, often harming themselves in the process (p. 77). This makes it all the more imperative for straight teachers and coaches "to take the lead in dealing with heterosexism" so that it will be easier for LGBTQ teachers and coaches who are more vulnerable to do the same (p. 92).

A fourth barrier to confronting harassment involves the reluctance of schools, parents, and families to recognize the special nature of gender-based bullying, or even the need to confront bullies. Perrotti and Westheimer (2001) relate how, when engaging a woman who was against anti-harassment policies in schools in a dialogue, she insisted that all kids get picked on so why make a special category of protection for sexuality or gender? Dolor's (2008) response might be, "I have never heard anyone suggest that a person was killed because he was heterosexual..." (p. 171). Opposition is also based on the unfounded belief that safe schools programs will lead K–12 students to a homosexual lifestyle. Another common tactic is denial. Many believe that there aren't many (if any) LGBTQ students at their school, so there is no merit to having a safe schools program (Forrest 2006). On the flip side, the need for anti-harassment education and policies can often feed the right wing propaganda machine that tries to portray LGBTQ people as miserable and suicidal, incapable of living a happy life (Perrotti and Westheimer 2001). This can also make teachers and administrators feel helpless to confront a problem they view as unsolvable.

Social and Psychological Factors

There are several social and psychological factors making it difficult to enforce anti-harassment policies. One is that the traits that bullies often display are the same ones admired in schools, especially when manifested by males, such as competitiveness, strength, toughness; what Meyer (2009) terms "hegemonic bullying" (p. 8). In these cases, bullying isn't technically an "antisocial" behavior. Other students often admire and support the perpetrators of gender-based bullying in order to gain favor or to avoid being a future victim. This has a radiating effect of contributing to sexual harassment, "...often done near the female students but not always directed at them, thus creating a space where women are targeted and objectified with no outlet for response or complaint of tangible harm" (p. 9). While straight males may be comfortable with lesbianism "because they believe they can control women's sexuality," the presence of gay males induces anxiety (Perrotti and Westheimer 2001, 11). When schools insist on using generic terms like "name calling," they avoid dealing with larger homophobic (and often racist and sexist) power dynamics, while an overall climate of "euroheteropatriarchy" is maintained (p. 17). As Brooks's (2005) account of his own high school experience revealed, "...because I was perceived as a 'faggot,' I could be ignored, or even worse, physically and verbally challenged" (p. 242).

A second factor reducing the likelihood of confronting homophobic bullying is the uncertainty of K–12 teachers and administrators when it comes to addressing gender-based forms of harassment. As with racism, school administrators tend to equate sexual orientation harassment with personal flaws or failings on the part of individuals, thereby perceiving their schools as having low levels of structural or institutional discrimination based on sex or race (Meyer 2009). Forrest's (2006) review of research with teachers found that many of them were complicit in the homophobic actions displayed by their students, such as laughing at anti-gay jokes. They also framed homosexuality solely in terms of "sex" rather than a more complex identity, resulting in further marginalization of LGBTQ students.

A third factor interfering with sexual rights is the fear that many LGBTQ teachers, administrators, and students have about coming out of the closet. Nicolari (2005) describes how the closeted teacher becomes part of the enforced heteronormativity, even when he or she can tell that a student might be questioning his or her own sexuality:

> I had another student stop by my room several times after school; she was struggling to understand her sexual orientation...As much as I wanted to be a support for her, she raised my anxiety level with her strong sense of self. Consequently, I always pretended to be busy, shuffling papers and avoiding her eyes. The reality that I couldn't be a support system for all of my students—especially the LGBTQ students—was taking its toll on me.

or limited public forum, including denying such access or opportunity or discriminating for reasons based on the membership or leadership criteria or oath of allegiance to God and country of the Boy Scouts of America or of the youth group listed in title 36 of the United States Code (as a patriotic society) (Boy Scouts of America Equal Access Act, 2002, sec.9525).

Though the act does include a "voluntary sponsorship" provision which states that no one can require schools to sponsor the Scouts, one can sense that the burden of proof would be on those who seek to maintain a setting free from sexual orientation discrimination. In terms of speculation, it doesn't take much to imagine a scenario where the Klan would be able to sponsor an organization that meets on public school grounds.

Homophobia through Gender-Based Bullying and Harassment

The problem of homophobia within K–12 schools is much deeper than often portrayed, requiring a more aggressive approach than the typical recommended remedy of "tolerance" training that seeks to change hearts and minds (Meyer 2009; Perrotti and Westheimer 2001). As Wolf (2009c) asserts:

> If we lived in a truly free society in which material and social constraints were removed, people would be neither oppressed nor even defined by their sexual or gender identities. Only then can we see how a liberated human sexuality could evolve and express itself. But in a class society that requires behavioral norms to discipline its workforce and ideology to justify the nuclear family, reactionary sexual ideas—including gender norms—are means of stoking division and repressing society as a whole. (p. 11)

As an example, Meyer's (2009) extensive study of gender-based bullying in K–12 settings presents an in-depth analysis of the psychology involved in community enforcement of nuclear family norms. This includes the general level of acceptance of anti-gay and sexist harassment as part of school culture. According to Meyer, "gender harassment" refers to a wide range of behaviors meant to enforce particular norms of masculinity and femininity. This differs somewhat from "bullying," in that the bully intends to harm someone physically while a harasser contributes to a larger climate of hostility. This means that the effect of the harassing behaviors is much more widespread since the perpetrators are more subtle and do not draw attention to themselves to successfully avoid detection. In order to fit in, heterosexual (along with closeted gay students) often have to display a willingness to engage in verbal insults of those deemed as violating gender norms. Other related forms of harassment that Meyer identifies include sexual harassment and sexual orientation harassment.

Council, but they pretty much exist to create programs and oversee policy. Most of the leadership is local in nature. The majority of troops are sponsored by religious organizations (65 percent) with one quarter being other kinds of private groups, and the rest publicly sponsored (p. 168). The policy on homosexuality mirrors the military's Don't Ask, Don't Tell:

> Official Scouting materials addressed to the boys do not refer to homosexuality or inveigh against homosexual conduct; rather they teach family-oriented values and tolerance of all persons...the handbooks for boys do not catalog immoral behavior for Boy Scouts. It cannot be inferred that unmentioned misconduct is consistent with Scouting's moral code." (p. 169)

Because the membership of the Scouts has become broader as society itself becomes more multicultural, it is more difficult to enforce vague concepts like "good citizen" or to claim that the presence of someone like Dale threatens the Scouts' ability to operate their organization as they see fit. In the Dale ruling, the nuances of group versus individual expression were not made manifest, instead, the ruling assumed that a private group had the right to exclude gay males. Yet the Scouts have a collection of general beliefs—other than the edict "morally straight," no rationale for banning gays exists in print. It is as if everyone is supposed to know that the scouts are not about promoting homosexuality. Only when Dale dared to violate this unspoken norm did he get kicked out of the Scouts. He wouldn't have had any problems if he had remained closeted.

Rosenbury (2007) analyzes Dale from a family law perspective, finding that a contradiction emerges. On the one hand, if we use the established precedent of parents having (with few exceptions) absolute authority over the rearing of their children, then the ruling in Dale conforms to the expectations of parents to be able to raise their children as they see fit (including the right of racists to teach children to hate minorities). But on the other hand, if one considers rulings concerning schools, where there are limits to the ability of groups—including private schools—to discriminate, then this runs counter to the ruling in Dale. Further complicating matters is this exception written into the No Child Left Behind Act, which essentially overrules the ability of schools to enforce anti-discrimination policy:

> Notwithstanding any other provision of law, no public elementary school, public secondary school, local educational agency, or State educational agency that has a designated open forum or a limited public forum and that receives funds made available through the Department shall deny equal access or a fair opportunity to meet to, or discriminate against, any group officially affiliated with the Boy Scouts of America, or any other youth group listed in title 36 of the United States Code (as a patriotic society), that wishes to conduct a meeting within that designated open forum

about radical figures in history such as Harvey Milk. When combined with media censorship of leftist LGBTQ perspectives, the entire radical history of the sexual rights movement is obscured (D'Emilio 1992).

The Significance of Dale

Boy Scouts of America v. Dale (2000) is an important case to examine in light of Petrovic's (2002) recommendation of positive systematic inclusion for K–12 public schools. James Dale, a long time scout, scout leader, and college student, was interviewed at a seminar he attended on gay and lesbian health issues. As a result of the interview being published, he was expelled from the Boy Scouts of America even though, as Rosenbury (2007) points out, he didn't broadcast his sexual identity to the troop nor did he advocate for LGBTQ issues while doing scouting activities. Dale challenged his expulsion, but the Court ruled that the Scouts, being a private organization, had the right to establish membership limitations, despite New Jersey's state law prohibiting discrimination on the basis of sexual orientation. The Scouts argued that they had the right to maintain "moral straightness" as a requirement of membership and therefore Dale had to go. As Seidman (2008) put it, Dale was deprived of his property rights. Despite the Court's ruling, even if there hadn't been the New Jersey law protecting against discrimination on which to base the defendant's argument, Dale should have had the right to remain in the Scouts.

In the ruling, Dale was referred to as a "gay rights activist" even though he was not officially "out" to his troop, equating Dale's sexual orientation with open advocacy of gay issues. As McGowan (2001) states, "to say that personal characteristics are speech would require a second look at many anti-discrimination statutes" (p. 122). Part of what makes the Dale decision interesting is that other cases in which the Court ruled that women had to be admitted to all-male golf organizations, for example, did not refer to the rights of private organizations to set membership requirements based on gender. Instead, the ruling in Dale seemed to get around this by claiming that Dale was an "activist" and that was the real problem, not his homosexuality per se. It was what he was doing with it. After Dale, one has to wonder if other protected categories like race can be overruled because some groups do not want to associate with different minorities. Lurking behind the Court's decision is the notion that being gay is a "choice" unlike gender or race. It suggests that being gay can be dismissed as mere activism rather than biologically based.

In actuality, the Scouts do not have a uniform belief concerning homosexuality (McGowan 2001). They are a decentralized organization, consisting of units called "troops," an assortment of 15 to 30 boys led by a Scoutmaster and his Assistant. These are usually adult males who themselves were Scouts as youngsters. Scoutmasters are selected by whoever sponsors a troop, and the troops reports to a local council. The highest authority is the National of the opposition to LGBTQ rights, as if it were a matter of different "viewpoints" about sexuality and not outright efforts at silencing and censorship on the part of the conservative side. Petrovic's (2002) approach demands a more radical analysis:

> My interpretation of non-oppression vis-à-vis the issue of same-sex orientation in the schools leads me to defend positive systematic inclusion. This requires the positive portrayal of same-sex orientation in the curriculum and forbids teachers from speaking out against same-sex orientation, regardless of their personal beliefs. (p. 146)

For Petrovic, the right of LGBTQ students to experience K–12 environments not only free of harassment, but ones that include positive and realistic portrayals of gays and lesbians, is critical. Petrovic came to this conclusion after working with education majors who believed that they had a right to openly disagree with homosexuality in front of K–12 students. If they couldn't oppose homosexuality in the classroom, then they felt they should banish its presence entirely from the curriculum in the name of fairness and neutrality. Petrovic linked this all or nothing mentality to key censorship efforts, pointing to a limitation of traditional liberal approaches to free speech in how it assumes equal distribution of power:

> Some people are simply at greater liberty to express their thoughts and engage in (read: control) discussion than others. This being the case, the liberal tenet of freedom of expression can serve to undermine the other liberal tenet of individual choice by concealing other views of the good life or by prejudicing the discussion against them, which consequently often makes their pursuit far too burdensome (p. 147).

Far from being neutral, the lack of representation within the curriculum has larger implications. For one, access to information has a lot to do with where one resides. If a student happens to live in the South, she or he is less likely to have the following in K–12 schools: a harassment policy with an anti-gender orientation, a gay-straight student alliance, LGBTQ representation in curriculum or texts, information via school libraries or the Internet on school computers, and access to supportive teachers and administrators (Kosciw, Diaz, and Greytak 2008). Students with gay or lesbian parents can often interpret a lack of curricular representation or activities that assume the two-parent, heterosexual household as the norm as a form of exclusion (Kosciw and Diaz 2008). The lack of representation also emboldens opponents of LGBTQ rights by contributing to the stereotypes of gays and lesbians as perverts—after all, if they were "normal," then why don't we see them more often in textbooks? (Aarons 1995). Goldstein (2003) feels that we should not be surprised when many LGBTQ people turn to more visible conservative gay and lesbian authors because they are not taught

Censorship

According to Moeller (2002), "dominant economic interests continue perniciously to shape education and media to the detriment of free and critical thinking" (p. 155). In the case of younger elementary students, most cannot take on homophobia on their own, even in the face of their families being castigated (Perrotti and Westheimer 2001). Teachers can also feel vulnerable, as Perrotti and Westheimer outline. The constant attack on LGBTQ individuals by a highly organized right wing can often lead to feelings of isolation. As Warnick (2009) describes, it is difficult to defend the right of students to have access to important information because rights to free speech exist in a more limited format in K–12 settings. Things have changed dramatically since the landmark Supreme Court ruling in favor of student expression in *Tinker v. Des Moines* (1969). Today, the courts are leaning toward a more narrow range of acceptable speech, often citing both the age and developmental level of K–12 students along with the compulsory nature of education, which means that "…the courts have found in the special institutional characteristics of schools only reasons to limit student rights" (Warnick 2009, 200). Indeed, Aarons (1995) describes how the implementation of a rainbow curriculum in Massachusetts was never considered. Instead, they focused on generic bullying and safety policies, so as not to rock the boat.

The majority of LGBTQ censorship instances (whether official or unofficial) revolve around not introducing important curricular topics because of student age, disruption, or safety. It is also the result of opposition to the mere existence of LGBTQ people as a group, or a belief that some topics should be avoided because they could offend others (Perrotti and Westheimer 2001). For Warnick (2009), this highlights a need for a special category of consideration, known as developmental rights, in which children should have access to information in light of "the theoretical existence of the future adult who will someday exercise individual liberties" (p. 202). As far as compulsory education goes, the Supreme Court has ruled that referring to the "captive audience" status is no reason to censor speech that is considered offensive (p. 204). Yet Meyer (2009) concludes that the conflict over K–12 curriculum boils down to difference between individuals who view the role of the school as existing to inculcate traditional values and those who see critical thought and democratic values as primary. An extreme example of this was the implementation of Section 28 in England, which barred teachers from "intentionally promoting homosexuality" in any way (Forrest 2006, 125). Though eventually repealed, the impact of state-sanctioned censorship had far-reaching effects, including continued self-censorship on the part of teachers (Ellis and High 2004).

Ellis and High (2004) find limitations to the existing moral neutrality surrounding LGBTQ issues in the curriculum. The sole focus on general terms like "equality" tend to gloss over the reactionary component behind much physical education and sport curricula, making it difficult to overcome the heterosexism involved. Since many physical education teachers also teach a health unit, the combination of hyper-masculinity along with the pathology of AIDS can create a lethal environment not only for LGBTQ students, but for straight students who may not fit in with established gender norms. The elevated visibility of sport and athletics in most schools also introduces an element of hierarchy and elitism, which can further silence students who do not conform. Indeed, Perrotti and Westheimer (2001) identify coaches, administrators (who were often once coaches themselves) and physical education teachers as key individuals for working to create an inclusive school environment since they often occupy more visible positions in both the school and community.

The inclusion of positive and complex portrayals of LGBTQ issues in the curriculum can serve several powerful purposes. Brooks (2005) and Moeller (2002) argue that until society fully recognizes the equality of sexual minorities, the contributions of LGBTQ people have to be featured in the curriculum. Record (2005) transforms the traditionally conservative character education curriculum into fuller definitions of citizenship, honesty, and tolerance regarding the LGBTQ community. Even in cases in which there is little to no coverage of LGBTQ material, Petr (2005) sees the figure of the LGBTQ teacher as an important form of curricular interrogation: "I wonder what my students see in this high school that gives them hope that they can have their chance to be gay...I wonder, and I worry, and then once in a while I remember that there's me" (p. 202). Meyer (2009) notices a positive relationship between LGBTQ curricula and increased motivation of teachers to become involved in enforcing anti-harassment policies. Perrotti and Westheimer (2001) consider the curriculum as providing an important support system to LGBTQ youth, as with exposure to life changing literature and realistic, relatable classroom lessons. D'Emilio (1992) makes a direct connection between curriculum and activism at the college and university level. The more that LGBTQ studies are visible on campus, the more that these ideas can become translated into overall demands for equality and human rights:

> Colleges and universities are important institutions. Large numbers of young people and adults pass through them. Universities also shape the public school system. Elementary and secondary school teachers were taught, trained, and certified through them. Textbooks are often written by college educators...institutions of higher education shape values, impart or withhold information, and define and describe social, cultural, and political reality...the purpose of education generally, and higher education in particular, is to offer and refine an accurate description of reality—a description of what is, of how things came to be, and how to make change. (p. 158)

sexuality as beyond the purview of the schools, students are already internalizing powerful norms about gender roles and regulations within K–12 settings (Forrest 2006; Luttrell 2003). Unfortunately, adults are often hesitant to address LGBTQ issues, thinking that they are automatically "about sex," whether it is due to their own hesitancy in dealing with sexuality, worries about being seen as a "recruiter" to the homosexual "cause," or negative community response (Forrest 2006; Perrotti and Westheimer 2001). The lack of large numbers of "out" LGBTQ faculty in K-adult educational settings makes it hard to push for a more diverse curriculum (D'Emilio 1992). Through a combination of denial, fear, and censorship on the part of key adults, K–12 students are unable to question the heteronormative school culture that is both hidden and obvious (Forrest 2006). The curriculum is a key vehicle for developing a historical materialist outlook, which is precisely why conservative groups have fought so desperately against the inclusion of LGBTQ issues. In light of this oppressive resistance to sexuality rights, Petrovic's (2002) recommendation that we move beyond classical liberalism to a more radical approach is advanced. This means considering the right of LGBTQ and straight students' access to information as more pressing than the right of teachers, parents, and administrators to oppose the presence of LGBTQ issues for religious or other reasons in the classroom.

The Prevalence of and Rationale for Including LGBTQ Issues in K–12 Curriculum

While exposure to LGBTQ issues in the curriculum is on the rise, the numbers are not encouraging. For example, roughly 11 percent of students surveyed indicated they had come across positive portrayals of the LGBTQ community, while nearly 15 percent noted an inclusion of LGBTQ information in textbooks and readings (Kosciw, Diaz, and Greytak 2008, xvi). When LGBTQ issues were mentioned, it was usually associated with pathology, such as AIDS (Aarons 1995; Ellis and High 2004). Fewer than 50 percent of students could locate information about gay and lesbian issues in their school library, while only 30 percent could use the Internet on school computers to find such information (Kosciw, Diaz, and Greytak 2008, 99). Close to 90 percent of students were not taught about gay history, people of note, or important events (p. 99). Representations of gay and lesbian families in the classroom were only reported by 31 percent of students (Kosciw and Diaz 2008, xviii). When LGBTQ issues were addressed, it was usually done in history, social studies, literature, or health classes (Ellis and High 2004; Kosciw, Diaz, and Greytak 2008).

School boards began approving the inclusion of lesbian, gay, and bisexual issues in health textbooks beginning in the late 1980s and early 1990s, mostly as a response to AIDS (Aarons 1995; Perrotti and Westheimer 2001). Sykes (2004) addresses the overwhelming message of masculinity that permeates

(D'Emilio and Freedman 1997). On the one hand, right-wing activists constantly point to the failure rate of condoms, using industry test statistics as "proof" (Chamberlain 2008; Mooney 2005). When used correctly, condoms can be as much as 98 percent effective (Feldt 2004). Yet on the other hand, the reliability factor could indeed be much higher if the industry made more durable condoms for anal sex, the most common way that AIDS is transmitted between gay men (Wolf 2009c). But manufacturers refuse to make these kinds of condoms because they don't want to be seen as promoting gay sex. This repressive—and deadly—approach contributes to a climate in which those who break so-called "virginity pledges" to remain celibate until marriage (88 percent eventually fail to do so) are not as likely to use contraception as those who do not participate in such rituals (Page 2006, 66). While these pledge-breakers have the same high STD rate of other teenagers, they tend to avoid confronting their situation. Embarrassment or denial prevents them from seeking medical care. As a result, they are often knowingly and unknowingly transmitting STDs to others. This doesn't seem to trouble abstinence-only proponents, as one spokesman stated: "...if you're talking about a person who is not going to keep the abstinence pledge anyway, whether or not they would use contraception isn't really something that concerns us" (p. 82).

Other instances of contributing to the already high teen abortion rate include conservative opposition (led by Bush appointee David Hagar) to over-the-counter status of the extra-dose birth control pill known as Plan B; links between abortion and mental illness; and the recent flap over the human papillomavirus vaccine. The FDA made an unprecedented move and refused to approve over-the-counter status for Plan B because conclusive data demonstrating that the drug would not be used differently by teenagers could not be provided—a demand that is scientifically unreasonable (Mooney 2005). Groups like the Medical Institute for Sexual Health, Concerned Women for America, and the conservative wing of the Republican Party (Oklahoma Senator Tom "no condom" Coburn in particular) bombard the media with "scientific" research that apparently uncovers deep secular conspiracies to recruit teenagers into sexual activity, which plays on public fears about sexuality in general. The results of these groups can only exacerbate the national problem of 50 percent of all pregnancies among adult and young women being unintentional (p. 229). Mooney relates how a book that Hagar once wrote suggested "...Bible readings for treating premenstrual syndrome" among other "medical" treatments, including relief from demonic possession (p. 228).

Curriculum of Silence

K–12 students spend a large part of their day in school, making issues related to sexuality imperative. Even if teachers, administrators, and parents view

Contraception and Abortion

Two key cases, *Planned Parenthood of Central Missouri v. Danforth* (1976) and *Carey v. Population Services International* (1977) established that adolescents are entitled to virtually the same rights as adult women when it comes to access to reproductive services. Both cases outlined the scope of parental consent regarding abortion and access to contraceptives. However, 31 states have laws on the books that require the consent of one or both parents in order for a minor to obtain an abortion (Rowland 2004, 548). On the flip side, every state in the union allows minors to agree to STD testing and treatment, along with full protection of a young woman's right to carry a baby to term. If a clinic accepts abstinence-only grant money, however, it is not allowed to dispense condoms or information about contraception, even to prevent the spread of HIV/AIDS. The continued assault on women's reproductive rights has resulted not in a reduction in pregnancy, but in a 20 percent increase in late-term abortions among adolescent girls (Page 2006, 84). The lack of availability of clinics performing these abortions increases the likelihood of compulsory childbirth (Rowland 2004).

The anticontraceptive movement is part of an overall assault on scientific knowledge, as Mooney (2005) addresses. A key strategy is to inject enough confusion into what should be noncontroversial issues in order to build support for reactionary ideas, or make people skeptical of scientific reason. Because uncertainty can never be fully disproven, conservatives use this line of thinking to advance ideas such as "don't use contraception because it doesn't work 100 percent of the time," a logic not applied to the military's use of so-called precision-guided weaponry, which has a failure rate far exceeding condoms. Based on percentages, comprehensive sex education is not controversial—an overwhelming majority of teachers (90 percent) support students learning about contraception in school—yet one-fourth of teachers are prohibited from addressing it, even to prevent HIV/AIDS (Feldt 2004, 68). Even the radical founder of Summerhill, A.S. Neill (1960), went against his conscience and did not distribute condoms at his school for fear of breaking the law and threatening the school with the possibility of closure. Nearly 50 years later, Neill's dilemma remains for many school districts. In contrast, the Netherlands begin their sexuality education in preschool and carry it through twelfth grade, with an emphasis on contraceptive use. This results in a Dutch teen abortion rate that is 33 percent lower and an AIDS case rate eight times less than those of the United States (p. 71).

Abstinence-only education forbids teachers and students from addressing contraception of any kind except for an emphasis on the failure rate of condoms (Murray 2008). The scare tactics involved in the anti-condom movement reveal a cold and calculating catch-22, reminiscent of the collective lack of response from the U.S. government in the wake of HIV/AIDS differently, social instability is brought about by globalization, economic restructuring, and diminished forms of social welfare are transposed into cultural symbols of vulnerability and dependency...made innocent in order to be seen as deserving of protection. (p. 36)

Federal funding for Title IX programming began to dry up at the precise moment social welfare was transformed from "entitlement" status to personal responsibility. Luttrell found that one of the key discourses used in this overarching deviance theory was that education was not a right, but a responsibility, a belief directly contradicting the original language of Title IX: "...the school curriculum reinforced images of teenage mothers are sexually irresponsible, likely to be bad mothers, and destined to become dependent, nonproductive citizens" (p. 22). The only way out of this cycle was for the girls to persist in the program, to "deliver themselves" from their teen mom status through compulsory attendance and assignment completion. This was difficult, of course, in that the classes the high school girls were required to take were not transferable to colleges and universities. The noncredit courses combined with segregation from resources (including outdated textbooks and classrooms in the basement) created a maternal ghetto of sorts—something Title IX was designed to avoid.

Luttrell (2003) also uncovered the ugliness of racism and social class in the framing of pregnant teens. For example, in the 1990s, several reports featured the "welfare-queen junior" image of predatory, older black men taking advantage of the pregnant African-American teen. This revived the classic racist stereotype of African-American males as rapists while resurrecting the notion of blacks as animalistic. At the same time:

...white girls...were treated as if they had a psyche—their deviance could be cured....Black girls, on the other hand, were not treated as if they had psyches or inner lives. Rather, their deviant motherhood was said to stem from their unruly and unredeemable sexual conduct. (p. 17)

Luttrell concluded that most schools have not created defined policies about teenage pregnancy and parenthood because over the past 30 years, teachers and other educational professionals have ceded ground to conservative policymakers. Current programs that deal with adolescent pregnancy refuse to examine the social contexts that contribute to sexual decision making (i.e., lack of resources, little to no economic or educational opportunity, no questioning of existing sexual stereotypes and expectations). The abstinence-only framework absolutely refuses to deal with emerging sexuality—especially female and LGBTQ, which creates a compulsory (masculine-centered) heterosexuality that is never questioned and assumed to be natural.

Using conservative groups' effective tactic of presenting religious dogma as science, pro-abstinence ideologues cloak discriminatory language in official-sounding statistics and concern for the health of young people (Chamberlain 2008; Mooney 2005). This enables those not directly involved with the religious Right to still accept the message of the abstinence-only movement, even though these same people may support comprehensive sex education. The Right frames sexuality as an "either/or" phenomenon (with a key focus on female virginity) rather than a human expression along a continuum during different points in one's life (Kinsey, 1998; Luttrell 2003; Neill 1960). This resulting confusion has created a dangerous backlash scenario in which young people distrust actual scientific information about HIV/AIDS, and perceive the warnings as yet another tactic by adults to keep them from having sex (Feldt 2004).

Teen Pregnancy

Sociologically speaking, as the median age of marriage rises—along with more people choosing alternative relationship arrangements or deciding to remain single—the promotion of abstinence is asynchronous with how many people live their lives (Coontz 2005). Without comprehensive sex education, it is no surprise that, despite a declining teen birthrate (117 pregnancies for every 1,000 adolescents between15 and 19 years old), the United States has one of the highest teen pregnancy rates in the industrialized world (Rowland 2004, 301). The U.S. rate is twice that of England, Wales, and Canada, and nine times higher than the Netherlands and Japan; countries that—not coincidentally—have excellent comprehensive sex education programs (p. 302). The high U.S. birthrate is due to a combination of a lack of access to and education about contraceptives along with early sexual intercourse that occurs when adolescent decision-making skills are not likely to be well established. For example, the majority of North Carolina high school students are exposed to abstinence-only education, yet that state's birthrate exceeds the national rate (Feldt 2004).

Luttrell's (2003) groundbreaking ethnography, *Pregnant Bodies, Fertile Minds*, exposes the hypocrisy of the public anxiety centered on teen pregnancy. She analyzed many Title IX programs created in the wake of the abstinence-only era to address pregnant teens in K–12 settings. In one of them, Luttrell found that both the school personnel and society at large subscribed to an overarching theory of deviance in their approach to adolescent pregnancy, including different "wrong" discourses: wrong girl (i.e., married women's decisions to have children are never viewed as irrational, while those of teen girls are); wrong family (a focus on African-American teen pregnancy as "the problem"); and wrong society (a more feminist viewpoint that takes into account social context more so than individual choice). The bulk of the research on teen pregnancy tends to fall within reactionary wrong-girl and wrong-family discourses:

> ...the campaign against teenage pregnancy serves to translate what is a political problem into a problem with individual dependency. Put slightly

Philosophy

The philosophy behind abstinence-only education has three essential points: (1) monogamy within marriage is the only acceptable outlet for sexual behavior; (2) anything outside of such monogamy is viewed as justly punished via natural causes (such as contracting STDs), psychological distress, or social shunning; and (3) more realistic safe-sex alternatives, such as condom use, are dangerous and flawed (Chamberlain 2008; Feldt 2004; Murray 2008). Page (2006) goes further to argue that for the Right, the issue isn't the effectiveness or ineffectiveness of abstinence policies—the point is that sin must be punished. It doesn't matter that education about contraception would reduce teen pregnancy or even lower the abortion rate. It is more important that a wayward girl be punished with the natural consequences of pregnancy or an STD; what Page calls "collateral damage for a good cause" (p. 65). Indeed, the discriminatory practices of such policies comes directly out of nineteenth- and twentieth-century purity movements in which young people were warned about their sinful natures and enrolled in educational programs in order practice mental hygiene (D'Emilio and Freedman 1997; Fone 2000). A modern manifestation of these same anxieties about youth eroticism includes the forced resignation of Surgeon General Joycelyn Elders in the 1990s due to her positive comments about masturbation (Perrotti and Westheimer 2001). Similarly, when conservative Surgeon General C. Everett Koop broke the silence in 1986 about AIDS prevention and the need for sex education in the early grades, he was vilified by prominent right-wingers as promoting sodomy (D'Emilio and Freedman 1997; Eisenbach 2006).

In response to this climate of repression, the About Face Youth Theatre (n.d.) based in Chicago uses the dramatic arts as a way to create a safe space for LGBTQ and straight young people. In addition to addressing social concerns in their traveling theater piece *Fast Forward*, the group speaks out against abstinence-only education, arguing that it promotes homophobia and violence against the gay community, as well as forcing poorer school districts to accept inferior health education to obtain federal funds that come with abstinence-only programming. Within the sex education curriculum (abstinence-only or comprehensive), a narrow range of heterosexual behavior is presented as the norm, automatically leaving out LGBTQ individuals who are not mentioned, as Forrest (2006) explains:

> …the centrality of the vagina and penis as penetrated and penetrating sexual organs effectively de-legitimizes some sexual acts…since same-sex sexual activity is characterized as a substitute for heterosexual sexual behavior, it is often portrayed as though it were an arrested sexual development…this may lie at the root of preconceptions that LGBT young people are failed heterosexuals who have either never had heterosexual sex or else were turned away by bad heterosexual experiences. (p. 128)

education classes—one for abstinence-only and the other for comprehensive instruction. The National Abstinence Education Association maintains a lobbying presence in Washington, resulting in over 900 funded programs that carry their own abstinence-based curricula. All claims of effectiveness come from one pro-abstinence company, the Institute for Research and Evaluation, headed by Stan Weed, who has been involved with faith-based funding efforts for close to 20 years (Chamberlain 2008). Bush also appointed several ideological figures, such as Dr. Alma Golden (an abstinence-only proponent) as deputy assistant secretary of population affairs (Feldt 2004). The U.S. Department of Health and Human Services created a Web site, 4parents.gov, under the guise of providing advice to parents when the government partnered with the National Physicians Center for Family Resources, a group that promotes a link between abortion and breast cancer (Mooney 2005). A similar initiative, Parents for Truth, spreads misinformation about HIV/AIDS prevention, and has links to the National Abstinence Education Association (Chamberlain 2008). Their latest project is lobbying to extend abstinence programming to 29-year-olds!

Feldt (2004) reveals stunning figures: 58 percent of all U.S. schools offer abstinence-only programming and 86 percent of public schools require heavy promotion of abstinence-only education. Of those public schools, more than 30 percent "require abstinence to be taught as the only option for unmarried people, and usually either prohibit any mention of contraceptives or limit discussion to their ineffectiveness" (p. 63). Despite these high numbers, there has been no research to date (other than anecdotal accounts or opinion surveys) showing definitive success of the abstinence-only approach compared to comprehensive programs (Berkowitz 2008). Recent investigations of abstinence-only curricula found that a whopping 80 percent of government funded materials "contained false, misleading or distorted information about reproductive health" (Murray 2009, 33). Some of the nuggets of wisdom Murray and Page (2006) lists as being funded by tax payer dollars include: a 43-day-old fetus is a "thinking person," half of gay teens have AIDS, and touching the genitals of the opposite sex can result in pregnancy. The National Coalition to Support Sexuality Education has attempted to counter such false claims (Chamberlain 2008), but they have a lot of work to undo. In Pennsylvania, since the implementation of abstinence-only curriculum, the rate of sexual activity among girls is at 42 percent, compared to the 27 percent with access to comprehensive sex education classes (Page 2006, 68). Texas, a leading abstinence-only state, is fourth in the nation for the percentage of people having HIV/AIDS, but even school janitors and support staff are not allowed to answer student questions about safer sex (p. 69). Southern states as a whole have much higher HIV/AIDS and STD infections, along with skyrocketing teen pregnancy rates (p. 79). Yet this is the region with the most abstinence-only programming, buttressed by extreme religiosity.

and just over 12 percent of urban gay and bisexual male teenagers have HIV (young African-American males are 30 percent HIV positive) (Perrotti and Westheimer 2001, 139). Indeed, Neill's declaration (1960) that "In most public schools, to tell the whole truth about love and birth would be to risk getting fired" holds true today (p. 219).

Though the Obama administration has begun to roll back some of the more repressive Bush-era policies concerning sexuality, laws, and funding remain intact—almost $95 million for grants (Omnibus Appropriations Bill 2009, 62). The abstinence-only movement had its legal genesis in the reactionary religious efforts that supported Reagan's Adolescent Family Life Act (1981), which was later found to be unconstitutional and subsequently edited. This was followed by the Clinton administration funding abstinence-only sex education in the Personal Responsibility and Work Opportunity Reconciliation Act (1996). Britain employed similar measures with Section 28, much like the U.S. Defense of Marriage Act (Forrest and Ellis 2006). According to Murray (2008), abstinence curricula for schools were not designed by doctors or teachers, but by the Family Research Council and the Christian Coalition, along with other conservative allies. Making matters worse, these programs were, for a few years, not even required to produce data on their effectiveness.. This is remarkable in the wake o No Child Left Behind's (NCLB) requirement that any program receiving federal funds has to show data driven improvements.

Page (2006) and Rowland (2004) outline how the Bush administration upped the ante in terms of making funding for schools contingent upon their following abstinence-only curriculum. When James Waggoner, a community organization leader, asserted that these approaches were tantamount to censorship, his group was hit with three Health and Human Services audits within one year (p. 72). While groups of parents have successfully challenged the religious content of abstinence-only programs (Feldt 2004), censorship persists. Perrotti and Westheimer (2001) write about conservative spies infiltrating an HIV/AIDS prevention program workshop where adolescents were posing private questions to medical experts. The education commissioner subsequently fired those involved in conducting the workshop, providing fodder for the Right wing.

The United States is the only nation to make funding for sexuality education contingent on abstinence-only curriculum (Feldt 2004). Aided by the recent wave of elected conservative school board members, over $1.5 billion has been spent on abstinence-only education, with specific grants going to religious organizations as part of Bush's faith-based initiatives (Chamberlain 2008; Feldt 2004). A $100 million grant went to the Compassion Capital Fund, with other grants going to Metro Atlanta Youth for Christ ($363,936), and the Catholic Diocese of Helena ($14,000), to name a few (Feldt 2004; Rowland 2004). Chamberlain (2008) describes how it has cost the Osseo, Minnesota school district over $100,000 to maintain two separate sex

Mormon Church played a key role in funding the passage of Proposition 8 in California in 2008.

The relationship of the LGBTQ community to the church has been problematic. This is most evident in the controversies surrounding membership and clergy in the Methodist, Presbyterian, and Episcopalian churches (The Episcopal Church, n.d.). The most conservative denominations within these churches have stated unequivocally that gays and lesbians cannot be members or clergy because they are living a life of sin. More liberal congregations see homosexuality as yet another facet of sexuality. Recent controversies in the Anglican Church center on whether gays and lesbians can be ordained as priests or bishops and whether or not to bless same-sex unions. In 2003, the Church voted "yes" to both. In some Anglican congregations, a form of Don't ask, Don't Tell became official policy—one is allowed to be gay in thought, but not in action (Forrest and Ellis 2006). However, a backlash erupted in 2006 with several congregations breaking off and becoming more conservative in their interpretations of what Religioustolerance.org calls the "clobber passages," the fewer than 12 Bible verses that condemn homosexuality (par. 4). It is likely that this controversy will continue to shape the Church over time, especially as the potential for federally protected same-sex marriage rights nears, and churches must decide whether or not to recognize such marriages.

Abstinence-Only Education

Prevalence and Funding

According to Forrest (2006), the evolving reactionary approaches toward K–12 sexuality instruction has to be viewed within the larger context of "bringing under increased control the whole system of maintained education" (p. 119). Perrotti and Westheimer (2001) connect resistance to comprehensive sex education to anti-LGBTQ movements, which have begun an organizing arm within public schools. The struggle persists, despite the fact that an overwhelming majority of parents (over 90 percent according to a combination of well-respected surveys) prefer comprehensive sex education curricula with 67 percent of U.S. voters supporting instruction about contraception (Chamberlain 2008, 21; Feldt 2004, 67; Murray 2008, 32). Comprehensive sex education is continually presented as a controversial issue, even when eight out of ten evangelicals endorse it in high school settings (Feldt 2004, 68; Murray 2008, 33). With 67 percent of adolescents having had sex by the age of 18, and marriage being delayed (if entered into at all) until they are roughly 25 years old, abstinence-only sex education is drastically irrelevant and irrational (Feldt 2004, 64; Rowland 2004, 302). The systematic denial of information to young people has dangerous consequences. For example, LGBTQ students are not as likely to use condoms than heterosexual students,

Carter's desk. Instead, more discharges were made during his administration than under post WWII Republican administrations (Wolf 2009c). Other nations, including right-wing dictatorships, allow LGBTQ people to serve in the military. But public attitudes are shifting in the United States. In 2008, three-fourths of Americans indicated support for LGBTQ people openly serving in the military—over 50 percent for both Democrats and Republicans (p. 67). Even in the hyper-masculine world of the military, 50 percent of its personnel support gays and lesbians serving openly (p. 68). The importance of LGBTQ rights in the context of the military is critical. If Don't Ask, Don't Tell falls, it paves the way for same-sex marriage rights and a chance to increase the scope of the social safety net within the United States.

The Church

A quick tour of the Internet reveals many websites devoted to religious gays and lesbians: Lesbian and Gay Christian Movement (http://lgcm.org.uk/); The World Congress of Gay, Lesbian, Bisexual, and Transgender Jews (http://www.glbtjews.org); Eye on 'Gay Muslims' (http://gaymuslims.org/); Evangelicals Concerned Inc (http://www.ecinc.org/); and DignityUSA (http://www.dignityusa.org/) to name a few. Gaychurch.org provides a database of same-sex friendly congregations worldwide, which number 5,301 and are increasing. Yet polling has revealed that the more a community is identified as highly religious, the less likely its residents are to list that community as a good place for LGBTQ people to live (Pelham and Crabtree 2009). There has been a seismic debate brewing within most Christian denominations over the inclusion of LGBTQ parishioners and clergy, mirroring important societal shifts:

> Urban churches are discovering large numbers of lesbians and gays in their midst as these people come out of the closet in today's more tolerant climate. Many demand greater acceptance, with the support of liberal minded straight congregants. These are churches permanently touched by the civil rights movement, the women's movement, and other social currents. (Aarons 1995, 260)

Aarons pinpoints the controversy to interpretations of fewer than 12 verses out of the entire Bible, primarily the Old Testament and the writings of Paul in the New Testament. Jesus never mentioned homosexuality in his teachings. Due to racist stereotypes of African-Americans being anti-gay, the African American church is often presented by conservatives as the gold standard for homophobia, especially in response to AIDS (Wolf 2009c). Yet majority white churches, like the Catholic Church, ignored AIDS as a social issue, making them the targets of protests in the 1990s. As Wolf points out, the practices. This means that if one is lesbian or gay and in the armed services, one has to essentially remain celibate or risk being discharged (Kotulski 2004). In 1975, Air Force Officer Leonard Matlovich challenged military policy by outing himself and appearing on the cover of *Time* after his discharge, motivating him to file a high-profile lawsuit (Eisenbach 2006). Matlovich was a problematic figure for straight America because he served three tours in Vietnam and was highly decorated. How could one be so patriotic and so...gay? To try to tamp down the growing activism surrounding Matlovich's lawsuit, the Air Force offered him a monetary settlement, which he accepted in lieu of reinstatement. Though activists were upset at his decision, Matlovich felt that the climate of the Supreme Court would have made reinstatement unlikely.

The military is the largest employer in the United States (Eskridge 2008). This makes it a key player in the fight for LGBTQ rights, which creates a dilemma for anti-war activists as Wolf (2009c) explains:

> Some progressives who oppose U.S. military operations around the world ask whether the left ought to support the right of sexual minorities to serve openly in the military...in essence, allowing the U.S. government to continue to discriminate on the basis of sexual and gender behavior in its military workforce of nearly three million people gives a green light to persistent social and legal restrictions on LGBT people and continued bigotry. There is nothing incompatible with demanding an end to draconian laws barring open LGBT folks from serving in the military while opposing armed forces recruitment and U.S. imperial actions all over the world. (pp. 71–72)

For example, current military policy dictates that even if same-sex marriages were recognized at the federal level, married gays and lesbians would still be barred from serving (Kotulski 2004). This should make overturning the current policy a priority for all activists.

The biggest barrier for LGBTQ people within the military is "Don't Ask, Don't Tell," a Clinton administration "compromise" in which gays and lesbians are "allowed" to serve as long as they don't raise suspicions about their sexual orientation. Each day four servicemen and women are discharged from the military with no benefits because of this policy, bringing the number to 12,500 since 1994 (Kotulski 2004, 173; Wolf 2009c, 146). Many of those discharged served key hard-to-fill positions, such as Arabic linguists, pilots, intelligence gatherers, and medical personnel. Wolf (2009c) outlines the situation that transgender service members face, including being turned away from Veteran's Administration benefits.

The issue of allowing gays and lesbians within the military has been raised repeatedly since Matlovich's challenge. Yet the Democrats, supposed allies of LGBTQ individuals, have been the least helpful. Legislation that would have provided relief for gays and lesbians in the military never crossed President in housing and employment (Goldstein 2003). As of 2003, there were only 14 states and 251 towns and cities with workplace anti-discrimination policies (p. 111). In 2009, the tally rose to 20 states along with Washington, D.C. (Wolf 2009c, 147). One's sexual orientation can also be raised in a criminal trial, even if it has no bearing on the case (including death penalty cases) (Goldstein 2003). Nearly one-fourth of LGBTQ people have problems obtaining housing and employment because of their sexual orientation, including conservative gay writer Bruce Bawer, who was fired when his sexual identity was discovered (p. 18).

Social class has a major impact when it comes to HIV/AIDS, with the bulk of cases happening in majority world countries, sites of dire poverty (Wolf 2009c). No longer a "gay" disease, the overwhelming majority of U.S. AIDS cases now consist of African-Americans and Latinos. Treatment for AIDS can be prohibitively expensive. Burroughs Wellcome, the manufacturer of AZT (an HIV/AIDS medication), put an annual price tag of $8,000 on the drug. The price of the latest medication, Norvir, increased five-fold to an annual cost of $7,800 (p.165). Though lawsuits against the pharmaceutical industry are important and understandable in light of these extortion-level prices, the fight against AIDS must include a broad offensive for universal health care, not just for access to these highly-priced drugs.

Anti-discrimination laws protect everyone, not just members of minority groups. Currently, the latest effort at including sexual orientation as a federally protected category at the workplace is the Employment Non-Discrimination Act (ENDA) of 2009. It includes transgendered people under the category of sexual minorities, which would make it illegal to discriminate against LGBTQ people in the hiring process and at the workplace. The inclusion of transgendered individuals is a milestone, considering that this group is among the most vulnerable of the LGBTQ community (Kuban and Grinnell 2008; Spade 2008). San Francisco, considered one of the most gay-friendly cities, has 60 percent of its transgendered residents earning less than $15,300 per year, with only a quarter employed full time (Wolf 2009c, 147). If ENDA passes during the Obama administration, a legacy of anti-gay discrimination—if not brought to an end—will be much harder to justify. This will make distant memories of the 1978 Briggs Initiative, which proposed firing public school teachers if they were gay or lesbian, and Oklahoma's past efforts in the 1970s to dismiss teachers and teacher aides for suspected or actual homosexual conduct (Eisenbach 2006; Eskridge 2008). It wasn't that long ago when public school teachers were in the crosshairs—including straight teachers who were suspected of being gay or lesbian.

The Military

Even though all sodomy statutes have been overturned in the United States, the military continues to outlaw anal sex—-along with a wide range of sexual charity-driven values rather than being led by the young people who need these services as a matter of human rights.

Even though the presence of a GSA can help push forward the safe schools movement, resistance is still felt among teachers, administrators, parents, and even students who view such groups as promoting sex (GLSEN and Harris Interactive 2008; Harris Interactive and GLSEN 2005; Ognibene 2005; Perrotti and Westheimer 2001). Often in their quest to gain acceptance and appear like a harmless social club, GSAs are pressured to become a "diversity" club instead of directly building gay-straight alliances. Or, as Perotti and Westheimer (2001) describe, schools will request "clean cut" and gender typical gay and lesbian speakers so as not to put off more conservative elements of the community. Transgendered students are rarely considered for speaking roles.

The existence of GSAs can be a critical source of support during the most vulnerable adolescent years when young people are beginning to experiment sexually and form relationships. D'Emilio (1992) asserts that despite the many gains of LGBTQ organizations, the majority of schools and universities still do not have gay student organizations. Even though the courts have consistently upheld the First Amendment rights of LGBTQ organizations, those who want to form clubs have to fight for this right while experiencing student-led backlash and administrative indifference. It often takes attending college for LGBTQ students to finally attain some level of self-acceptance. But what about the majority of the population that never sets foot on a university campus?

Employment and Health

The media is replete with images of affluent gays and lesbians, which the right wing uses as a rationale for not needing to grant "special rights" to sexual minorities. It is quite interesting that the "average salary" of $80,000 is regularly attributed to gays and lesbians along with unionized auto workers. This is part of a strategy to over-inflate these salaries in order to build resentment and demobilize activism among the working class. In reality, compared to straight men, gay men earn less overall, while lesbians and straight women are comparable in terms of salaries (Goldstein 2003). However, compared to heterosexual households, lesbian households earn much less. Those with networks of support that enable them to come out tend to be the most visible. They are the subjects of most magazine and consumer product surveys, which skews the perception of upper-middle class groups of gays and lesbians (Wolf 2009c). Meanwhile, transgendered, intersex, and queer people are more likely to live below the poverty line, especially if they are also members of ethnic minority groups.

LGBTQ individuals are not covered by federal sexual minority job protections, which makes it easier for this group to be targeted for discrimination

November 3, 2009. This highlights the futility of framing the fight for marriage rights as a state issue.

Gay-Straight Alliances

Parents and Friends of Lesbians and Gays (PFLAG) was formed in 1972 by a couple from New York whose gay son was violently attacked. PFLAG currently has over 200,000 members across 500 U.S. affiliates (PFLAG Website, n.d.).. Soon after its formation, clubs in secondary schools and universities multiplied to provide fellowship and socialization opportunities for gays and lesbians, even though these kinds of clubs existed in the 1950s and 1960s, though under much more hostile conditions (D'Emilio 1992). Eventually, PFLAG's approach was combined with these student clubs to include straight allies. The concept of a gay-straight alliance (GSA) was born. On the faculty front, the Gay, Lesbian, and Straight Education Network (GLSEN) was originally formed by a few hundred closeted teachers in order to protect their jobs. It has evolved into an "out" membership with 40 chapters nationwide (GLSEN Website, n.d.). Currently, there are more than 3,600 GSAs registered with GLSEN and every state in the union has at least one GSA (Jennings 2005, xiv). The growth of the GSA might have to do with more students coming out at younger ages than ever before (Ognibene 2005).

The average GSA has a majority white female and straight membership and numbers 10–12 people. In any given GSA, one is likely to find 6–8 females, 1 or 2 of whom identify as bisexual or lesbian, 1 male who identifies as gay, and possibly 1 male who identifies as straight (Perrotti and Westheimer 2001, 65). It is difficult to build membership among straight males because of the stigma attached to homosexuality and its perceived threat to masculinity in K–12 settings. Perrotti and Westheimer theorize that the majority female membership has to do with girls relating homophobia to abuses suffered under sexism. However supportive the GSA structure can be, there are many LGBTQ youth who are not being reached—even the GSAs in the more affluent suburban schools are at the mercy of budget fluctuations (Aarons 1995). There remains a severe lack of outreach to trans-youth and LGBTQ youth of color who live on the streets (O'Brien 2008; Rosado 2008). Kuban and Grinnell (2008) maintain that the large portion of queer youth who have dropped out of school or who attend rural schools make the traditional GSA model inadequate for kids who need basic health services and a safe place to stay off of the streets. Rosado (2008) estimates that 35 percent of New York City's 22,000 homeless youth population are out LGBTQ young people (p. 319). The fewer than 200 LGBTQ oriented programs in existence are clustered in New York and California and are dependent on outside funding (Aarons 1995). The minimal services that are available for LGBTQ youth on the margins of society are often based on paternalistic, Judeo-Christian

In addition to being denied marriage protections, LGBTQ people are barred from receiving the benefits of divorce, including not being evicted from one's home if the partner meets someone else (Kotulski 2004). Current immigration policy also forbids gays and lesbians from sponsoring a partner from another country. Transgendered people cannot officially declare transformation into the opposite sex unless proof of surgery- the surgical process which is itself expensive- is presented,. Full marriage rights would overturn these policies and automatically allow all adults, regardless of sexual identity, equal protections. Oddly enough, intersex people have the most flexibility under the DOMA, since they have the genetic makeup to marry either sex and still be considered in a heterosexual marriage. Goldstein (2003) sees the automatic widening of definitions of marriage and family if the DOMA falls and same-sex marriages are finally recognized. This is what the right wing fears most, for reasons that will be explored in chapter two.

Nate Silver (2009) of the blog FiveThirtyEight, provides a tart, tongue-in-cheek analysis of Maine's recent anti-gay Stand for Marriage campaign in support of Question 1. In a bulk email sent to Silver, the group urged conservative radio listeners to chime in on an upcoming broadcast debate with a pro-gay marriage advocate, listing the following talking points instructions:

> Remember...they had every opportunity to expressly prohibit gay marriage from being taught in schools but they did not; the State Department of Education admits there is nothing prohibiting schools from teaching about same-sex marriage; there is already the infrastructure to advance gay issues in the schools...and if Question 1 fails, homosexual marriage will not become equal to traditional marriage because traditional marriage will be totally eliminated. Marriage will be genderless. It will totally exist for the benefit of adults. (par. 5)

In his brilliant response, Silver neatly challenges these nonsensical arguments against same-sex marriage:

> So, paraphrasing somewhat, the arguments that the Yes on 1 campaign seems to be making are as follows: (1) The new law won't make gay marriage equal to straight marriage. Instead, it will create a new kind of marriage in which gay people and straight people are equal. (2) Although we may not have proven any connection between gay marriage and public education, our opponents haven't disproven the connection, and it's their fault that the subject came up. (3) If gay marriage is upheld, then marriage will exist solely to make people happy. (par. 7)

Unfortunately, the combination of Silver's wit and widespread activism failed to prevent Maine voters from overturning the right to same-sex marriage on was limited to immediate family and spouses—not partners...though Langbehn had documents declaring her Pond's legal guardian and giving her the medical power of attorney, Jackson officials refused to recognize her or the kids as family (par. 1–15).

In states that do not recognize same-sex marriages, it doesn't matter if you have been with your partner for several years. If he or she passes away without a will, directive, or other legal document (all of which are cumbersome and expensive to obtain), someone from your partner's family has more legal rights to determine a place of burial than you do (Kotulski 2004; Wolf 2009c). With marriage between partners of the opposite sex, these rights are automatically granted to spouses.

In the 1990s, same-sex marriage efforts centered on obtaining domestic partnerships and civil union status for LGBTQ people, with significant victories in Hawaii, California, and Vermont (Kotulski 2004; Wolf 2009c). The majority of people who benefit from domestic partnerships are heterosexual and unmarried, so much of the legislative efforts to extend domestic partner benefits have come from this demographic, and not "activist judges." Close to half of a company's employees are unmarried, which reflects larger societal shifts. As a result, more Fortune 500 companies offer partner benefits, partly because they cost less than full marriage benefits packages (Coontz 2005; Wolf 2009c). Kotulski (2004) outlines the lengthy process of applying for domestic partner benefits in California. Those applying for domestic partnerships must go to their City Hall. In some cases, the couple ends up registered for only that city, and not the whole state. City Hall might not have the statewide partnership forms or the basic information on how to obtain them. Once they have the forms, the couple has to have them notarized and mailed to the Secretary of State. By contrast, heterosexual married couples are able to do all of this with just one visit to City Hall with no notarization steps.

Marriage includes nearly 1,300 rights and legal protections, while civil unions and domestic partnerships only offer a fraction of these rights (Kotulski 2004; Wolf 2009c). The push for full same-sex marriage recognition accelerated in the wake of HIV/AIDS, when LGBTQ people began to see the consequences of the lack of protections offered through the health-care system (Eisenbach 2006). The marriage rights movement also gained inspiration from the overturning of interracial marriage bans in the 1960s—bans which were often justified using the same lines of reasoning as today's anti-gay marriage backlash (Coontz 2005). As Johnson (2009) rightly points out, most LGBTQ people are in the working class, not affluent, and desperately need the federal rights that come with marriage recognition. Gay seniors are especially vulnerable as they often have few surviving family members to oversee financial and health matters (Wolf 2009c). Setting aside all leftist debates about the appropriateness of marriage, the working class as a whole would gain from the DOMA being overturned.

being "hidden recruiters" of school-age children to the gay lifestyle Kosciw and Diaz report that LGBTQ parents have an overall level of satisfaction with their children's school, and that they experience minimal harassment. The main problems tend to come from other parents and classmates.

Same-Sex Marriage

On April 3, 2009, the Iowa Supreme Court ruled that laws against same-sex marriage are unconstitutional (Colson 2009). Four other states—Connecticut, New Hampshire, Massachusetts, and Vermont—also allow for full marriage equality for lesbians and gays. In addition, the District of Columbia recognizes same-sex marriage rights performed in these other states (Johnson 2009; Thweatt and Rose 2009). One of the Iowa justices, Joyce Kennard, declared "the argument is between the inalienable right to marry and the right of people to change the constitution as they see fit" (Colson 2009, par.12). Even with these important victories, same-sex couples in these states still do not have access to federal protections. These include social security and disability benefits for surviving partners or the freedom to move to another state and still have their marriages recognized (Kotulski 2004; Thweatt and Rose 2009; Wolf 2009c).

The chokehold on same-sex marriage rights is the 1996 Defense of Marriage Act (DOMA), which federally recognizes marriages only between one man and one woman. DOMA has enabled conservative groups to successfully use the state referendum strategy to further limit same-sex couples' access to even minimal protections—43 states passed limitations by the end of 2004 (Coontz 2005, 274). The closest DOMA came to being seriously challenged via judicial scrutiny was in California during the potential—but unsuccessful—overturning of ballot initiative Proposition 8 (Schulte 2009). The current status of same-sex marriage rights does not reflect public opinion, with 52 percent opposing DOMA, and large majorities supporting the extension of health care benefits and visitation rights to partners (Wolf 2009c, 16). At the national level, 55 percent of Americans agree with federal recognition of same-sex partnerships (p. 257) and 42 percent in support of full marriage recognition (Johnson 2009, 3).

Schulte (2009) describes key legal cases in which partners were denied health-care benefits and hospital visitation rights. Though DOMA wasn't directly harmed in the rulings, some of which were in favor of LGBTQ plaintiffs, these kinds of cases move closer to a formal challenge. Figueroa (2009) exposes the consequences of bigotry as long as DOMA stands:

> As her partner of 17 years slipped into a coma, Janice Langbehn pleaded with doctors and anyone who would listen to let her into the woman's hospital room...staffers advised Langbehn that she could not see Pond [her partner] earlier because the hospital's visitation policy in cases of emergency

Gay and Lesbian Families

There has been a steady growth in the number of lesbian and gay families, much to the chagrin of religious fundamentalists and traditional familialists (Hardisty 2008). The high-profile celebrity of Thomas Beatie, the female-to-male transgendered "pregnant man," has challenged the idea that only women can give birth (Wolf 2009c). Contrary to the stereotypes of childless gays and lesbians, roughly 33 percent of lesbian households and over 20 percent of gay households contain biological children under 18 years old (Coontz 2005, 275; Kotulski 2004, 91). Translated into numbers, there are over seven million LGBTQ parents with school-age children in the United States today (Kosciw and Diaz 2008, vii). It is not only the LGBTQ community that is challenging the norms of what it means to be a family. For example, more than 25 percent of all households in the United States consist of only one person (p. 276).

Though not a majority, four out of ten U.S. adoption agencies have made placements in gay and lesbian households (Coontz 2005, 275). This low rate has much to do with state prohibitions against gays and lesbians adopting children (Kosciw and Diaz 2008). For example, Utah, Arkansas, Mississippi, and Florida do not allow lesbian or gay people to act as foster parents, even though there are currently 134,000 children nationwide in foster care awaiting adoption (Dave Thomas Foundation 2002). Marriage status also makes a difference. California requires the non-biological parent in a registered domestic partnership to go through the time and expense of adopting the child But in a civil marriage, biology makes no difference—the child is considered automatically adopted (Kotulski 2004). These second-parent adoptions are often used by gay and lesbian couples as an additional precaution since they are currently denied marriage rights. This allows same-sex domestic partners to become stepparents to each other's children so that the children cannot be seized by other family members in the event of a partner's passing.

Even though a majority of secondary school principals report that students at their schools have been harassed for having a lesbian or gay parent, nearly half indicate that these same students would feel "very safe" at their school (GLSEN and Harris Interactive 2008, 11). Only one in six principals believes that a gay or lesbian parent would feel hesitant to join school activities, like participating in parent-teacher associations, being a classroom helper, or assisting on field trips (p. 11). Despite these perceptions, in the areas of parent-teacher conferences and volunteering, gay and lesbian parents tend to be more involved in their child's school than heterosexual parents (Kosciw and Diaz 2008). They also have more overall contact time with school staff than heterosexual parents—68 percent of LGBTQ parents indicated they contacted their child's school versus 38 percent of parents nationally (p. xv). LGBTQ parents are also forthright about their sexual orientation with teachers and administrators, contradicting the stereotype of gays and lesbians cope with the inevitable feelings of inadequacy. These feelings stem from the overwhelming social problems and fears of admitting they are letting their students down (p. 174). One common defense mechanism is to defer authority to administrators, while decreasing their own competence and agency. Meyer (2009) describes how these defense mechanisms create barriers against teachers openly confronting homophobia because they lack a materialist analytical frame where they could comprehend the functions of schooling within a capitalist society beyond just a reactionary position:

> They are frustrated by the limitations, but none critique the formal structures of the school that cause them to feel overwhelmed. They spoke of it as if it were simply the reality that must be dealt with. When asked to suggest changes that could improve their ability to address bullying and harassment, no teacher mentioned reducing class sizes, limiting the number of class preparations, or adding educational assistance to alleviate some of these demands on their time and energy...by dealing with behavior issues only within the microstructures of their classrooms rather than addressing the macro-structures of the school, they are extremely limited in what they can do to improve student safety and school climate. (p. 27)

By relying exclusively on current institutional practices to determine their strategies, teachers quickly become cynical and apathetic. They are less likely to confront homophobia.

LGBTQ individuals who struggle with the decision to reveal their sexual orientation to other faculty and students experience a hyperawareness of their movements and speech and its potential impact on others (D'Emilio 1992; Freedman 2005; Goldstein 2003; Petr 2005; Record 2005; Roberts 2005). Transgendered educators face even more challenges as they undergo the lengthy surgical procedures of sex change operations (Roberts 2005). D'Emilio (1992) focuses on how remaining closeted makes it impossible for people to organize and form important solidarities with other workers—precisely why the Right clamors for locking the closet door as a remedy:

> ...the fear that compels most gay people to remain hidden exacts a price of its own. It leads us to doubt our own self-worth and dignity. It encourages us to remain isolated and detached from our own colleagues and peers, as too much familiarity can lead to exposure. And it often results in habitual patterns of mistrust and defensiveness. (p. 151)

When teachers decide to come out of the closet, it creates a "ripple effect" which can send an important message to LGBTQ and straight students (Nicolari 2005, 24). The presence of an out teacher interrupts a homophobic school climate—business isn't as usual.

of a conspiracy of one-world government (Gandossy 2009). Regardless of the specific labels applied to those who are not exclusively heterosexual, there are 8.8 million people who identify as lesbian, gay, or bisexual (Gates 2006, 1; Wolf 2009c, 241). There are now more than 594,000 same-sex households in the United States, and the numbers are growing (Rowland 2004, 303). These households exist in large concentrations even in unlikely states: West Virginia, Montana, Wyoming, and Arkansas. According to Goldstein (2003), across all categories of social class and ethnicity, LGBTQ people as a group tend to be more left-leaning politically than straight constituents. Gay males' voting records resemble straight women's when it comes to gender issues. There is growing discontentment, however, with the Democratic Party, which has continually failed to confront discriminatory policies, both in the workplace and with marriage rights (Wolf 2009c). A lack of coherent action concerning LGBTQ issues on the part of the Democrats persists, even though public support for relations between consenting LGBTQ adults has risen to 59 percent, and workplace equality is supported by almost 90 percent of Americans (p. 128).

One of the major shifts within society has been increased contact between out LGBTQ and straight people. Within 25 years, the number of Americans who report knowing someone who is openly gay or lesbian has increased from less than one quarter of the population to more than half of the country (Eskridge 2008, 267). In K–12 schools, the likelihood of students knowing an openly gay or lesbian student is more than 50 percent, while 20 percent have a close friend who is gay or lesbian (Harris Interactive and GLSEN 2005, 13). Nearly 50 percent of teachers know of a gay or lesbian coworker, though the higher percentages are in public schools than private religious school settings (p. 87).

Experiences of LGBTQ Students and Teachers

K–12 schools are part of a society that currently denies several key rights to LGBTQ people, creating a climate of legal alienation (D'Emilio 1992; Fone 2000; Wolf 2009c). Students and teachers exist within a framework in which only 10 states, along with Washington, D.C., prohibit sexual orientation discrimination in public schools. Four of those states prohibit gender identity discrimination (Kosciw and Diaz 2008, xix; Perrotti and Westheimer 2001, 36). LGBTQ students are less likely to report instances of harassment to teachers and administrators because doing so might reveal their sexual orientation to parents and friends (Kosciw, Diaz, and Greytak 2008). It wasn't that long ago when LGBTQ teachers didn't have anti-discrimination clauses in their contracts, making the closet the logical choice for self-preservation (Beach 2005).

According to Luttrell (2003), schools are a key site for "managing anxiety," where teachers have to create several defense mechanisms in order to tones is involuntary, yet racism still persists. As Wolf (2009c) states, "tying civil rights to biology avoids the central argument that human beings deserve humane and equitable treatment regardless of what others think of their sexual preferences" (p. 216). Anders (2008) injects a sense of humor to counter the assertion that no sane LGBTQ person would "choose" their present identity: "because, of course, if it weren't for that one difference, queer people would all be just like Republican hatemonger Orrin Hatch" (p. 88).

The evolution of sexual orientation has brought with it a myriad of contemporary terms. The word "queer" had negative and homophobic connotations in the 1960s, so it was replaced with "gay" and "lesbian." More radical factions of contemporary movements have co-opted the word queer to indicate someone who eschews any type of gender or orientation label (Fone 2000). Queer can refer to people who are gay, lesbian, bisexual, transgendered, intersex, or any of these categories combined with race and social class differences, though there is debate about the appropriateness of its use in liberation movements (Wolf 2009c). Transgendered refers to those who choose to identify as the opposite of his or her biological sex at birth. Bisexuals are those who are attracted to either sex. Both of these terms are more flexible categories of sexual orientation and gender. Someone who is transgendered can self-identify as homosexual or heterosexual, while bisexuals consider themselves as a separate orientation category instead of strictly gay or lesbian. However, some bisexuals may select gay or lesbian as a primary identity and bisexual as a secondary one. Other labels include the Q, for questioning and two-spirit, which references Native American beliefs that homosexuals possess both male and female souls (Goldstein 2003).

Even within the broad coalition of LGBTQ, transgendered individuals and intersex people (hermaphrodites who are born with both male and female sex organs) have suffered the most discrimination and violation of their civil rights. There are five million intersex people in the United States today, and many of them are not aware of it (Wolf 2009c). Because of the intense pressure to conform to gender norms, parents of intersex children often secretly choose which set of genitals to remove. Sometimes doctors perform the surgery without the parents' full consent, which is a shocking violation of the basic human right of self-definition. Activist groups like the Intersex Society of North America and Hermaphrodites with Attitude have done much to raise awareness of this marginalized group. They argue that despite the presence of both sexual organs, intersex individuals should have the right to select their own gender.

Because sexual orientation is a self-identified category with several potential definitions, it can be difficult to accurately track numbers since some categories like transgender and intersex have not been well-researched (Sexual Minority Assessment Research Team 2009). For example, 2010 is the first year that the U.S. Census is systematically tracking the marital status of same-sex couples, causing an uproar in conservative groups that already view the census as part defiance and/or are actively depriving students as a whole of valuable information. Historically, schools have not been places where anti-sexist or anti-racist practices happen—the subjects aren't even broached as topics of conversation (Perrotti and Westheimer 2001). D'Emilio (1992) reminds us of the fact that capitalism has shaped the nuclear family, thereby pushing people, regardless of their particular orientation, into this particular template for heterosexual relationships. This template—along with gender role expectations—is internalized, and people of all ages are constantly comparing themselves to the template's requirements. As a result, most of the population is just beginning to gain an understanding of the LGBTQ community, particularly within a K–12 context. As Record (2005) states, "heterosexual privilege has significant power to blind people to the reality in front of them" (p. 255).

A Matter of Identity

Sexual orientation as an integral feature of identity is a relatively recent invention from an anthropological standpoint. The rise of industrial capitalism in the mid-1800s provided the specific historical conditions for sexual identity to become a social framework (Cotter 2001; D'Emilio 1992; Fone 2000; Forrest and Ellis 2006; Wolf, 2009c). The professionalization of the medical field, an emerging discipline called sociology, and the growth of psychiatry and psychology brought about an interest in categorizing different sexual practices. The term "homosexual" first appeared in the text of a German letter in the late 1860s. In 1880s England it was used in an early pro–gay rights essay. The first American use of the term was in a medical article written in the 1890s (Fone 2000). Popular use of the terms "homosexual" and "heterosexual" began in the 1920s and 1930s. When examining the phenomena of sexual orientation, it is important to distinguish between sexual behaviors, which have been a part of humanity since day one, and the notion of a fixed sexual identity, which is a modern concept (D'Emilio 1992; Eisenbach 2006; Wolf 2009c). Sexuality ranges along a continuum of exclusively heterosexual to exclusively homosexual, proposed by Kinsey (1998) during his ground breaking research where he discovered that few expressed preferences on either end of the scale. While human beings have always had sex, the meanings attached to sex have been directly influenced by economic and social factors.

Both D'Emilio (2003, 2009) and Wolf (2009c) make the compelling point that focusing on the biological-versus-choice debate when it comes to homosexuality is a zero-sum game. This essentialist struggle will be explored more fully in chapter four. Suffice it to say that justifying LGBTQ rights by trying to find a biological basis for homosexuality has not advanced the liberal perspective much. In the 1930s many progressive authors argued that homosexuals should not be persecuted because they were "born that way." But this did little to prevent gays and lesbians from being rounded up and exterminated. The same is true with more obvious biological features. Having darker skin sexuality, people are reluctant to bring up these issues in K–12 settings. In particular, the LGBTQ activist community has steered clear of school settings because of unfounded associations with pedophilia and other forms of external stereotyping (Perrotti and Westheimer 2001). The overall contraction of the public sector has also contributed to a culture of "bare bones" budgeting, making LGBTQ and sexuality programming seem frivolous (Luttrell 2003).

This chapter will present information examining the existing sexuality and LGBTQ issues in K–12 classrooms. The main goal of this chapter is to present the reader with a portrait of sexuality in today's classroom, highlighting abstinence-only education and discrimination. In addition, laws and policies related to gay marriage will be reviewed throughout the chapter, as these have a direct impact on K–12 students, teachers, and parents or guardians. The first section addresses the contemporary situation facing LGBTQ students and their families within an educational context. It begins with a demographic portrait of sexual minorities, the marriage rights movement, workplace discrimination, and composition of families, while addressing services currently offered in schools (gay and lesbian organizations and counseling services). The second section presents the censorship and discrimination inherent in abstinence-only education, a movement with a large funding base coming from the religion-industrial complex. Meacham and Quinn (2007) describe this religion industry as being comprised of non-profits, college and university research departments, think tanks, direct marketing firms, and corporate sector investments which back policies like abstinence-only education. The third section examines the lack of LGBTQ representation within the curriculum and the resulting enforced censorship that contributes to students' and teachers' reluctance to come out of the closet or to step up as straight allies. The fourth section looks at the current climate of enforced heteronormativity in K–12 settings. Examples range from teasing and name calling to violent manifestations of murder and suicide, as represented by the well-known tragedies of Matthew Shepard and Bobby Griffith. These point to the need for, at minimum, and expansion of existing hate crimes legislation to include protections for LGBTQ people and the addition of sexual orientation to protected categories of school discrimination policies.

The LGBTQ Community and K–12 Schools

Dolor's (2008) study of teachers' attitudes toward LGBTQ students and families revealed that most educators feel that it is quite easy to pinpoint which of their students are gay because they assume that LGBTQ students display non-normative gender cues such as males being effeminate. When juxtaposed against Dolor's other findings—that teachers tend to avoid LGBTQ topics in the curriculum—it doesn't take much to conclude that they are most likely overlooking a large portion of students who are part of a sexual minority group because they are basing their assumptions on stereotypes of more visible gender

Chapter One

The Existing Climate of K–12 Classrooms

In the spring of 2009, 11-year-old Carl Joseph Walker-Hoover hanged himself at home with an electrical cord. He had been actively involved in sports, scouting, and church. What led him to suicide was daily verbal and physical assaults centered around his perceived sexual orientation—even though Walker-Hoover was not gay (Kornegay 2009). His mother had made continual attempts to contact the school to ask them to intervene, but to no avail. Kornegay also describes a similar case of 17-year-old Eric Mohat, who shot himself in response to constant harassment at the hands of classmates. Like Walker-Hoover, Mohat did not consider himself gay. This corresponds with the findings of a 2005 study by the Gay, Lesbian, and Straight Education Network (GLSEN) that the second-most-common cause of bullying in school is "perceived or actual sexual orientation and gender expression," with appearance being the primary cause (National School Climate Survey 2005, p. 4). What is clear is that it is no longer acceptable to marginalize gay and lesbian issues within K–12 education. All students are potential victims of homophobic actions. Just being perceived as violating gender norms is enough to make one a target (Meyer 2009).

According to Cudd (2002), "freedom of expression and the right to equal opportunity often conflict. There may be no place where the conflict is more immediate and more wrenching than in our colleges and universities" (p. 217). One could also argue that K–12 settings are also sites of this conflict. This is especially true when it comes to sexuality rights: the right to information about sexuality and the right to exist as a gay, lesbian, bisexual, transgender, queer/questioning, intersexed, or straight individual. Meyer (2009) likens the struggle for sexuality rights to destroying the very walls of our childhood home. For some, these walls are comforting and familiar, thus the loss is mourned. But we often forget that for others, the walls are confining and exclusionary. Because of the heightened shame and sensitivity surrounding issues of

Chapter 6, "Building Effective Educational and Resistance Movements in K–12 Classrooms," culminates in a revolutionary message of how we can use issues of sexuality, including addressing issues important to working-class families, to build activist tactics and strategies. A discussion of common barriers to activism sets the stage for possible solutions.

beyond the usual frame of "tolerance," homophobia is presented as a form of false consciousness that is deliberately used by capitalists to divide the working class. Many corporations use diversity training and include sexuality as a protected minority. However, when times get tight, big business often plays off of religious conservative objections. The industry benefits from rallying consumers around "family values." This chapter opens with an analysis of the defeat of California's "No on 8" campaign, which resulted in overturning same-sex marriage rights. Homophobia is then defined and presented from a historical vantage point. This is followed by a systematic presentation of various current sources of homophobic ideology.

Chapter 3, "The Family: Conservative, Psychoanalytical, Anarchist, and Materialist Readings," presents a variety of theories about the significance of how humans have arranged themselves within the framework of relationships. In the 1960s and 1970s, activist scholars paid particular attention to materialist interpretations of the family. But lately, these viewpoints have been missing in action, so to speak, so they are revived here to shed light on the current instability of the nuclear family. Conservatives often present the nuclear family as an unchanging entity and source of morality, but the historical record reveals a complex picture of humans adapting to material conditions. Today's nuclear family is unsustainable, as evidenced by the alienation that impacts two-earner and single-parent families. Of particular interest to K–12 educators is the current discussion of children as "super-products" of reproduction.

Chapter 4, "Identity Politics: Limits of Postmodernism and Queer Theory," critiques postmodernist and poststructuralist approaches to sexuality. The chapter opens with a brief overview of the philosophical foundations and impact of postmodernism, queer theory, and identity politics. Instead of being treated as a subject apart from context, postmodernism is viewed as an ideological movement stemming from neoliberalism and global capitalism. As a result, this view of postmodernism does not acknowledge Marxism's insistence on placing class at the center of sexuality. Several harmful aspects of postmodern theories are presented, including privileging localized identities, prioritizing desire over material need, valorizing consumerism, and promoting libertarian-style rootlessness and individualism.

Chapter 5, "The Socialist Feminist Message: Sexuality, Work, and Liberation," presents alternatives to post-al theories. Women's issues are first outlined with connections to LGBTQ oppression. This includes the family, unpaid labor, sexual harassment, and reproductive freedom. Common myths about the family and patriarchy are exposed, including the notion that sex divisions are eternal and insurmountable. Contemporary anthropological research has much to say about pre-class societies and their social and work arrangements. These are quite illuminating as we attempt to craft an alternative vision of human society. An extensive discussion of reproductive rights, along with a review of conservative feminist philosophies, will be useful to educators trying to distinguish viewpoints among competing ideologies.

While much of the anti-gay sentiment is couched in discussions of family values, activist responses have been limited to queer theory, identity politics, or consumerism as solutions, which enhance capitalism. When social class is removed from sexuality as a central point of analysis, we are reduced to relying only on accounts of experience. Nowlan (2001) outlines how a dependence on identity politics sustains the fiction of the lone individual who has agency and choice. This propagates ahistoric thinking, in which realities are presented as malleable "texts" that can be changed at whim, including relations involving exploitation such as the worker under capitalism (Cotter 2001). With no history in which to frame oppression, the transformation of society for the betterment of everyone is jettisoned for more local, desire-based responses and performances. Theory becomes "a means to justify one's own legitimacy in the marketplace without interrogating the structural exploitation upon which this legitimacy and class privilege is founded" (p. 172).

Instead, what is required—and what is presented in this volume—is a revival of dialectical materialist analysis applied to sexuality. As Hill, Sanders, and Hankin (2002) urge, class is not a monolithic concept; it needs to be analyzed in a multidimensional fashion. The postmodern equation should be inverted where sexuality is shaped by class, not the other way around. Ebert's red cultural studies (2001) support this volume's attempt to go beyond desire and the body in order to effectively confront capitalism. This is done by addressing oppression's choke point: the relation of the worker to the means of production which undergirds the functioning of capitalism, allowing it to shape all relations, including the family. Hickey (2006) explains that historical materialism "was the first systematic attempt to indentify structures that constrain human behavior" (p.187). It radically broke with traditional approaches to sexuality, including philosophies of natural rights, human nature, empiricism, individual will, romantic desire, and religious explanations. We are facing a world in which concerns once viewed as "private," such as the family, are now part of public policy debates, which further exploits the working class (D'Emilio 2002).This has impacted both social conservatives and Marxists. By focusing on dialectical materialist sexuality, we gain much in our efforts to understand and change the world, both in K–12 settings and beyond.

Chapter 1, "The Existing Climate of K–12 Classrooms," examines the current situation regarding those who identify as LGBTQ. On a K–12 level, this chapter addresses abstinence-only education, LGBTQ harassment in schools, the significance of equal access, the safe schools movement, and curricular concerns. In addition, there is an examination of larger, related social issues, including demographic portraits of LGBTQ people, LGBTQ families, same-sex marriage, religion, military policies, and workplace discrimination. Focus is placed on the significance of high-profile tragedies, such as Matthew Shepard.

Chapter 2, "Sources of Opposition to Sexuality and LGBTQ Rights in the Schools," attempts to uncover the roots of homophobic ideologies. Going

Louisiana State Representative John LaBruzzo proposed a "solution" to the problem of poor people reproducing: pay low-income women a $1,000 incentive to get sterilized (Waller 2008). A few months into 2009, just as unemployment applications were skyrocketing, proposals were presented, requiring welfare, food stamp, and unemployment recipients to submit to random drug testing in order to obtain aid. Citing the common reactionary canard that no one is forced to use these benefits, supporters of the testing claim to be "motivated out of a concern for their constituent's health and ability to put themselves on more solid financial footing once the economy rebounds" (par. 4). As Warren Buffett famously said, "*There's class* warfare, all right, but it's *my class*, the rich *class*, that's making *war*, and we're *winning*" (Stein 2006 par. 6)

As the legalization of same-sex marriages gained momentum in Washington, D.C., the Catholic Church responded by threatening to withdraw all of its social services (Craig and Boorstein 2009). While the Church isn't the only provider of any specific services, it does supplement the struggling city's budget by $10 million. One of the council members, a Democrat, equated denial of marriage rights with the right of a Christian caterer to not have to bake a cake with "two grooms on top" (par. 14). This ignored the larger implications of the church's threat—denying food and health care to the poorest in the city.

This short sampling of news events illustrates, in the words of D'Emilio and Freedman (1997), that "sexuality has become central to our economy, our psyches, and our politics...it is likely to be vulnerable to manipulation as a symbol of social problems and the subject of efforts to maintain social hierarchies" (p. 360). Much has been written about sexuality per se, but not in the context of K-12 education. When sexuality is addressed in any setting, it is from the standpoint of experience-based, identity-politics accounts or presented as a postmodern romance with popular-media-as-liberation. For example, the proliferation of gay- and lesbian-themed entertainment is often used as evidence that society is now more open-minded, a notion that Kotulski (2004) critiques:

> The phenomena of straight folks enjoying *Boy Meets Boy*, *Will and Grace*, and *Queer Eye for the Straight Guy*, doesn't mean LGBT people have arrived any more than whites owning Sambo dolls and watching Amos and Andy signaled equal rights for African-Americans. (p. 178)

Indeed, Cloud (2001) notes that homophobic rhetoric and actions have only increased in the wake of changing state domestic partnership and marriage laws. This is tied to the challenge of people not recognizing that their own oppression is a result of false consciousness and interpellation, in which anti-LGBTQ ideology is embraced by sections of the working class (Cole 2009; Moeller 2002).

inability to sustain itself once work disappears. In 2008, a desperate widower of ten children took advantage of a state Safe Harbor law that allows a parent to drop off a child at designated sites with no questions asked. He left nine children—ranging in ages from 1 to 17—at a local hospital, stating, "I couldn't do it anymore" (Ross 2008). His wife had died in the past year, and he was unemployed. Redmond and Colson (2008) report at least six other cases in which parents from other states drove to Nebraska to take advantage of the law's lack of upper age limits of children eligible for Safe Harbor.

One of the other parents had dropped off her eleven-year-old son who had bipolar disorder and uncontrollable outbursts. When her health insurance turned down coverage for her son's mental health, the Safe Harbor law was the only option:

> The lawmakers' amendment was accompanied by a blame the victim mentality among politicians and the media that smeared parents who surrendered their children as lazy or irresponsible... insurers discriminate against both adults and children with mental illness and find ways to avoid covering them. When they do provide coverage, care is restricted. It's simply not profitable for insurers to treat chronic mental illnesses that aren't curable. And for the estimated 47 million Americans who are uninsured, it's almost impossible to get ongoing care for mental illness. (Redmond and Colson 2008, par. 18–23)

Instead of viewing these situations as an indication that urgent help is needed for struggling families, the Nebraska legislature spent a total of $80,000 to hold a special session to quickly amend the age limits. Now infants no older than 30 days can be left at designated sites.

Amidst a backdrop of ideological violence and the abandonment of families in a collapsing capitalist economy, in August 2008, Republican presidential candidate John McCain selected an unknown Governor from Alaska as his vice presidential candidate. It didn't take long for the media to discover that Sarah Palin—a vocal supporter of abstinence-only education and a contraception opponent—had a 17-year-old daughter who was five months pregnant and unmarried. The impromptu shotgun engagement was announced on television at the conclusion of the Republican National Convention. As the camera zoomed in on the bewildered groom, viewers were unaware that just weeks before, on his MySpace profile, Levi Johnston had put "children: don't want any." The Palin family spectacle revealed how conservative ideologies about the family and sexuality could often seem contradictory, yet very powerful in their persistence. Larger initiatives such as federal funding for abstinence programs in schools while limiting access to contraception and abortion illustrate how sexuality impacts public policy while being marketed as something to be handled privately.

Introduction

Over the past two years, intersecting ideologies of family, sexuality, public policy, and capitalism were evident in a series of news events great and small. In 2008, on a hot July day in Knoxville, Jim D. Adkisson walked into Tennessee Valley Unitarian Universalist Church with a rifle and an intent to kill liberals. According to his manifesto found shortly after the attack, these liberals "are a pest, like termites, millions of them...the only way we can rid ourselves of this evil is to kill them in the street where they gather" (Adkisson 2008). Adkisson selected a Unitarian church precisely because of their publicized support for LGBTQ rights, along with their antiwar and multicultural platforms.

Adkisson's recommendation for right wing foot soldiers to confront the leftists where they spent a majority of their time was misguided as few liberals seemed inclined to gather in large enough groups these days. For example, United for Peace and Justice is the most powerful anti-war coalition in the United States. But the most "street action" they could muster for their annual assembly in 2008 was 248 attendees. And most of these people were uncertain about challenging newly elected president Barack Obama's position on escalating military action in Afghanistan (Suber 2009).

Police later found books in Adkisson's house by Bill O'Reilly, Michael Savage, and Sean Hannity, premiere hate talkers in the culture wars (Havenworks n.d.). The irony of Adkisson—whose food stamps were being cut—consequently blaming liberals, the very people who support funding the very assistance programs conservatives were cutting—appeared lost on the major cable news networks. The killing of three police officers in Pittsburgh has also been attributed to the ideological anti-government promptings of talk show hosts Glenn Beck and Alex Jones. In April 2009, Richard Poplawski opened fire on the police because he feared a socialist new world order ushered in by the Obama administration (Blumenthal 2009).

While instances of violence make a dramatic and often immediate impact in the media, the long-standing deterioration of the nuclear family under economic pressures loom large. Recent issues stemming from Nebraska's Safe Harbor laws exemplify the privatization of the nuclear family and its

This page intentionally left blank

Contents

Introduction	1
Chapter One The Existing Climate of K–12 Classrooms	7
Chapter Two Sources of Opposition to Sexuality and LGBTQ Rights in the Schools	49
Chapter Three The Family: Conservative, Psychoanalytical, Anarchist, and Materialist Readings	79
Chapter Four Identity Politics: Limits of Postmodernism and Queer Theory	119
Chapter Five The Socialist Feminist Message: Sexuality, Work, and Liberation	151
Chapter Six Building Effective Educational and Resistance Movements in K–12 Classrooms	189
Conclusion	223
References	225
Index	243

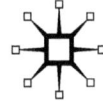

MARXISM AND EDUCATION BEYOND IDENTITY
Copyright © Faith Agostinone-Wilson, 2010.
Softcover reprint of the hardcover 1st edition 2010 978-0-230-61608-0
All rights reserved.
First published in 2010 by
PALGRAVE MACMILLAN®
in the United States—a division of St. Martin's Press LLC,
175 Fifth Avenue, New York, NY 10010.

Where this book is distributed in the UK, Europe and the rest of the world, this is by Palgrave Macmillan, a division of Macmillan Publishers Limited, registered in England, company number 785998, of Houndmills, Basingstoke, Hampshire RG21 6XS.

Palgrave Macmillan is the global academic imprint of the above companies and has companies and representatives throughout the world.

Palgrave® and Macmillan® are registered trademarks in the United States, the United Kingdom, Europe and other countries.

ISBN 978-1-349-37976-7   ISBN 978-0-230-11355-8 (eBook)
DOI 10.1057/9780230113558

Library of Congress Cataloging-in-Publication Data

Agostinone-Wilson, Faith.
   Marxism and education beyond identity : sexuality and schooling / Faith Agostinone-Wilson.
      p. cm.—(Marxism and education)
    1. Sex—Social aspects—United States. 2. Homosexuality and education—United States. 3. Marxian school of sociology—United States. 4. Families—United States. I. Title.

HQ23.A36 2010
306.70973—dc22   2010001972

A catalogue record of the book is available from the British Library.

Design by Newgen Imaging Systems (P) Ltd., Chennai, India.

First edition: October 2010
10 9 8 7 6 5 4 3 2 1

Marxism and Education beyond Identity

Sexuality and Schooling

Faith Agostinone-Wilson

fathers, 77, 81, 83, 93, 95, 96, 97, 98, 101, 151, 192
fathers' rights movement, 97, 151
Federal Equal Access Act, 1984, 42
Federal Marriage Amendment, 8, 69, 181
Feinstein, Diane, 197
feminism, 71, 129, 153, 154, 157, 164, 176, 213
  and anthropology, 157
  and colonialism, 163
  and conservative feminism, 5, 156, 161
  and essentialism, 151
  and lesbianism, 127, 196, 199
  and LGBTQ rights, 124, 136, 196
  and patriarchy, 5, 84, 86, 132, 156, 157, 158, 160, 161, 164, 165
  and postfeminism, 130, 151
  and postmodernism, 127, 143, 149, 167
  and power feminism, 98, 151, 156, 160, 161, 162, 163, 164
  and race, 124, 151, 159, 162
  and research, 213
  and social class, 116, 124, 127, 135, 151, 159, 162, 165
  and socialism, 86, 124, 127, 141, 151, 159, 165, 166, 167, 168, 186
feminist, 26, 42, 84, 124, 127, 141, 148, 196
  critique of Marx, 85, 151, 156
Fenrich, John, 44
Fenrich, Steen, 44
fertilization, 172, 179, 180
  and in vitro, 100, 179
fetus, 174, 178
  and embryo, 176, 178
  laws protecting, 175, 179
  as person, 24, 176, 178, 180
FIERCE, 215
First Amendment, 18
  and religious freedom, 72, 208
Flores v. Morgan Hill, 2003, 42
Focus on the Family, 57, 58
Foxx, Virginia, 55
free love movement, 108, 110, 112, 141, 167, 170

free speech, 32, 187
free will, 89, 97, 139, 146
Freidan, Betty, 177
Freud, Sigmund, 103
Frist, Bill, 172
fundamentalism, 45, 172, 195
  and homophobia, 13, 53, 55, 74, 75, 196
  and procreation, 53, 75, 182
  and the Bible, 74

gag rules, 28, 48
Garden Cities, the, 167
gay, 7, 10, 17, 18, 20, 21, 23, 25, 29, 31, 33, 34, 35, 36, 37, 38, 39, 40, 41, 42, 43, 44, 46, 47, 48, 51, 52, 54, 56, 57, 60, 61, 62, 63, 64, 65, 66, 72, 73, 75, 77, 89, 101, 106, 120, 122, 123, 124, 125, 130, 131, 133, 136, 140, 141, 143, 147, 156, 166, 185, 186, 190, 195, 196, 197, 198, 199, 200, 201, 202, 203, 204, 205, 206, 207, 212, 214, 215, 217, 218, 221
Gay Activists Alliance, 204
gay agenda, 55, 56, 68
Gay Lesbian and Straight Education Network (GLSEN), 7
  number of, 17
Gay Liberation Front, 133, 197, 204, 220
gay rights legislation, 21, 43, 48, 59, 85
  repeal of, 52, 53
gay-straight alliances, 7, 17, 18, 33, 40, 46, 48, 134, 197, 203, 209, 215
  and First Amendment rights, 18
  and lack of outreach to poor and minority youth, 17, 215
  membership of, 17, 199, 202, 204, 205
  number of, 17
gender, 10, 34, 37, 83, 90, 101, 124, 127, 128, 131, 143, 152, 162, 185, 190, 193, 196, 222, 223
  and evolutionary biology, 74, 77, 114, 157, 161

gender—*Continued*
  and identity, 7, 36, 39, 40, 41, 43, 61, 78, 126, 147
  and Marxism, 87, 126, 165
  and non-normative, 8, 10, 30, 36, 56, 87, 98, 102, 105, 106
  and roles, 30, 62, 74, 77, 84, 85, 87, 98, 103, 115, 148, 152, 154, 158, 159, 160, 166, 184, 185
  and sociobiology, 9, 129, 165
  and traditional roles, 43, 50, 52, 62, 75, 89, 90, 102, 105, 154, 166, 167, 184, 185
gender-segregated education, 87, 154
GI Bill, 95, 184
Gingrich, Newt, 96, 162
Giuliani, Rudolf, 215
Golden, Alma, 24
Goldsmith, Stephen, 96
Griffith, Bobby, 8, 44, 45, 47, 48
Griffith, Mary, 44, 45, 47
Griswold, Estelle, 171
Griswold v. Connecticut, 1965, 50, 171, 175

Hagar, David, 29
Hagee, John, 63
Haggard, Ted, 55, 208
Hammil, Peggy, 172
Hannity, Sean, 1
harassment, 12, 73
  gender-based, 11, 36, 37, 38, 40, 41, 42, 48, 215
  and sexual orientation, 4, 7, 11, 33, 36, 37, 38, 39, 40, 41, 42, 218
Hatch, Orrin, 10, 42
hate crimes legislation, 8, 42, 43, 44, 47, 48, 56, 60, 73
Hay, Harry, 202
Health and Human Services (HHS), 23, 24
health care, 14, 17, 51, 67, 69, 71, 89, 112, 118, 131, 134, 137, 138, 140, 142, 146, 152, 155, 163, 167, 169, 174, 176, 177, 178, 179, 180, 183, 186, 193, 194, 196, 197, 198, 200, 206, 209, 213, 214, 215, 216, 217, 218, 220, 222, 223
  and women, 81, 152, 168
health insurance, 186
  and women, 2, 168, 169
hegemony, 37, 81, 129, 135, 156, 212
Henkle v. Gregory, 2001, 42
Heritage Foundation, 57, 182
hermaphrodites. *See* intersex
Hermaphrodites with Attitude, 10
heteronormativity, 8, 10, 25, 30, 33, 37, 189, 206, 221
heterosexism, 31, 38, 56, 157
heterosexual, 9, 22, 25, 33, 36, 42, 43, 52, 55, 62, 63, 68, 73, 78, 89, 127, 147, 153, 154, 167, 197, 212, 217
  likelihood of know LGBTQ people, 11
heterosexuality, 27, 46, 120, 122, 132, 214
  origin of term, 9, 121
high school, 22, 26, 27, 31, 37, 45, 46, 84, 91, 119, 152, 189, 190, 192, 214
Hill, Anita, 162
Hill, Paul, 57, 156, 176
Hispanics, 19, 162, 163
historical materialism, 4, 30, 142, 148, 165
HIV/AIDS, 23, 24, 26, 28, 42, 47, 54, 55, 60, 73, 77, 131, 136, 137, 149, 196, 205, 212, 213, 215, 216, 221, 222
  and desire vs. need, 138
  and legal impacts, 15, 61, 123, 198, 206
  and prevention programs, 23, 25, 61, 72, 75, 172, 173, 198, 206
  and social class, 19, 127, 154, 198
  and youth rates, 24, 206
homework, 100
homocons, 43
homophile, 76, 202, 204
homophobia, 4, 10, 12, 17, 25, 36, 37, 38, 42, 44, 47, 48, 51, 53, 57, 61,

70, 72, 75, 78, 112, 114, 126, 131, 132, 134, 140, 143, 144, 194, 196, 199, 200, 202, 204, 207, 208, 209, 210, 211, 212, 217, 220, 221, 223
and African-Americans, 21
and backlash, 3, 18, 49, 56, 57, 65, 77, 205
and capitalism, 42, 59, 78, 130, 207, 214
definition of, 5, 49
and economy, 54
and family, 49, 67, 68, 75, 166, 167
history of, 5, 49, 54, 55
and lesbians, 61, 166
and nuclear family, 32, 36, 60, 62, 67, 166, 167
personal impact of, 49, 61, 190
and pseudo-science, 49, 75, 76
and racism, 60, 189, 190
and religion, 21, 46, 49, 54, 72, 73, 75
and violence, 4, 25, 38, 39, 40, 42, 44, 56, 60, 196
omosexual, 9, 10, 19, 42, 45, 46, 48, 53, 58, 74, 75, 102, 106, 114, 118, 147, 201, 202, 203, 205, 208
number identifying as, 11
percentage identifying as, 78
omosexuality, 17, 21, 22, 30, 33, 34, 35, 37, 45, 50, 52, 58, 59, 62, 63, 72, 74, 75, 129, 131, 134, 208, 211
attitudes toward, 8, 73
and behaviors, 8, 25, 29, 55, 62, 63, 76, 77, 124, 128
as biological, 9, 34, 53
and capitalism, 42, 62, 122, 207
as choice, 9, 34, 49, 70, 76
definition of, 120, 122
and identity, 34, 61, 65, 121, 123, 124, 128, 147
origin of term, 9, 121
as pathology, 30, 48, 76, 77, 201
psychiatric definition of, 49, 204
and socialism, 5, 64, 117, 141, 151
as threat to masculinity, 17, 54, 102, 143

Horowitz, David, 125
hospital visitation rights, 14, 51, 206
households, 44, 83, 88, 91, 107, 158, 166, 167, 181, 183
and single, 93, 95, 98, 100, 167, 183, 188, 223
and single parent, 74, 95, 97, 98, 100, 182, 183, 188, 223
and two-parent, 33, 70, 72, 73, 81, 100
and unmarried, 57, 67, 68, 182
Hudson, Rock, 61, 206
human capital, 153
Hurricane Katrina, 63, 106, 134, 145
husband, 69, 85, 89, 103, 115, 154, 164, 168, 169, 170, 171, 176
Hyde Amendment, 1976, 175, 176, 178

Icarian Movement, 107
identity, 9, 94, 118, 122, 124, 126, 128, 132, 133, 144, 146, 152, 157, 160, 162, 165, 202, 208, 212, 221
and localized, 5, 37, 120, 121, 129, 132, 196
identity politics, 3, 4, 5, 114, 120, 122, 123, 124, 129, 130, 151, 195, 202, 205
and activism, 134, 135, 202, 205, 221
and capitalism, 123, 124, 134, 165, 207
and feminism, 160
and localized, 52, 221
and racism, 121, 128, 196
within-group, 124, 133, 162, 196, 211
identity tribalism, 158, 191, 195, 196
illegitimate children, 70, 82, 93, 94, 98, 100, 114, 118, 181, 182, 188
illiteracy, 195
implantation, 172, 179
in vitro fertilization, 100, 179
incest, 71, 115, 174, 179
Independent Gay Forum, 61
individualism, 5, 57, 145, 146, 160, 164, 167, 188, 197, 217

individualism—*Continued*
  Marxist critique of, 97, 121, 142
  and sexuality, 57, 144, 191
interpellation, 3
intersex, 7
  definition of, 10
  and marriage, 16
  number of, 10
Intersex Society of North America, 10

Jackson, Lulia, 200
Jane, underground abortion movement, 112, 168, 199, 217, 219, 220
Jefferson, Thomas, 64
Jesus, 21, 44, 52, 75, 102, 103, 194, 208, 210
Johnston, Levi, 2
Jones, Alex, 1
Jones, Cleve, 206
Jones, Paula, 162

King, Martin Luther, 147, 177, 207
Kinkaid, Bob, 72
Kline, Phill, 177
Knights of Columbus, 58
Kopp, James, 176
Kucinich, Dennis, 112
Kuo, David, 58

La Reunion, 107
labor power, 86, 140, 157, 161, 184, 212
LaBruzzo, John, 3
Lambda Legal Defense and Education Fund, 204
Laramie Project, the, 189
Latino/as, 19, 162, 163
Lawrence, Gary, 58
Lawrence v. Texas, 2003, 65, 67, 71, 171
leadership, 57, 199, 207, 211, 217, 218, 219, 220
Lee University, 210
legitimacy, 4, 82, 83
leisure, 167, 184, 185
lesbian, 7, 10, 17, 18, 20, 21, 33, 38, 41, 42, 44, 46, 51, 54, 57, 61, 71, 72, 73, 75, 96, 106, 119, 120, 122, 123, 124, 125, 130, 133, 136, 140, 141, 156, 166, 185, 190, 196, 197, 198, 199, 200, 201, 202, 203, 204, 205, 206, 212, 214, 215, 217, 220
lesbianism, 37
  and feminism, 127, 199
LGBTQ, 1, 4, 7, 22, 25, 27, 31, 34, 41, 42, 47, 58, 65, 67, 78, 93, 94 106, 122, 123, 124, 128, 129, 132, 139, 143, 144, 160, 162, 175, 185, 193, 195, 197, 198, 199, 203, 205, 209, 214, 220
  and adoptions, 13, 53, 67, 140
  and bullying, 7, 12, 36, 37
  and censorship, 8, 32, 33, 75, 189, 207
  and clergy, 21, 22
  closeted teachers, 8, 12, 17, 19, 30, 31, 37, 39, 44, 46, 52, 56, 60, 61 66, 73, 119, 123, 136, 205
  and conservative, 30
  and curriculum, 8, 16, 30, 31, 32, 33, 40, 48
  and divorce, 16
  and employees, 123
  and families and adoption laws, 13, 53, 67, 68, 71, 85, 97, 140, 166, 182, 183, 185, 186, 212, 222, 223
  and families participating in school functions, 4, 13, 30
  and feminism, 124, 136, 196
  and HIV/AIDS, 15, 30, 60, 73, 138, 196, 205, 206
  and hospital visitation rights, 14, 51, 206
  and immigration policy, 16
  and legal rights, 4, 15, 48, 50, 51, 52, 57, 59, 67, 133, 134, 136, 140, 147, 167, 186, 200, 206, 208, 212
  and military, 21, 65, 66, 205
  and number of discharges from military, 20, 65, 66
  and number of parents, 13

and organizations, 13, 18, 52, 201, 202, 204, 205, 206, 207, 209, 211, 212, 213, 215
and parents, 13, 33, 44, 45, 46, 76, 77, 140, 207, 210, 218
and religion, 4, 21, 22, 46, 60, 64, 218
and religious groups, 21, 114
and resistance movements, 6, 198, 203
and retirement, 186, 209
and schools, 9, 12, 37, 46, 52
and senior citizens, 15, 68, 141, 186, 205, 209
and social class, 51, 134, 136, 157, 205
and socialism, 5, 64, 117, 141
stereotypes and suicide, 7, 38, 45
stereotypes of, 3, 8, 13, 33, 59, 61, 66, 136, 147, 203, 204
and suicide, 8, 212
and teachers, 73
teachers' attitudes toward students, 8, 38, 39, 190
transgendered teachers, 37, 38, 52, 56, 60, 61, 66, 205
and transgendered youth, 17, 42, 43, 60, 68, 133, 215, 218
and youth dropout rate, 17, 39, 46
and youth suicide rate, 39, 42
GBTQ studies, 40, 77, 130, 134, 147
bertarian, 64, 65, 112, 116, 121
bertarian anarchism, 107, 116
bertarianism, 5, 107, 142, 146
iberty University, 208, 209
iddy, G. Gordon, 163
imbaugh, Rush, 48
calized, 120, 121, 128, 129, 131, 132, 148, 156, 191, 196, 204
og Cabin Republicans, 52
nching, 65, 105

arch on Washington for Lesbian and Gay Rights, 205, 206, 213
arriage, 39, 62, 63, 67, 73, 82, 83, 85, 89, 90, 91, 95, 96, 97, 99, 102, 108, 109, 111, 115, 116, 117, 132, 140, 141, 154, 157, 167, 175, 181, 184, 188, 193, 212, 214, 217, 222
interracial bans on, 15, 70
legal benefits of, 13, 15, 67, 69, 74, 95, 114, 152, 184, 185, 223
and Marxism, 223
median age of, 26, 181, 182
and procreation, 53, 60, 64, 69, 74, 75, 83, 85, 123, 168, 169, 171, 212
and religion, 3
rights and legal protections of, 3, 50, 67, 69, 95, 110, 118, 133, 152, 184, 185, 223
marriage equality, 3, 51, 67, 110, 112, 133, 140, 141, 142, 186
Marxism, 81, 111, 112, 113, 118, 120, 121, 125, 126, 137, 146, 148, 156, 190, 197, 200, 201, 202, 204, 223
and commodification, 86, 132
and consumption, 86, 224
critique of anarchism, 112
critique of Stalinism, 127
and gender, 85, 223
HIV/AIDS, 131
and labor power, 86, 123, 157, 161
and marriage, 223
and women, 79, 85, 131, 157
masculinity, 17, 30, 36, 37, 61, 62, 98, 102, 103, 143, 185
masturbation, 25
maternity leave, 93, 153, 179, 183, 188, 200, 217
Matlovich, Leonard, 20
matriarchy, 115, 116, 159
Mattachine Society, 76, 197, 199, 201, 202, 204, 213, 218
Matthew Shepard and James Byrd Jr. Act, 2007, 43
McCain, John, 2
McCarthy, Joseph, 57, 60, 65, 66, 202, 213
McCarthyism, 57, 60, 65, 66, 202, 213
McGovern, George, 59
McIlvenna, Ted, 218
Mead, Margaret, 159

media, 3, 18, 32, 34, 48, 49, 50, 51, 53, 57, 65, 66, 71, 72, 77, 84, 89, 90, 92, 99, 106, 113, 120, 136, 137, 151, 155, 156, 160, 163, 164, 177, 182, 185, 187, 189, 194, 195, 196, 198, 199, 201, 202, 204, 205, 207, 209, 210, 211, 212, 213, 217, 218
Medicaid, 96, 141, 175, 179, 186, 193
Medical Institute for Sexual Health, 29
Meese, Edwin, 162
men, 84, 89, 90, 101, 102, 109, 115, 131, 132, 152, 153, 157, 163, 166, 168, 183, 199, 208, 221
metanarrative, 121, 148, 149
micropolitics, 128, 130, 156
midwives, 82, 174
military
  and LGBTQ, 20, 21, 65
Milk, Harvey, 34, 197, 205, 206, 216, 221
minorities, 18, 31, 34, 35, 51, 52, 55, 56, 60, 65, 66, 68, 70, 93, 94, 101, 106, 113, 123, 128, 129, 131, 133, 134, 135, 140, 141, 143, 154, 156, 162, 163, 177, 178, 180, 184, 186, 189, 193, 194, 195, 196, 202, 203, 205, 214, 220
miscegenation, 83
modernism, 128, 195
modernist, 128, 195
monogamy, 70, 74, 78, 115, 141, 154
  and abstinence-only education, 25
Montgomery v. Independent School District No. 709, 2000, 42
Moody, Howard, 217
morality, 5, 34, 62, 67, 104
  nation-as-family, 107
  strict-father, 107
Mormon church, 22, 58, 72
morning-after pill, 172
Moscone, George, 205
motherhood, 84, 90, 91, 92, 94, 116, 117, 153, 168, 174
Motherhood Protection Act, 179

mothers, 77, 79, 80, 83, 84, 90, 91, 98, 101, 115, 168, 173, 175, 183, 184, 188, 192, 196, 220
movements, 34, 84, 98, 101, 112, 124, 125, 127, 130, 137, 147, 149, 156, 157, 164, 176, 182, 197, 202, 207, 213, 214, 217, 219, 221
  authority in, 202, 218, 220
  organization of, 181, 191, 196, 198, 202, 204
multiracial, 99, 180

Nabozny v. Podlesny, 1996, 42
Nagin, Ray, 134
Names Project, 206
NARAL Pro-Choice America, 176, 177
Nashoba, 111
National Abstinence Education Association, 24
National Anarchism, 113
National Anarchists, 113, 114
National Association of Scholars, 57
National Coalition to Support Sexuality Education, 24
National Endowment for the Arts, 189
National Gay Task Force, 204
National Physicians Center for Family Resources, 24
National Right to Life Committee, 178
National Woman Suffrage Association, 171
Native Americans, 10, 64, 83, 116, 158, 160, 166
neoliberalism, 5, 49, 121, 124, 135, 145, 152, 168, 191, 192, 193, 214
Nixon, Richard, 223
No Child Left Behind Act, 2001, 23, 35, 192
North American Man-Boy Love Association, 71
North Central University, 211
nuclear family, 36, 55, 57, 68, 69, 70, 71, 74, 75, 81, 82, 83, 84, 85, 87, 89, 92, 93, 95, 100, 101, 102, 114, 115, 117, 122, 141, 146,

154, 158, 159, 160, 164, 167, 181, 182, 183, 188, 193, 197, 199, 205, 212
and legal protection for, 1, 5, 9, 36, 60, 70, 71, 78, 81, 83, 95, 133, 184, 186
origins of, 115
and patriarchy, 70, 71, 157
and women's oppression, 78, 153, 166, 184

Obama, Barack, 1, 19, 23, 43, 50, 51, 52, 72, 99, 107, 137, 155, 198, 223
objectivism, 126, 142, 146, 147
Oklahoma Baptist University, 211
Omnibus Appropriations Bill, 2009, 23
Oneida, 111
oppression, 3, 4, 36, 55, 56, 70, 78, 81, 103, 104, 111, 112, 115, 116, 122, 123, 124, 125, 127, 128, 129, 130, 131, 132, 134, 138, 139, 140, 142, 143, 144, 147, 148, 149, 151, 152, 153, 156, 157, 158, 160, 165, 166, 170, 190, 191, 195, 196, 200, 202, 205, 207, 208, 212, 214, 215, 220, 221, 222
O'Reilly, Bill, 1, 156
outing, 34, 52, 58, 199, 200, 210
out-of-wedlock births, 70, 93, 96, 98, 100, 102, 118, 170, 175, 179, 181, 182, 188

Paglia, Camille, 140
Palin, Bristol, 2
Palin, Sarah, 2, 164
Parents and Friends of Lesbians and Gays (PFLAG), 46, 204, 218
history of, 17
number of members, 17
Parents for Truth, 24
"partial birth abortion," 176, 178
"partial-birth abortion," 168
patriarchy, 37, 87, 98, 114, 116, 117, 120, 122, 131, 132, 133, 143, 154, 156, 157, 158, 159, 160, 163, 165, 167, 214, 222
conservative views of, 158
and nuclear family, 70, 71, 115, 157
and religion, 171
patriarchy theory, 151, 158, 160
critique of, 157
Paul, Ron, 112
pedagogy, 216
and anti-homophobic, 9, 38, 194
pedophilia, 53, 62, 65
and homosexuality, 8, 47, 64, 76, 114, 127
and recruiting, 14, 48, 52, 54, 64, 65, 76, 120, 127
Personal Responsibility and Work Opportunity Reconciliation Act, 1996, 23, 175
Petersen, Laci, 179
Pharmacists for Life International, 172
Phillips, Will, 190
physical abuse, 101, 106, 152, 163, 177, 203
Plan B, 29, 172
and FDA, 172
Planned Parenthood, 28, 155, 171, 177, 196, 215
Planned Parenthood of Central Missouri v. Danforth, 1996, 28
Planned Parenthood v. Casey, 1992, 180
Plato, 62, 63
politically correct (PC), 58, 67, 147
polygamy, 67, 70, 71, 74, 75, 108
Poplawski, Richard, 1
pornography, 140, 162, 189
Posse Comitatus, 112
postfeminism, 151, 161
postmodernism, 3, 4, 125, 135, 137, 148, 149, 157, 205, 223
as anti-solidarity, 78, 123, 131, 142
and capitalism, 113, 123, 127, 135
and closet, the, 122, 123
and consumerism, 5, 123, 124, 135, 224
critique of, 113, 127

postmodernism—*Continued*
  critique of Marxism, 5, 121, 126, 137, 156, 195
  and deconstruction, 121, 122, 125, 143, 146, 198
  and difference, 123, 133
  and feminism, 124, 127, 143, 149, 167
  and identity, 118, 128, 129, 132
  origins of, 121
  and personal experience, 129, 130
  and power, 121, 130, 144
  and queer theory, 5, 119, 122, 123, 127, 132, 141, 142, 209
  and rejection of Marxism, 120, 122, 156, 195
postpartum depression, 183
poverty, 3, 18, 44, 51, 68, 71, 91, 92, 93, 94, 95, 96, 97, 98, 101, 115, 116, 132, 134, 139, 140, 154, 160, 161, 163, 165, 169, 170, 174, 175, 179, 180, 182, 183, 186, 192, 193, 194, 195, 196, 204, 206, 215, 218, 220, 221
Powell, Colin, 129
power, 53, 99, 101, 105, 108, 113, 118, 119, 121, 124, 127, 130, 134, 139, 143, 144, 145, 147, 162, 168, 170, 185, 199, 216, 221
power feminism, 120, 151, 156, 160, 161, 162, 163, 164
preclass societies, 116, 163, 166
prefigured societies, 111, 112
pregnancy, 29, 82, 84, 153, 154, 168, 169, 171, 172, 173, 174, 176, 179
  and compulsory, 82, 156, 170, 171, 176
  cost of, 152, 169
  prevention of, 25, 84, 155, 168
prejudice, 61, 145, 201
  hysterical, 54, 66, 120
  narcissistic, 54
  obsessional, 54
Prince, Erik, 58
prison, 93, 104, 106, 109, 111, 192, 212, 221

privilege, 4, 9, 57, 73, 95, 120, 123, 131, 133, 134, 136, 139, 140, 197, 214, 223
pro-choice, 155
procreation, 75, 110, 123, 146, 171, 179
  and abstinence, 2, 69
  and fundamentalism, 74, 172
  and marriage, 60, 64, 84, 171, 212
production, 82, 85, 91, 131, 132, 135, 142, 153, 156, 184
projection, 102, 106, 107
Pro-Life Action League, 170
Pro-Life Wisconsin, 172
promiscuity, 46, 76, 78, 104, 114, 139, 144, 170, 175
Promise Keepers, 101
property, 34, 62, 64, 68, 74, 78, 82, 83, 97, 104, 110, 114, 115, 116, 117, 145, 158, 165, 166, 181, 188, 206, 214, 223
Proposition 8 (CA), 5, 14, 22, 42, 49, 50, 51, 52, 58, 59, 72, 143, 194
  and African-Americans, 133
  and funding, 191
Proposition 21 (CA), 133
pseudo-science, 49, 75, 177
  and evolutionary biology, 77, 114, 157
  procedures of, 76, 77
  and sociobiology, 77
punishment, 25, 103, 106, 111, 183
purity movement, 108
  19th century, 25, 94
  20th century, 25, 65, 84
  21st century, 94

queer, 7, 43, 44, 52, 114, 119, 123, 126, 127, 136, 138, 139, 141, 186, 193, 203, 209
  meaning of term, 10, 53
queer theory, 122, 123, 124, 125, 127, 139, 141, 146, 205
  and identity politics, 4, 5, 120, 122, 123, 130, 132, 144
  and postmodernism, 5, 119, 120, 130, 132, 144

rejection of Marxism, 120, 144, 197
Question 1 (ME), 16, 191
questioning, 7, 37
  meaning of term, 10

race, 10, 34, 64, 65, 99, 121, 124, 126, 127, 128, 151, 162, 193, 196, 223
racism, 10, 27, 35, 37, 39, 44, 53, 56, 57, 60, 67, 72, 83, 92, 99, 105, 111, 113, 121, 130, 131, 134, 140, 144, 145, 146, 156, 157, 159, 162, 177, 187, 190, 193, 194, 196, 212, 217, 220, 221
  and capitalism, 94, 130
  and homophobia, 190
Rand, Ayn, 90, 112
rape, 27, 47, 65, 74, 99, 105, 133, 140, 152, 157, 161, 162, 174, 177, 179, 180
reactionary populism, 12, 55, 191, 193, 194, 195, 200
Reagan, Ronald, 23, 50, 52, 61, 93, 127, 205, 216, 223
reform, 194, 197, 204, 205, 212, 213, 216
Regent University, 209
religion, 39, 47, 54, 57, 64, 69, 70, 73, 78, 97, 103, 104, 109, 134, 148, 154, 159, 165, 193, 195, 207, 209
  attitudes toward LGBTQ people, 8, 21, 44, 46, 49, 64, 72, 75, 93, 207, 208, 217, 218
  and capitalism, 154, 207, 208
  and patriarchy, 171
religion industrial complex, 58
remarriage, 82, 96, 100, 183, 186
reproduction, 5, 53, 85, 116, 136, 171, 208
  physical, 86
  social, 86
reproductive freedom, 155, 160, 167, 169, 171, 175, 176, 181, 183, 196, 197, 215, 219
  and equality, 28, 112, 173
  and socialist feminism, 5, 151, 168
reproductive rights, 5, 28, 118, 149, 151, 152, 153, 154, 155, 168, 169, 170, 171, 173, 180, 183, 196, 215, 217, 223
Republic Windows and Doors occupation, 220, 221
Republican party, 21, 29, 94, 221
Republicans, 21, 52, 58, 221
research, 8, 58, 129, 177, 188, 213
resistance movements, 6, 88, 189, 190, 198, 202, 204, 207
retirement, 209
  and Medicaid, 141
  and medical coverage, 141
  and nursing homes, 141, 186
  and women, 152
revolution, 81, 97, 117, 143, 147, 148, 157, 198, 199, 201, 205, 213, 216
Revolutionary People's Constitutional Convention, 220
Rice, Condoleezza, 129, 151
Rihanna, 152
Roe v. Wade, 1973, 50, 154, 155, 168, 171, 173, 174, 175, 176, 178, 179, 180, 181
Roeder, Scott, 176
Röhm, Ernst, 200
romantic love, 46, 83, 108, 109, 212
Romer v. Evans, 1995, 66, 67
Rose, Lila, 177
Rose, Tricia, 137
Rudolf, Eric, 176
Rumsfeld, Donald, 119
rural, 52, 118, 165, 192, 193, 194, 198
Rushdoony, R.J., 58
Russian Revolution, 118, 127, 153, 165, 223
Rust v. Sullivan, 1991, 178
Rustin, Bayard, 219

Safe Harbor law, 1, 2
Safe Schools programs, 4, 18, 38, 39, 40, 41, 43
Sagarin, Edward, 76, 202

same-sex households, 44, 62, 197
  demographics of, 11
  earnings of, 18
  number of, 11, 13
same-sex marriage, 3, 5, 10, 16, 20, 22, 51, 59, 67, 68, 70, 71, 72, 73, 75, 78, 85, 93, 133, 141, 167, 197, 213, 217, 220
  as civil rights, 3, 14, 59, 73, 140, 152, 190, 202, 208, 209, 210, 212, 214
  legalization of, 3, 14, 50, 78, 133
  percent of people approving of, 14, 50, 58
  states allowing, 14, 43, 50, 58, 59, 66, 213
  and states' rights, 14, 50, 53, 66, 67, 70, 213, 222
  as threat to the family, 16, 50, 51, 52, 60, 62, 67, 70, 73
same-sex relationships, 33, 57, 59, 61, 62, 63, 64, 75, 78, 89, 120, 141
  legalization of, 3, 14, 58, 67
  support for, 11, 21, 33, 46, 51, 54, 58, 133
Sanford, Mark, 93, 96
Santorum, Rick, 71
Sarria, José, 203
Savage, Michael, 1, 125
Scalia, Antonin, 66, 67
scarcity, 142, 143, 145, 154, 160, 163, 191, 223
Scheidler, Joseph, 170
school, 56, 57, 82, 90, 92, 111, 118, 119, 145, 150, 153, 164, 184, 188, 190, 192, 195, 207, 210, 215, 216, 219, 220, 223
  and administrators, 12, 37, 38, 40, 210
Section 28 (Britain), 23, 32, 42
self-help books, 90
senior citizens, 15, 141, 186, 188, 205, 209, 222
sex, 37, 39, 61, 90, 91, 94, 108, 116, 140, 149, 157, 168, 170, 181, 189
  as desire, 104, 120, 138, 169
  and marketplace, 84
  as need, 109, 120, 138
sex drive, 82, 83, 116, 170
sex work, 46, 62, 63, 65, 76, 84, 99, 116, 118, 137, 141, 144, 218
sexism, 56, 57, 105, 130, 132, 140, 143, 144, 145, 158, 159, 161, 194, 196, 212, 220, 222, 223
  and capitalism, 55, 87, 130, 187
  and homophobia, 17, 37, 55
sexual abuse, 101, 119, 120, 152
sexual harassment, 4, 37, 87, 99, 119, 120, 131, 140, 145, 149, 151, 161, 162, 163, 168, 187, 196, 218, 223
  conservative feminist response to, 128, 161
  and free speech, 32, 48, 187
  institutional, 36, 128, 187, 188
  quid pro quo, 146, 187
  and teachers, 31, 119
sexual identity, 9, 19, 34, 36, 56, 73, 76, 94, 106, 118, 122, 126, 128, 212, 221
  born, 10, 77, 125, 129, 206
  and capitalism, 9, 62, 122, 153
  choice, 7, 10, 70, 125, 153, 162, 191, 213
  definitions of, 10, 125
  history of, 9, 62, 124
  and homosexuality, 7, 62
  scientific rationales for, 121, 125
sexual orientation, 7, 8, 9, 10, 11, 12, 13, 19, 20, 34, 36, 37, 38, 39, 41, 42, 43, 44, 46, 47, 52, 60, 61, 66, 67, 76, 78, 119, 144, 193, 239
Sexual Orientation Non-Discrimination Act, 2002 (NY), 61
sexual revolution, 95, 98, 139
sexual violence, 55, 58, 64, 67, 105, 133, 140, 149, 157, 196
sexuality, 1, 3, 4, 6, 23, 36, 58, 62, 68, 70, 78, 99, 103, 116, 117, 118, 121, 122, 124, 126, 127, 129, 134, 136, 138, 143, 146, 150, 171, 178, 188, 190, 198,

199, 200, 201, 203, 204, 206, 207, 213, 215, 216, 217, 222, 223
conservative views about, 2, 22, 73, 76, 77, 78, 94
and individualism, 9, 57, 97, 121, 140, 146, 149, 191, 213
and procreation, 60, 62, 63, 64, 84, 168, 174, 179
sexuality education, 3, 49, 109, 189, 190, 213
and abstinence-only, 2, 22, 23, 24, 25, 171
and contraception education, 2, 22, 26, 28, 168, 171
and curriculum, 8, 25, 189, 206, 220, 223
and percent supporting comprehensive, 22, 28, 155
resistance to, 8, 22, 25, 109
and STIs, 24, 25
Shepard, Judy, 47
Shepard, Matthew, 4, 8, 42, 43, 44, 47, 56, 60, 189
slavery, 42, 83, 104, 105, 111, 161, 166
slaves, 62, 82, 83, 99, 104, 111, 136, 181
Smith, Susan, 92
social class, 4, 10, 18, 27, 56, 57, 58, 68, 110, 111, 116, 121, 124, 128, 135, 138, 141, 149, 151, 156, 157, 158, 159, 162, 163, 165, 170, 175, 181, 185, 190, 192, 195, 196, 199, 203, 212, 213, 220
and historical materialism, 4, 64, 91, 94, 117, 135
social context, 26, 27, 46, 56, 99, 148, 156, 162, 199
social Darwinism, 99, 159, 161
social safety net, 21, 57, 79, 80, 90, 93, 99, 125, 142, 193, 204
Social Security, 14, 75, 140, 141, 186, 193, 209
socialism, 1, 80, 81, 90, 94, 97, 107, 109, 113, 116, 117, 118, 121, 141, 166, 196, 198, 212, 214, 223

and feminism, 5, 141, 151, 165, 167, 186
and homosexuality, 64, 201
and LGBTQ, 64, 167, 201
socialist feminism, 5, 124, 149, 151, 159, 165, 166, 168, 186
and family, 71, 141
and marriage rights, 8, 71, 141, 167, 213
socialization, 81, 84, 87, 111, 188, 212, 223
society, 26, 112, 212, 220, 224
sodomy laws, 53, 55, 61, 62, 63, 65, 196, 201, 204
and criminalization, 50, 60, 63, 64, 66, 77, 181
and HIV/AIDS, 60
and homophobia, 42, 60, 63
and racism, 65
repeal of, 19, 43, 59, 64, 65, 66, 67, 71, 196, 214
and teachers, 60, 205
Sodoni, George, 99
solidarity, 12, 117, 123, 131, 134, 136, 142, 144, 145, 146, 147, 161, 181, 192, 193, 196, 206, 209, 210, 211, 212, 216, 217, 221
Sotomayor, Sonya, 162, 163
Souder, Mark, 172
Soulforce Equality Ride, 207, 208, 209, 210, 211, 212
Southgate, Troy, 114
sports
and curriculum, 31
and health class, 31
and homophobia, 7, 38
and LGBTQ, 31, 38
and physical education, 31, 38
and sexuality, 7
Ssempa, Martin, 72
standardized testing, 92, 184, 192
states' rights, 53, 70
Steele, Michael, 93
stereotypes, 27
of African-Americans, 27
of LGBTQ, 13, 33, 66, 136

sterilization, 3, 77, 175
Stirner, Max, 143
Stonewall, 52, 66, 124, 136, 199, 201, 202, 203, 204, 205, 206, 213
straight, 7, 16, 18, 19, 20, 21, 25, 30, 31, 34, 35, 37, 38, 39, 40, 43, 45, 46, 48, 50, 51, 54, 57, 60, 61, 62, 65, 68, 77, 121, 157, 194, 196, 197, 199, 203, 204, 211, 214, 215
   definition of, 17
Student Homophile League, 202
Stupak Amendment, 214
Summerhill, 28, 77, 104, 170, 223
super-child, 82, 90, 91
supply-side economics, 96, 97
survival of the fittest, 159
Symonds, John, 64

talk radio, 163, 195
teachers, 92, 133, 205, 216
   and closeted, 8, 11, 12, 17, 30, 37, 38, 44, 56, 61, 66, 119, 123
   and homophobia, 11, 37, 38, 189, 205
   and LGBTQ, 66
   and sexual harassment, 31, 37, 40, 119
   and sexuality curriculum, 8, 189, 223
teen pregnancy, 29, 84, 96, 155, 169
   and education programs, 24, 25
   and racism, 26, 27, 177
   and Title IX, 26
   and U.S. birthrate, 26, 93, 154
   and welfare, 27, 93
Terry, Randall, 155
think tanks, 164
   and funding, 8, 57
   and media, 57
Third Position politics, 113
Third World, 163, 191
Tiller, George, 156
Tinker v. Des Moines, 1969, 32
Title IX, 26, 27, 41, 168
Title X, 168, 169, 172
torture, 105
transgendered, 7, 10, 13, 18, 19, 20, 41, 42, 55, 61, 75, 106, 199, 203, 204, 218, 221
   and discrimination, 12, 16, 18, 19, 118, 185, 203, 214, 215
   and economic disadvantage, 16, 68, 123, 136, 193, 215
   and marriage laws, 8, 16
   teachers who are, 12, 123
two-spirit
   meaning of term, 10
Tynggarden, 188

Ulrichs, Karl, 76
Unborn Victims of Violence Act, 2001, 179
unemployment, 2, 3, 38, 54, 85, 88, 93, 95, 96, 99, 106, 152, 163, 171, 176, 195, 202, 205, 217
unions, 18, 51, 145, 147, 194, 197, 206, 220, 221
   corporatization of, 111, 217
   membership in, 193, 200, 205, 217
   and organizing, 133, 157, 181, 216
unwed motherhood, 82, 83, 93, 94, 96
urban, 21, 23, 43, 52, 63, 64, 78, 83, 123, 188, 193, 194, 201, 206

virginity, 26, 74
virginity pledges, 29
visibility, 58, 136, 148, 214, 218

wage gap, 43, 57, 59, 71, 91, 94, 95, 131, 135, 145, 151, 152, 153, 168, 183, 184, 192, 194, 222
wages for housework movement, 81, 86
Wagner, Jodi, 172
war, 1, 3, 20, 57, 84, 95, 102, 104, 110, 112, 114, 118, 122, 123, 124, 128, 137, 141, 153, 159, 184, 194, 200, 201, 202, 221
Warren, Rick, 72
Weaver, Randy, 113
Webster v. Reproductive Health Services, 1989, 180
Weinberg, George, 54
welfare, 3, 71, 91, 94, 95, 96, 106, 175, 182, 201

and mothers, 27, 93, 155, 168, 175, 183, 184, 217
Weller, Susan, 172
White, Ryan, 77, 206
White Night Riots, 205
white supremacy, 113, 121, 122, 196
wife, 69, 74, 82, 84, 85, 102, 103, 109, 115, 166, 171, 181
   submission of, 82
Wilde, Oscar, 64, 65
Winfrey, Oprah, 151, 163
wives, 84, 101, 115, 154, 181, 183, 219
   submission of, 82, 83, 103
women, 77, 84, 85, 88, 90, 102, 103, 104, 105, 106, 109, 113, 115, 127, 131, 132, 137, 140, 147, 148, 152, 163, 165, 167, 169, 175, 189, 208, 220
   and double standard, 83, 115, 152, 169, 170, 184
   and exploitation, 87, 151, 153, 160, 184
   and family, 79, 80, 82, 84, 98, 152, 153
   and health care, 79, 80, 81, 84, 112, 152, 155, 163, 168, 169, 172, 174, 175, 186, 214, 216, 220
   and health insurance, 2, 152, 168, 186
   and marriage, 82, 152, 182, 183, 186
   and Marxism, 87
   and rape, 64, 65, 152, 157, 161, 162
   and reproductive rights, 84, 112, 118, 149, 151, 152, 153, 160, 168, 169, 170, 171, 173, 174, 175, 176, 177, 179, 183, 197, 214, 215, 219, 220
   and retirement, 152
   and sexual harassment, 37, 145, 149, 151, 152, 161, 162, 187
   and sexually active, 37, 149, 152, 169, 172, 176, 215
   and wage gap, 84, 151, 152, 153, 163, 183, 184, 185, 217
   and wages, 81, 84, 86, 87, 151, 152, 163, 170, 183, 184, 185, 217
   and work, 81, 84, 85, 86, 87, 98, 101, 117, 118, 135, 140, 145, 153, 159, 161, 165, 169, 176, 181, 183, 184, 185, 186, 187, 196
   in workforce, 57, 81, 84, 117, 118, 135, 152, 163, 168, 176, 181, 183, 184, 185, 186, 187, 196
women's oppression, 78, 115, 116, 122, 128, 132, 134, 143, 147, 151, 153, 156, 157, 158, 160, 165, 166, 170, 179, 196, 214, 216
work, 84, 96, 116, 123, 157, 167, 181, 182, 195, 207, 220, 222
workforce, 84, 85, 117, 118, 122, 134, 137, 146, 151, 153, 154, 156, 158, 165, 167, 168, 184, 185, 194, 212
   and earnings, 18, 85, 89, 135, 145, 152, 183, 184, 192, 194, 195, 219
working class, 3, 4, 5, 15, 18, 64, 78, 88, 97, 127, 128, 131, 134, 136, 138, 141, 146, 149, 151, 152, 153, 154, 156, 157, 159, 160, 161, 165, 167, 168, 170, 171, 176, 184, 187, 188, 190, 192, 193, 194, 195, 198, 200, 203, 212, 213, 214, 215, 223
   as conservative, 143
   and industrialization, 87, 111, 116
   as reactionary, 56, 144
workplace discrimination, 4, 8, 11, 19, 57, 61, 77, 145, 152, 161, 184, 185, 186, 187, 223
Works Progress Administration (WPA), 118
Wurzelbacher, Samuel, 53

Zapatistas, 111
zaps, 206, 221

GPSR Compliance

The European Union's (EU) General Product Safety Regulation (GPSR) is a set of rules that requires consumer products to be safe and our obligations to ensure this.

If you have any concerns about our products, you can contact us on

ProductSafety@springernature.com

In case Publisher is established outside the EU, the EU authorized representative is:

Springer Nature Customer Service Center GmbH
Europaplatz 3
69115 Heidelberg, Germany

www.ingramcontent.com/pod-product-compliance
Lightning Source LLC
LaVergne TN
LVHW011810060526
838200LV00053B/3729